Cell Cycle Regulation and Development in Alphaproteobacteria

Emanuele Biondi
Editor

Cell Cycle Regulation and Development in Alphaproteobacteria

Springer

Editor
Emanuele Biondi
CNRS
Institute for Integrative Biology of the Cell
Gif sur Yvette, France

ISBN 978-3-030-90620-7 ISBN 978-3-030-90621-4 (eBook)
https://doi.org/10.1007/978-3-030-90621-4

This Springer imprint is published by the registered company Springer Nature Switzerland AG.
The registered company address is: Gewerbestrasse 11, 6330 Cham, Switzerland

"To all hard-working young and not so young scientists that love bacteriology"

Preface

Cell cycle and cellular differentiation are fascinating biological phenomena that are highly regulated in all organisms. In the last few decades, many laboratories around the world have been investigating these processes in *Alphaproteobacteria*. This bacterial class comprises important bacterial species, studied by fundamental and applied research. The complexity of cell cycle regulation and many examples of cellular differentiations in this bacterial group represent the main motives of this book. Hopefully, this ensemble of excellent contributions will fascinate new generations of biologists to embark on the investigation of the biology of bacteria.

Starting from the model species *Caulobacter crescentus* and proceeding to important models, such as *Agrobacterium tu*mefaciens, *Brucella* species, or *Sinorhizobium meliloti,* this book has the ambition to give a fairly complete overview of how the cell cycle is mechanistically regulated and what are the main regulators of cellular differentiation in the class *Alphaproteobacteria.*

After the first chapter that discusses the regulation of cell cycle in alphaproteobacterial species by a system biology perspective, the following chapters specifically focus on *C. crescentus* multiple layers of regulation, from transcriptional cascades to proteolysis and dynamic subcellular regulation of cell cycle regulators. Moreover, several chapters describe in detail the cell division process, chromosome segregation, and growth of the cell envelope mostly using *C. crescentus* as an example, but also other interesting and emerging model *genera* such as *Asticcacaulis* and *Hyphomonas*. The complexity of cell cycle is also described using mathematical modeling in a specific chapter.

Finally, the last three chapters talk about three well-studied examples of non-*Caulobacter* alphaproteobacterial models, such as *A. tu*mefaciens, *Brucella* species, and *Sinorhizobium meliloti.* These cell cycle models, although less known than *C. crescentus*, indeed represent new frontiers in the investigation with the goal of using the knowledge on this important species for the development of society, such as new human and veterinary antibiotics (*Brucella*) or anti-plant pathogens molecules for agronomy (*Agrobacterium*) or improving nitrogen fixation for agriculture (rhizobia).

Gif sur Yvette, France — Emanuele Biondi

Contents

Toward a Comparative Systems Biology of the Alphaproteobacterial Cell Cycle

Antonio Frandi, Francesco Pini, Wanassa Beroual, Andrea Bianchetti, Alice Chiodi, Elia Mascolo, Lorenzo Miano, Greta Petazzoni, Emanuele G. Biondi, and Matteo Brilli

Abstract

This chapter outlines how important properties of the bacterial cell cycle arose during evolution, and how it has been integrated to sustain different lifestyles in different niches. The circuits controlling the cell cycle not only set the pace of cell division but also actively influence the global response of the bacterial cell to its environment. Possibly because of this key role in cellular organization, certain mechanisms have evolved to specifically respond to different needs dictated by the different ecological niches occupied by the alphaproteobacteria. In addition, RNA and protein synthesis act in a balance to control essential cell cycle functions and differentiation.

A. Frandi
Department of Fundamental Microbiology, University of Lausanne, Lausanne, Switzerland

F. Pini
Department of Agronomy, Food, Environment and Forestry (DAGRI), University of Florence, Florence, Italy

W. Beroual · E. G. Biondi
Laboratoire de Chimie Bacterienne, CNRS, Marseille, France

A. Bianchetti · E. Mascolo · L. Miano · G. Petazzoni
Department of Biosciences, University of Milan, Milan, Italy

A. Chiodi
Department of Biosciences, University of Milan, Milan, Italy

Department of Earth and Environmental Sciences, University of Pavia, Pavia, Italy

M. Brilli (✉)
Department of Biosciences, University of Milan, Milan, Italy

Paediatric Clinical Research Center "Romeo ed Enrica Invernizzi", University of Milan, Milan, Italy
e-mail: matteo.brilli@unimi.it

E. Biondi (ed.), *Cell Cycle Regulation and Development in Alphaproteobacteria*,
https://doi.org/10.1007/978-3-030-90621-4_1

1 Introduction

Evolution often generates novelties by recycling pre-existing objects; a well-known example concerns the proteins of the vertebrate eye lens, that evolved from proteins devoted to unrelated functions (Slingsby et al. 2013). Rarely, similar events result in the *emergence* of a brand-new feature. The evolution of the alphaproteobacterial lineage witnessed such an event, when a eukaryotic-style cell cycle was assembled from pre-existing proteins. Their recruitment allowed the set-up of a *biological oscillator*, and its coupling with the DNA replication and cell division machineries originated a peculiar cell cycle, with no known parallel in Bacteria except within the same taxonomic group.

In this chapter, we will update the comparative genomics analysis of this system, we will provide a summary concerning the transversal knowledge about it across several alphaproteobacterial species, and we will conclude by providing a well-supported hypothesis about the evolutionary path leading to this exquisitely precise biological oscillator.

The main task of the chapter, however, is to introduce a new approach to the study of biological systems that we named *Comparative Systems Biology*, which is based on the observation that Evolution is blind to the inner structure of a certain system and it can only work on the output produced by the system. Therefore, a full understanding of regulatory circuits from an evolutionary perspective requires an approach where their structure in different species is compared considering their dynamical properties.

To maintain genomic integrity, cells have to coordinate DNA replication, DNA partition, and cell division. Eukaryotic cells possess regulatory circuits enabling precise progression across the distinct phases of the cell cycle. The orderly path from one cell cycle phase to the next is mediated by the timed activation of distinct cyclin/cyclin-dependent kinase complexes, and it is regulated by monitoring the replication status, DNA integrity or chromosome alignment, and other physiological/structural aspects of the cell, at well-defined checkpoints (Li and Nicklas 1995).

Likewise, in bacteria, regulatory networks coordinate cell cycle progression with cell growth, temporal and spatial control of DNA replication and cytokinesis, to ensure equal partitioning of chromosomes into daughter cells. However, the cell cycle of most prokaryotes is much simpler than the eukaryotic counterpart: in *E. coli* and other well studied bacterial species, DNA replication is partially independent from cytokinesis and the daughter cells are precise copies of the mother cell. Conversely, stunning functional similarities to the eukaryotic cell cycle are found in members of the class Alphaproteobacteria, notably in the aquatic alphaproteobacterium *Caulobacter crescentus*. This bacterium, akin to eukaryotic cells has evolved an intricate regulatory network, which couple DNA replication with cell division and morphological differentiation.

Recent discoveries have substantially improved our understanding of the genetic circuits controlling cell cycle progression and differentiation in *C. crescentus*. In addition to the ones characterized in *C. crescentus*, similar genetic circuits have been identified in other alphaproteobacteria, such as the plant symbiont *Sinorhizobium meliloti*, the plant pathogen *Agrobacterium tumefaciens*, and the facultative pathogens of the *Brucella* genus. Proteins involved in this system, traced across the Alphaproteobacteria class, were shown to be mostly conserved; some of them are specific to this lineage (Brilli et al. 2010) and appeared at well-defined points of the Alphaproteobacteria tree. However, several cell cycle proteins that accordingly to a role in the cell cycle are essential in *C. crescentus* and closely related organisms are instead dispensable in other alphaproteobacteria. In the latter, they are involved in cellular activities other than the cell cycle and associated differentiation, especially in motility. Therefore, the evolutionary history of the cell cycle regulatory network is particularly intriguing as its assembly happened during the radiation of the alphaproteobacteria.

This provides the unprecedented possibility of studying how the cell cycle circuit and asymmetric cell division emerged. This is one of the very fundamental questions in biology that cannot be properly studied in Eukaryotes, as the involved circuits are mostly the same in all members of the kingdom.

As precisely orchestrated differential spatial localization plays a fundamental role in the division and differentiation process of *C. crescentus* and other alphaproteobacteria, a detailed understanding of the evolutionary paths leading to the present situation might also shed some light on how spatial regulation originated. Dimorphism—the ability of a mother cell to originate different cell types—is indeed widespread in alphaproteobacteria, in more or less evident degrees depending on the species.

In the crescent-shaped alphaproteobacterium *C. crescentus* the cell cycle is spatiotemporally coordinated with the dimorphic differentiation through a peculiar regulatory network.

This network allows a mother *Caulobacter* cell to produce two genetically identical cells expressing distinct morphological features and regulatory programs: a swarmer daughter, which has a single flagellum and several pili; this form is the bacterial equivalent of the eukaryotic G1 cell in that chromosome replication is silenced. The idea is that the swarmer cell scavenges in search of nutrients, and when it finds them, differentiates into the sessile stalked cell that is responsible for the colonization of the niche, but the exact signal stimulating the differentiation, is still unknown.

The second cell type generated at each cell division is called *stalked* for the presence of a cylindrical polar structure (the *stalk*) capped with an adhesive polysaccharide. Additionally, swarmer cells at some point also differentiate into this form through a complicated developmental program and become competent for division. During the swarmer to stalked cell transition, the flagellum is shed, pili are lost and growth of the envelope is redirected to build the stalk. Coincident with these morphological changes, the differentiating cells acquire the ability to initiate chromosome replication, and enter a pre-divisional phase during which they complete chromosome replication and prepare the division septum. They also execute a developmental program that creates a new swarmer pole opposite to the stalk, where flagellum and pili will soon appear to propel the future swarmer daughter cell.

Despite *C. crescentus* showing a prominent dimorphic life cycle, the latter might be less evident in other alphaproteobacteria in terms of morphological or physiological/functional features that enable to discern among the two different progenies. This is for instance the case of *S. meliloti* whose daughter cells have cell sizes/shapes that are only marginally different, and internal differences are also less evident than in *C. crescentus*.

The molecular circuits responsible for this behavior have been studied in depth in *C. crescentus* and in *Sinorhizobium meliloti* (Kobayashi et al. 2009; Pini et al. 2013, 2015; Penterman et al. 2014); moreover a variety of studies has begun to accumulate for additional species (Bellefontaine et al. 2002; Belas et al. 2009; Mercer et al. 2010; Bird and MacKrell 2011; Cheng et al. 2011; Greene et al. 2012; Kim et al. 2013a; Zan et al. 2013; Wang et al. 2014; De Bolle et al. 2015; Francez-Charlot et al. 2015) (see below) providing a wide picture of how the entire system might have been assembled and improved under different selective pressures in different taxonomic lineages.

A few top-level master regulatory proteins organize and coordinate the many functions that constitute the cell cycle. CtrA is the best characterized one, and it was identified and studied in other alphaproteobacteria (Hallez et al. 2004; Belas et al. 2009; Mercer et al. 2010; Cheng et al. 2011; Greene et al. 2012; Kim et al. 2013b; Zan et al. 2013; Wang et al. 2014; De Bolle et al. 2015; Francez-Charlot et al. 2015). CtrA-regulated functions include synthesis of the flagellum, control of DNA replication and cell division (Laub et al. 2002). The two-component signaling protein CtrA is a critical regulator of asymmetry through its dual roles as an inhibitor of the initiation of DNA replication—in *C. crescentus* by binding and blocking the origin of replication—and by acting as a classical transcription factor controlling many genes (Laub et al. 2002). Because of the central role of CtrA in cell cycle progression and differentiation, there are multiple layers of regulation of its activity, including differential synthesis, activation by phosphorylation, and timed degradation.

Everything is finely interlaced to an astonishing degree of synchronization such that each daughter cell not only inherits a single copy of the genome but also a carefully selected set of proteins that determine the asymmetry in cell division and therefore the daughter cells' fate.

Swarmer cells are characterized by a high level and activity of CtrA. DNA replication starts after CtrA proteolysis by the ClpP-ClpX protease during the G1-S transition (Joshi and Chien 2016); its degradation frees bacterial replication origins allowing DnaA binding and induces differentiation from swarmer to stalked cells. In predivisional cells, *ctrA* transcription initiates thanks to the sequential activation of promoters P1 and P2. During this phase, CtrA reaches its peak of activity (Domian et al. 1996), being phosphorylated in a cell cycle-dependent fashion by an essential phosphorelay. The phosphorelay is formed by a hybrid histidine kinase (CckA) and a histidine phosphotransferase (ChpT), which transfers the phosphate from CckA to both CtrA and CpdR (Biondi et al. 2006), a protein involved in CtrA degradation

(Joshi and Chien 2016). Phosphorelay activity is under control of the response regulator DivK, in its phosphorylated form, DivK blocks the phosphorelay thus promoting cell cycle progression by CtrA inactivation. CtrA promotes transcription of *divK*; in its unphosphorylated form, DivK is not able to sequester DivL. The binding of DivL with CckA promotes the auto-phosphorylation of the latter, promoting CtrA activity. Two histidine kinases control DivK phosphorylation, DivJ and PleC, which are respectively the principal kinase and phosphatase of DivK. These two proteins are spatial and temporal localized during the cell cycle: DivJ localizes to the stalked pole at G1-S transition; PleC at the flagellar pole in swarmer and predivisional cells (Wheeler and Shapiro 1999; Subramanian et al. 2015). CtrA presence and phosphorylation is then opposite in stalked and swarmer cells, while in predivisional cells there is a gradient of phosphorylated CtrA from the swarmer (high) to the stalked (low) pole, which allows the formation of two morphologically different daughter cells.

Proteins central to the cell cycle are often conserved in most alphaproteobacterial species (Brilli et al. 2010), but regulatory interactions seem to be much less conserved as inferred by the bioinformatics predictions using the available information about the CtrA binding site (Fig. 1). By exploiting this information, we derived five variants of the regulatory circuits driving cell cycle progression, with those of *Rhizobiales* and *Caulobacterales* representing the most complicated and the most studied ones; partial or weak evidences of reduced regulatory circuits were found in other alphas. Several predictions, reported in (Brilli et al. 2010), where validated in the next years by many independent research labs. For instance, Greene et al. showed that CtrA is dispensable in *Magnetospirillum magneticum*, where it regulates motility and not the cell cycle (Greene et al. 2012).

The spatio-temporal regulation of CtrA seems to be particularly well adapted to the aquatic lifestyle of *C. crescentus* but it appears to be surprisingly conserved in other alphaproteobacteria with very different lifestyles. In *Brucella abortus*, the core circuit controlling CtrA activity is conserved with some variation compared to that of *Caulobacter* and this might reflect the bacterial requirement to precisely regulate its cell cycle depending on its intracellular environment. Infection and cell cycle are strictly connected; *Brucella* cells in the G1 phase are more infective than cells in other cell cycle stages (Deghelt et al. 2014); moreover, in the first 6 h after infection cells are blocked in the G1 phase, chromosome replication and cell cycle are unlocked just before they reach the intracellular proliferation compartment (De Bolle et al. 2015). In *B. abortus*, CtrA binds upstream of several promoters regulating cell cycle genes and also cell envelope biogenesis, which could potentially have an impact on the bacterial fitness when inside host cells (Francis et al. 2017). This type of regulation performed by CtrA is not only exclusive of *Brucella*; in fact, the intracellular pathogen *Ehrlichia chaffensis* is thought to regulate the expression of a major outer-membrane stabilizing protein through CtrA. In addition to the intracellular pathogens *Brucella* and *Ehrlichia* also the plant symbiont *S. meliloti* and the plant pathogen *A. tumefaciens* do share some similarities in their potential CtrA targets. However, to determine if the functions regulated by CtrA have really been shaped by evolution to match the lifestyle of these bacteria

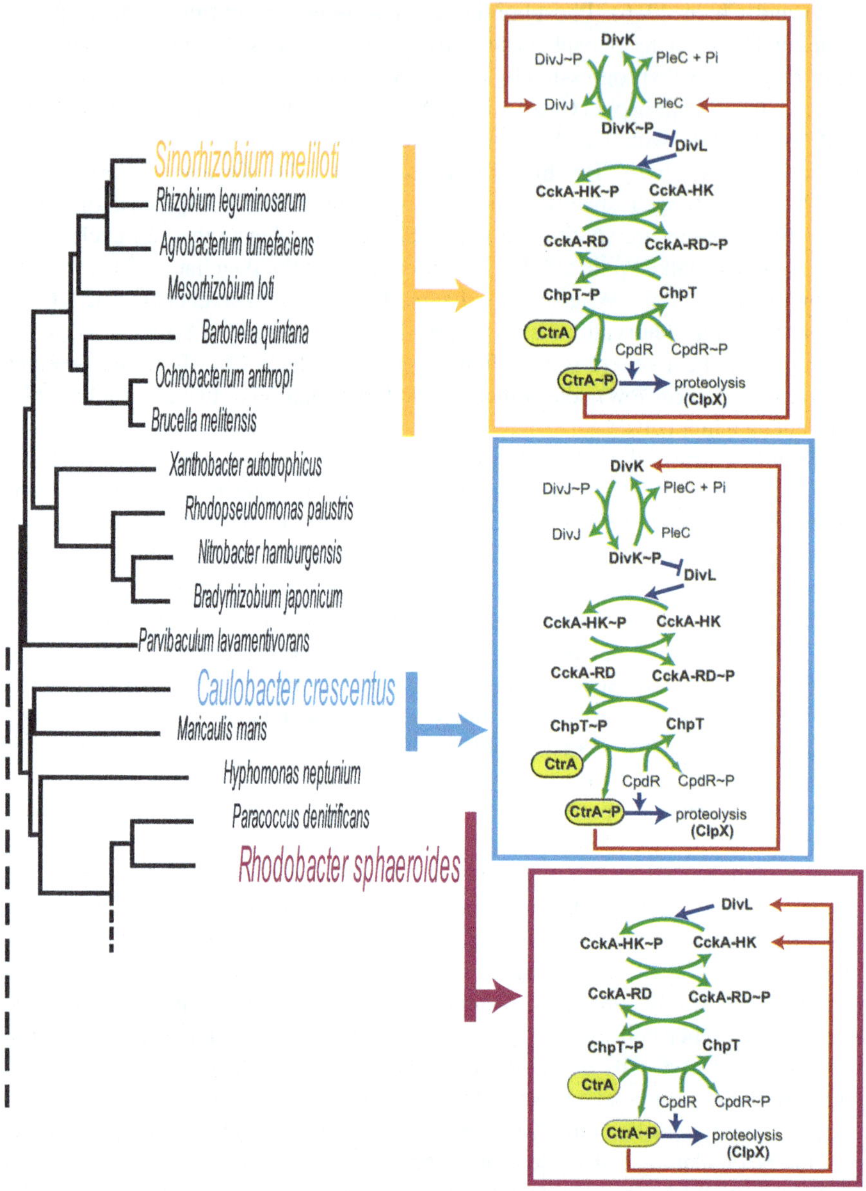

Fig. 1 Examples of alternative arrangements of cell cycle regulatory networks reconstructed in (Brilli et al. 2010) and focusing only on CtrA regulatory sub-networks. A common core of proteins is interconnected by slightly different regulations. Comparative interpretation of the models built for the different systems can elucidate the impact of these differences on system's functionality

would require more experimental data from other bacteria with different lifestyles and niches. On the other hand, not much is known about CtrA essentiality and functions in other groups of strictly intracellular alphaproteobacteria as *Wolbachia* and *Rickettsia*. Few reports have shown that also in these intracellular pathogens CtrA controls essential functions as cell division and chromosome replication. In *Rhodobacter sphaeroides*, two sets of flagellar genes are present, called *fla1* and *fla2*, which are responsible for producing two different flagellar structures. Binding sites for CtrA of these promoters were identified in silico, tested by site-directed mutagenesis and confirmed by Chromatin Immunoprecipitation (IP) supporting the notion that CtrA directly controls the expression of *fla2* genes (Rivera-Osorio et al. 2018). *Rhodovolum sulfidophilum* produces a gene transfer agent-like particle (GTA-like particle) (Nagao et al. 2015). GTAs were first discovered in *Rhodobacter capsulatus* but are present in various prokaryotic species (Lang and Beatty 2006); GTAs are shaped like bacteriophage particles, but they contain short DNA fragments randomly sampled from the bacterial genome instead of the bacteriophage genome. In *Dinoroseobacter shibae* (Wang et al. 2014), *R. capsulatus* (Lang and Beatty 2001), and *Rhodovolum sulfidophilum* (Komatsu et al. 2018), these systems are controlled by CtrA and the latter, also produces extracellular nucleic acids in a CtrA-dependent way. Loss of CtrA in *R. capsulatus* has pleiotropic effects and influences about 6% of the genes, including flagellar motility genes and a number of other putative regulatory proteins, but does not appear to include any genes involved in the cell cycle.

2 Comparative Genomics of the Cell Cycle

The basis for cell cycle studies in many different alphaproteobacteria was provided thanks to bioinformatics analysis on alphaproteobacterial genomes, obtaining a map with the presence of cell cycle-related proteins (Brilli et al. 2010); predictions of regulatory interactions based on the presence of well-studied binding sites were also performed. Since then, several papers have shown that CtrA or other *C. crescentus* cell cycle proteins are not essential throughout all alphaproteobacteria, which strongly suggest involvement in processes other than cell cycle regulation. A trait, which is evolutionary conserved, is the control exerted by CtrA on motility, chemotaxis, membrane/cell wall/envelope biogenesis, and signal transduction, suggesting these may be the ancestral roles of CtrA, before recruitment for cell cycle regulation. Our predictions of the promoters likely regulated by CtrA in other alphaproteobacteria, together with their distribution with respect to the alphaproteobacterial phylogenetic tree, reinforced this view. CtrA was later experimentally shown to be dispensable and involved in the regulation of motility (and often quorum sensing) in *Rhodospirillum centenum* (Bird and MacKrell 2011), *Ruegeria* sp. (Zan et al. 2013), *Dinoroseobacter shibae* (Wang et al. 2014), *Rhodobacter capsulatus* (Mercer et al. 2010), *Sphingomonas melonis* (Francez-Charlot et al. 2015), *Silicibacter* TM1040 (Belas et al. 2009) and *Magnetospirillum magneticum* (Greene et al. 2012). In most of these organisms, CtrA often has additional and important roles; in *Ehrlichia chafeensis*, CtrA coordinates the development of the stress resistance for the passage from a host cell to the next one (Cheng et al. 2011). Motility and chemotaxis are also part of the CtrA regulon in *Alphaproteobacteria* species such as *C. crescentus*, *Brucella* spp., *S. meliloti* (Bellefontaine et al. 2002), and *A. tumefaciens* (Kim et al. 2013b), where CtrA is essential and involved in cell cycle regulation. From an evolutionary/ecological point of view, this might be important as motility and chemotaxis are often connected to sense environmental changes, which could be important to coordinate DNA replication and cell division to the availability of nutrients. These properties provide a way to partition the alphaproteobacteria into two large groups; the first one includes the *Rhizobiales* and the *Caulobacterales*, where CtrA controls DNA replication and cell division, in addition to motility; the second group comprises all the species where CtrA mainly controls motility and is not involved in cell cycle regulation. The partition in these two groups is congruent with the ancestral presence of CtrA and its recruitment for cell cycle control in one of the ancestors of the *Rhizobiales* and *Caulobacterales* lineages.

Since 2010, thousands of new genomes have been sequenced, many of which belong to the *Alphaproteobacteria* class. Additionally, the many papers published so far have revealed important roles for proteins previously not implicated in the cell cycle, meaning that we can today provide a more complete picture of the system. We first obtained the phylogenetic profiles for cell cycle-related proteins starting from the *C. crescentus* protein sequences and a selection of alphaproteobacteria.

This allowed reconstructing the presence–absence profile matrix shown in Fig. 2. Since sequenced genomes are often draft and therefore missing genes are quite common, we merge the information concerning different species of the same genus and plot one line per genus. We are aware that in this way it is possible to lose the information about genes with true heterogeneous profiles within genera, but for the purpose of the present analysis, confidence is preferable in respect to precision. Moreover, as we are using the profiles to get information about proteins sharing function, it is worth noticing that biological sequence databases are rarely well balanced from a taxonomical point of view. This can bias the calculation of similarities among profiles; for instance, pathogen species are over-represented with respect to organisms from peculiar environments. This bias tends to diminish when one considers taxonomical categories (e.g., Species, Genus, Family) instead of the actual genomes present in the database that correspond to strains. Collapsing the profiles at the level of genera, therefore, allows reducing the distortion caused by the uneven organism sequencing.

The study of the cell cycle began on *C. crescentus* and for a few years the focus was on this only organism. The scientific community started to address the same questions in different species only later, and today this additional information has started to accumulate. If we continue studying the cell cycle with reference to *Caulobacter* we may end up missing important features that it does not possess but that could be important in other lineages. Some of these comprise species—such as *S. meliloti*, *A. tumefaciens* and *Brucella* spp.—that engage in complicated associations with plants and metazoans and may have genomes with additional

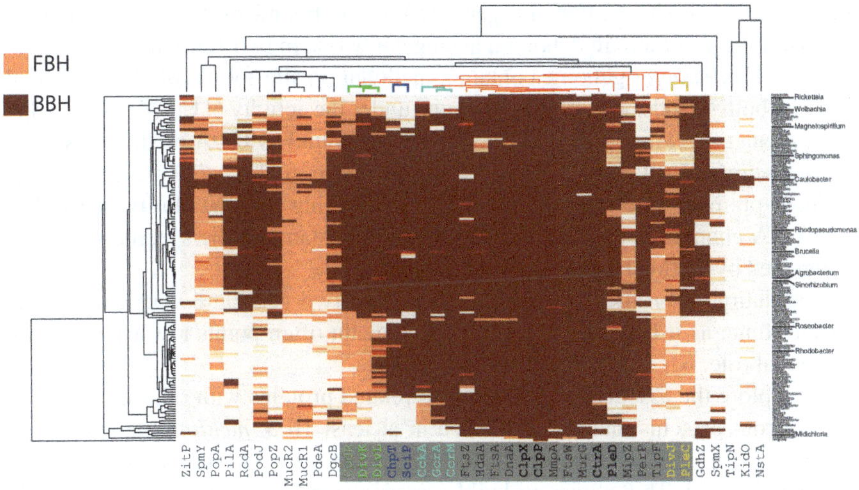

Fig. 2 Clustering of the phylogenetic profile matrix obtained when using proteins listed in Table 1 as query for a BBH or FBH search against a selected sample of Alphaproteobacteria covering all major taxonomic groups within the class. The most studied cell cycle genes all cluster together and at least 3–4 functional units are recognizable: in yellow, DivJ and PleC working antagonistically on DivK activity through phosphorylation/dephosphorylation. In green DivK and DivL, the former negative regulator of the second, who also "feeds" the phosphorelay. The presence of CpdR in this group of co-occurring proteins is unclear. However, it interacts with ClpX, which becomes phosphorylated by the phosphorelay controlled by DivK-DivL. The functional association among ChpT and SciP could be explained in the following way: ChpT is part of the phosphorelay and therefore it is involved in the regulation of the activity of CtrA; data indicate that SciP stabilizes the binding of CtrA to DNA, which, in turn, leads to the stabilization of CtrA against degradation by ClpXP (Gora et al. 2013). To conclude, the light blue proteins also tend to co-occur often; concerning GcrA and CcrM, it is well known that GcrA activity is dependent on the methylation status of the DNA it binds to (Brilli et al. 2010; Fioravanti et al. 2013), while the co-occurrence with also CckA is not easy to explain but might be indirect on CcrM through the phosphorelay and CtrA

levels of complicacy with respect to the free-living *C. crescentus*. For this reason, we decided to start a new co-occurrence analysis in a selection of alphaproteobacterial genomes starting from the whole proteome of *S. meliloti*. By using the profiles of known cell cycle-regulated proteins, it is possible to identify proteins with strongly correlated patterns, thereby providing testable hypothesis about their role in the cell cycle. As known since long, correlated profiles of the presence/absence of genes in genomes are indicative of the involvement of the corresponding proteins in common functional processes and therefore this method has been used to fish for proteins involved in interesting processes with only partial characterization. In Brilli et al. (2010) for instance, based on less than 100 alphaproteobacterial genomes, we inferred the association between GcrA, whose function is still somehow controversial but clearly belongs to the cell cycle at least in *Caulobacter*, and the CcrM methylase, also essential in this organism and several other alphaproteobacteria. At the time, this association was unclear from a functional point of view, but since then it has been shown that the genome regions bound by GcrA often contain methylation sites; more in detail, on average only one in two 1000 nt long windows from the *Caulobacter* genome contain a methylation site for CcrM (GAnTC, with the A on both strands becoming methylated under the activity of CcrM), while when considering only windows centered on positions bound by GcrA on the basis of the accumulation of ChIP-seq reads, the average increases to 2 sites/1000 nt. Stimulated by these premises, we then succeeded in establishing that the ability of GcrA to bind DNA changes depending on the methylation status of the specific sequences recognized by CcrM (Fioravanti et al. 2013).

Correlation in profiles can also happen when two genes are very close on the chromosome, as it happens in operons; this, again often points toward a common functional role.

To explore the association of cell cycle-related proteins with other proteins, we started from proteins that are conserved in at least 10 *S. meliloti* genomes to get orthologs in selected alphaproteobacterial genomes by exploiting the Bidirectional Best Hit method. This allows to build a matrix storing the information about the presence/absence of each seed protein in the other genomes, and we can ask what are the proteins whose profiles are similar to those of proteins that have been previously implicated in the cell cycle. By measuring distances among different profiles using the Euclidean distance and an arbitrary threshold of 4.5, chosen as a compromise among the size of the result and stringency, we were able to retrieve over 350 proteins that tend to co-occur with at least one of the "seed" cell cycle proteins.

We can assume that if this set of proteins is indeed enriched in proteins with functions related to the cell cycle, it should be enriched in essential proteins (as cell cycle-related proteins are often essential) and eventually in proteins encoded by transcripts that change significantly their expression level during cell cycle progression. To check if this is indeed the case, we used experimental data obtained in *S. meliloti* and more precisely: essentiality data (diCenzo et al. 2018), differential gene expression analysis during the cell cycle (De Nisco et al. 2014), and differential gene expression in a CtrA depletion strain (Pini et al. 2015). Among the 367 proteins retrieved, 67 are also essential, which is a significant enrichment ($p = 1.4\text{E-}05$) with respect to random sampling genes in the genome. Additionally, 27 are in common with the set of cell cycle-dependent genes, and 5 belong to the set of genes whose expression level is significantly affected by the depletion of CtrA. Overall, we found that 86 of the genes that were found by our strategy belong to at least one of these three categories, and they are reported in Table 1.

Among these 86 genes we also retrieved 4 out of 14 cell cycle genes used as seeds (*cck*A, *gcrA*, *ccrM*, and *cpdR1*; Table 1 shaded in gray), and another well-known cell cycle gene, *sciP* coding for a CtrA inhibitory protein. Strictly connected with cell cycle regulation are those genes required for cell division, cell wall, and cell envelope biosynthesis. Peptidoglycan cell wall is fundamental for morphogenesis and survival of bacteria, and several genes which encode for enzymes required in peptidoglycan synthesis were retrieved (*murA,C,D,G* and *mrcA1*). FtsI and FtsW are involved in cell division: FtsI catalyzes cross-linking of the peptidoglycan cell wall at the division septum (Typas et al. 2012); FtsW is

Table 1 Proteins co-occurring with cell cycle proteins. In light gray proteins that are also seeds in this search. In the header of the table, Pini15 means the protein is also in the list of differentially expressed genes after CtrA depletion (Pini et al. 2015), diCenzo18 means the protein is also in the essential genes list from (diCenzo et al. 2018), DeNisco14 indicates overlapping with the list of genes that significantly change their expression level across cell cycle progression (De Nisco et al. 2014). Gray shading corresponds to proteins used as seed that are also recovered by their proximity with other seeds

Gene Code (Smeli1021)	Protein name	Pini15	diCenzo18	DeNisco14	Gene name	Description
sma0126	AAK64724			x	*cspA8*	CspA8 cold shock family protein
sma0244	AAK64787			x	–	Dehydrogenase, FAD-dependent
smc00007	CAC45473		x		*aroC*	Chorismate synthase
smc00016	CAC45484		x		*ispH*	4 hydroxy 3 methylbut 2 enyl diphosphate reductase
smc00021	CAC45498		x	x	*ccrM*	Adenine DNA methyltransferase
smc00034	CAC45525	x			–	Putative quinone oxidoreductase
smc00077	CAC45493		x		*thrC1*	Threonine synthase
smc00118	CAC45572		x		–	Hypothetical protein
smc00155	CAC46432		x		*aroF*	DAHP synthetase prtein
smc00161	CAC46425		x		*nadE*	NAD synthetase
smc00232	CAC46236		x		*glmU*	Bifunctional N acetylglucosamine 1 phosphate uridyltransferase/ Glucosamine 1 phosphate acetyltransferase
smc00333	CAC41690		x		*aroA*	3 phosphoshikimate 1 carboxyvinyl-transferase
smc00394	CAC41751		x		*guaA*	GMP synthase
smc00408	CAC41765		x		*uppP*	UDP pyrophosphate phosphatase
smc00471	CAC46381		x		*cckA*	Sensor histidine kinase transmembrane protein

(continued)

Table 1 (continued)

Gene Code (Smeli1021)	Protein name	Pini15	diCenzo18	DeNisco14	Gene name	Description
smc00522	CAC46324			x	*rhlE1*	Putative ATP-dependent RNA helicase
smc00530	CAC46314		x		–	Cysteine desulfurase activator complex subunit SufB
smc00531	CAC46313		x		–	ABC transporter ATP binding protein
smc00532	CAC46312		x		–	Hypothetical protein
smc00580	CAC45729		x		*pdxA*	4 hydroxythreonine 4 phosphate dehydrogenase
smc00582	CAC45731		x		*LptD*	Hypothetical protein
smc00583	CAC45732		x		*LptG*	Hypothetical protein
smc00637	CAC47313		x		*glmM*	Phosphoglucosamine mutase
smc00643	CAC47307		x		*purA*	Adenylosuccinate synthetase
smc00657	CAC47294	x		x	*sciP*	Conserved hypothetical protein
smc00696	CAC47255		x		*aroB*	3 dehydroquinate synthase
smc00701	CAC47250		x		*cobT*	Cobalamin biosynthesis protein
smc00723	CAC47228		x		*lysA*	Diaminopimelate DAP decarboxylase
smc00825	CAC45342		x		*gsh1*	Glutamate cysteine ligase precursor protein
smc00862	CAC45416		x		*Ipk*	4 diphosphocytidyl 2 C methyl D erythritol kinase
smc00985	CAC45439		x		*pdxR*	Oxidoreductase
smc01040	CAC46035		x		*ispDF*	Bifunctional 2 C methyl D erythritol 4 phosphate cytidylyltransferase/2 C methyl D erythritol 2 4 cyclodiphosphate synthase

(continued)

Table 1 (continued)

Gene Code (Smeli1021)	Protein name	Pini15	diCenzo18	DeNisco14	Gene name	Description
smc01048	CAC46043		x		*hfq*	RNA-binding protein Hfq
smc01109	CAC41848		x	x	*metK*	Probable S-adenosylmethionine synthetase
smc01138	CAC41819		x		–	ABC transporter ATP binding protein
smc01161	CAC41795		x		*dfp*	Bifunctional phosphopan-tothenoylcysteine Decarboxylase/ phosphopantothenate synthase
smc01183	CAC46189			x	*lexA*	Putative LexA repressor transcription regulator
smc01209	CAC46161		x		*coaD*	Phosphopantetheine adenylyltransferase
smc01215	CAC46155		x		*carB*	Carbamoyl phosphate synthase large subunit
smc01301	CAC45943		x		*rpmC*	50S ribosomal protein L29
smc01334	CAC45909		x		*mrcA1*	Penicillin binding 1A transmembrane protein
smc01343	CAC45900		x		*aroQ*	3 dehydroquinate dehydratase
smc01344	CAC45899		x		*accB*	Acetyl CoA carboxylase biotin carboxyl carrier protein subunit
smc01362	CAC45881		x		–	Glycerol 3 phosphate acyltransferase PlsY
smc01407	CAC46665		x		*pdxJ*	Pyridoxine 5 phosphate synthase
smc01569	CAC46899		x		*carA*	Carbamoyl phosphate synthase small subunit

(continued)

Table 1 (continued)

Gene Code (Smeli1021)	Protein name	Pini15	diCenzo18	DeNisco14	Gene name	Description
smc01585	CAC46909			x	*cspA3*	Putative cold shock transcription regulator
smc01772	CAC45789		x		*ribD*	Riboflavin biosynthesis protein
smc01781	CAC45798		x		–	Hypothetical protein
smc01784	CAC45801		x		*plsX*	Glycerol 3 phosphate acyltransferase PlsX
smc01860	CAC46761		x	x	*ftsI*	Probable peptidoglycan synthetase FtsI
smc01864	CAC46757		x		*murD*	UDP N acetylmuramoyl L alanyl D glutamate synthetase
smc01865	CAC46756		x		*ftsW*	Cell division protein FtsW
smc01866	CAC46755		x		*murG*	UDP diphospho muramoylpen-tapeptide Beta N acetylglu-cosaminyltrans-ferase
smc01867	CAC46754		x		*murC*	UDP N acetylmuramate L alanine ligase
smc02114	CAC46061	x			–	Putative hydrolase
smc02137	CAC45090		x		*argF1*	Ornithine car-bamoyltransferase
smc02139	CAC45088		x		*gcrA*	Hypothetical protein
smc02141	CAC45086			x	*phoU*	Probable phosphate transport system transcriptional regulator
smc02143	CAC45084			x	*pstA*	Putative phosphate transport system permease ABC transporter
smc02144	CAC45083			x	*pstC*	Phosphate ABC transporter permease

(continued)

Table 1 (continued)

Gene Code (Smeli1021)	Protein name	Pini15	diCenzo18	DeNisco14	Gene name	Description
smc02147	CAC41945			x	*phoR*	Probable phosphate regulon sensor histidine kinase transmembrane protein
smc02163	CAC41920		x		*pgi*	Glucose 6 phosphate isomerase
smc02252	CAC45128			x	*galE*	Probable UDP-glucose 4-epimerase
smc02305	CAC45181		x		*murA*	UDP N acetylglucosamine 1 carboxyvinyl-transferase
smc02377	CAC45605		x		*etf*	Electron transfer flavoprotein ubiquinone oxidoreductase
smc02560	CAC41431		x		*chvI*	Transcriptional regulator
smc02644	CAC45659			x	–	Putative transcriptional accessory protein (Tex; transcriptional accessory protein)
smc02678	CAC46920		x		*rph*	Hypothetical protein
smc02686	CAC46927		x		*prsA*	Ribose phosphate pyrophosphokinase
smc02848	CAC41560	x		x	–	Conserved hypothetical protein
smc03778	CAC47741		x		*nadD*	Nicotinic acid mononucleotide adenylyltransferase
smc03783	CAC47745		x		*ctpA*	Carboxy terminal processing protease precursor signal peptide protein
smc03809	CAC47771		x		–	Hypothetical protein
smc03820	CAC47782		x		–	Transcriptional regulator

(continued)

Table 1 (continued)

Gene Code (Smeli1021)	Protein name	Pini15	diCenzo18	DeNisco14	Gene name	Description
smc03826	CAC47788		x		*argG*	Argininosuccinate synthase
smc03888	CAC47850		x		*ispG*	4 hydroxy 3 methylbut 2 en 1 yl diphosphate synthase
smc03978	CAC47341		x		*tkt2*	Transketolase
smc04024	CAC47387			x	–	Membrane-bound lytic murein transglycosylase precursor
smc04043	CAC47405			x	*hutG*	Conserved hypothetical protein (HutG; N-formylglutamate amidohydrolase)
smc04044	CAC47406		x	x	*cpdR1*	Single domain response regulator
smc04083	CAC47897		x		*cynT*	Carbonic anhydrase
smc04270	CAC46517		x		–	Dehydrogenase
smc04318	CAC46636			x	*cspA1*	Cold shock transcriptional regulator
smc04346	CAC46647			x	*ilvC*	Ketol-acid reductoisomerase
smc04384	CAC47868			x	–	Putative oxidoreductase: GlcD; FAD/FMN-containing dehydrogenases

a lipid II flippase but could also work as a peptidoglycan polymerase in complex with a class B penicillin binding-protein (Taguchi et al. 2018). UppP is one of the key enzymes in the lipid II cycle of wall biosynthesis. CtpA has been proposed as a regulator of cell division and peptidoglycan biosynthesis in *C. crescentus* (Shapland et al. 2011). In Gram-negative bacteria the envelope biogenesis requires the synthesis of lipopolysaccharides and their delivery to the outer membrane. LptG and LptD are two proteins that are essential respectively for LPS biosynthesis and their localization in the outer membrane (Braun and Silhavy 2002; Wu et al. 2006). Moreover, UDP-N-acetylglucosamine-1-P is a precursor of both LPS and peptidoglycan, and its biosynthesis from fructose-6-P and glutamine involves only three enzymes (Rodríguez-Díaz et al. 2012), two of which were also identified by this strategy (*glmM* and *glmU*). Additionally, we retrieved genes—in particular *pgi* and *galE*—that are involved in gluconeogenesis, which has a primary role in the production of precursors for cell envelope biogenesis (Sperber and Herman 2017). The gene *galE* has been moreover implicated in cell division regulation in *Bacillus subtilis* (Gamba et al. 2015).

Isoprenoids are essential metabolites for growth and viability (they have roles in electron transport, cell wall, and membrane biosynthesis, etc.) (Pérez-Gil and Rodríguez-Concepción 2015). In Bacteria, isoprenoids are mainly produced through the MEP (methylerythritol 4-phosphate) pathway; three enzymes out of four, which are required for the transformation of MEP into DMAPP (dymethylallyl diphosphate), are putatively connected to cell cycle genes (*ispDF*, *ispG*, and *ispH*). Cell cycle regulation has to be strictly associated with cell metabolism and should rapidly adapt to different environmental stresses (Beaufay et al. 2015). It is therefore not surprising to find 10 genes, which have a role in the SOS response (*lexA*), RNA chaperones proteins (CspA1, CspA3, CspA8, and Hfq), proteins required in phosphate metabolism (PhoR, PhoU, PstA, and PstC) and exopolysaccharide production (ChvI). Notably, *cspA1* and *cspA8* are among those genes that undergo down-regulation in *S. meliloti* cells exposed to peptide NCR335 (Tiricz et al. 2013); NCR peptides are produced by legume plants and induce a terminal differentiation process in *S. meliloti* altering its normal cell cycle. We conclude that the strategy used to generate the gene list that we briefly analyzed above can be considered a valuable approach for the identification of genes related to a given phenotype and provides useful information for a focused experiment in the lab.

3 Cell Cycle Involvement of sncRNAs

Small RNA-dependent regulations are one of the hot topics of recent years and cell cycle students are also starting to explore their involvement in their system. Small RNAs (sRNA) are extremely versatile regulators in terms of the input–output response curve, and indeed they are able to originate variegated dynamics. Recent works have for instance demonstrated that integrated circuits with both transcriptional and sRNA-mediated regulations are able to produce bilinear responses, ultrasensitive switches, and bistability (Mitarai et al. 2008, 2009; Semsey et al. 2009; Liu et al. 2011). The discovery of sRNA involved in the regulation of the cell cycle in these bacteria has therefore a relevance to understanding how their activity affects the properties of the cell cycle.

In *Alphaproteobacteria*, sRNAs have been studied in several model systems, notably *Sinorhizobium meliloti*, *Rhodobacter* spp., *crescentus*, *Brucella* spp. and *A. tumefaciens* (Ulvé et al. 2007; Landt et al. 2008, 2010; Robledo et al. 2018). They generally play a role in response to stress conditions and activating specific regulatory programs in certain physiological conditions (Storz et al. 2011; Gottesman and Storz 2011).

In *S. meliloti* for instance, NfeR1 (Nodule formation efficiency RNA) is expressed under salt stress conditions and during symbiotic interaction. Loss of function of NfeR1 induces problems in osmo-adaptation and misregulation of the genes responding to high salt concentration. Because this stress response is highly linked to plant infection, NfeR1 has a major role in the infection process being severely impaired in the colonization of plants (Robledo et al. 2017).

In *Brucella*, *S. meliloti* and *Agrobacterium*, AbcR1 and AbcR2 have been characterized as a conserved family of sRNAs, generally involved in nutrient uptake systems (Overlöper et al. 2014; Torres-Quesada et al. 2014). In *Brucella* the deletion of both results in attenuation of the virulence in mice (Wilms et al. 2011; Sheehan and Caswell 2017).

Despite being a model system for the investigation of cell cycle, only a few examples of sRNAs have been linked to regulation of cell cycle in alphaproteobacteria. sRNAs have been studied in *C. crescentus* too, connecting their activity to the peculiar cell cycle of this organism (Landt et al. 2008; Schrader et al. 2014; Beroual et al. 2018). Results suggested that ncRNAs are intimately connected to the known regulatory network, playing a crucial modulatory role in cell cycle progression (Beroual et al. 2018).

The cell cycle is connected to sRNA also in *S. meliloti*; here EcpR1, a sRNA conserved across *Rhizobiales* that belongs to the stringent response, and whose expression is induced by various stress factors and during the stationary phase. EcpR1 overproduction led to cell elongation and increased DNA content, negatively affecting the levels of two master regulators in alphaproteobacteria, GcrA and DnaA (Robledo et al. 2015).

In *Agrobacterium* spp. a sRNA, named PmaR controls growth, motility, and virulence. In particular, PmaR targets the gene *ampC* coding for a beta-lactamase involved in ampicillin resistance (Borgmann et al. 2018).

The abundance and complexity of global regulation in the class of *Alphaproteobacteria* strongly suggest, as hypothesized for *Caulobacter* (Beroual et al. 2018) that their important role in controlling cell cycle will be clarified in the near future.

4 Toward Comparative Systems Biology of the Cell Cycle

The major conclusion of the comparative genomics analyses performed so far is that the ancestor of present-day *Alphaproteobacteria* very likely missed a *Caulobacter*-like cell cycle, but (most of the) proteins implementing this circuit were likely present in its genome. Thinking in an evolutionary way, we imagine that at some point, a novel system able to produce sustained oscillations evolved from a simpler, pre-existing circuit; the most likely ancestral function of the system was the regulation of motility and other processes; the coupling of this oscillating system with DNA replication and division originated the cell cycle: a novel system *emerged* from pre-existing genes (the genes required for the CtrA *oscillator*, cell division genes and DNA replication genes).

We think that the most important evolutionary questions in an evolutionary perspective are: How the transition to the first oscillator took place? Which were the key innovations that allowed this breakthrough? How many of these innovations were strictly required and which were instead optional? Can we find alternative arrangements still able to drive cell cycle progression? What are the evolutionary meaningful differences of the alternatives we are describing in different lineages?

Three milestones took place during this evolutionary path: (1) the origin of a biological oscillator driven by CtrA, and CtrA recruitment for the regulation/coordination of (2) DNA replication and (3) cell division. These three steps are potentially independent and the order in which they appeared is not easily predictable, even considering the most recent experiments outside the *Caulobacter/Rhizobia*; at the beginning, CtrA could have controlled cell division by simply integrating environmental clues, without the need for the circuit to undergo sustained oscillations; it could have done so for instance by integrating environmental signals to decide if the conditions were appropriate for division and might also have worked uncoupled from DNA replication; as in other bacteria, some degree of polyploidy would have guaranteed genome repartition among daughter cells. The ecology of this hypothetical ancestor could also have played a role; the strategy of *Caulobacter* could be appropriate for the oligotrophic freshwater environments where it is found the most: at division, one daughter cell stay attached and can divide shortly after, while the other swims away eventually reaching another appropriate area to settle down. This in principle reduces competition among daughter cells and allows to colonize novel areas.

Comparative genomics analyses are fundamental to guide experimental design and provide the tools for transferring functional information across organisms. By carrying out a similar set of experiments in different organisms, we will understand how the activities change in time during the cell cycle. However, it is hard to imagine that we will be able to elucidate how the system evolved without a detailed understanding of its dynamical properties.

Evolution selects genetic systems on the basis of their activity; in the case of the cell cycle, the most essential ability is to sustain oscillations.

Mathematical modeling of the cell cycle circuits characterized in different organisms can provide information about how the different regulations/arrangements affect evolutionary meaningful properties of the circuits, whereas static reconstructions cannot. Additional dynamical features can also be evolutionarily important, such as the robustness of the output with respect to changes in physical or biological parameters. In models of biological systems, each parameter could be considered as an evolutionary volatile quantity, therefore subject to changes as a consequence of mutations; indeed, one important property of biological system is their robustness to changes in the parameters (for instance, the affinity of a transcription factor for a binding site on the genome can change by mutation of the genomic locus or the binding site on the regulator; the maximum velocity of enzymes can also change as a function of mutations in the coding sequence, etc.). This introduces a simplification in the modeling as we seldom have all the experimental data required to identify the parameters with precision: following the above reasoning, we might expect that a model well describing a biological system should be robust with respect to parameter changes, i.e., we do not need fully parameterized models to start working with them.

Modeling could for instance show how the redundant regulations of CtrA and other proteins affect the dynamic properties of the cell cycle: artificial modulation of the expression of *divJ* in vivo was shown to modulate the mean duration and the associated noise of the cell division in *C. crescentus* (Lin et al. 2010); similar properties might be attributed to additional players of the cell cycle, for instance, one natural question could be how different regulatory schemes affect the dynamical properties of the cell cycle. Answering this question means being able to extrapolate important evolutionary insights on the selective pressures that shaped the system. Are the alternative arrangements all similarly robust with respect to evolutionary or environmental variations? Can we understand which features are responsible for optimizing such robustness?

Systems biology is today considered the most promising way to understand how living systems behave, since it produces realistic simulations of the most disparate biological systems; by integrating different sources of information into meaningful mathematical models, it allows to study the dynamical properties with a level of detail that is impossible to obtain through wet-lab experiments. While the study of a single model allows understanding its specific properties, it does not allow to elucidate the evolution of a system's structure in terms of its dynamical properties. That is, redundant or simplified regulations in different organisms can be described through experiment and comparative genomics, but understanding the effects of these differences on the properties of the system can best be achieved through a deep characterization of dynamic models.

We propose an approach that we call *Comparative Systems Biology*, whose aim is to introduce evolutionary thinking in modeling approaches. In the present case, mathematical models of the cell cycle in different alphaproteobacteria may be interpreted as independent evolutionary outcomes of an ancestral system, therefore providing hints on the reasons why during evolution certain structures emerged in different lineages. Are those alternatives equivalent? How they differ? Are the differences correlated with different needs?

Basically, we are convinced that a full understanding of the system requires integrating classical comparative genomics with experiments and modeling techniques to translate the regulatory circuits into systems of differential equations allowing to explore their dynamical properties at a detailed level not achievable by the experiments.

The different evolutionary paths represented by the models, corresponding to organisms from different taxa however evolving from a common ancestor, could be compared at first by their ability to produce stable oscillations and then one could study more specific properties.

One main difficulty in modeling biological systems concerns their parameterization. The number of unknowns, the strong non-linearity of these models, and several other reasons make the identification of such models difficult even with large amounts of omics data, therefore this approach is fruitful for only a few systems, and mainly in model organisms. However, from an evolutionary point of view as discussed above, no biological system has a fixed set of parameters, as mutations changes them continuously. We may therefore imagine building models and then studying their properties in a wide and biologically meaningful range of parameter values.

One modeling approach allowing to avoid a full parameterization of the models is the so-called structural kinetic modeling (Steuer et al. 2006). This approach does not require explicit rate functions for reactions and therefore it enables a thorough exploration of the parameter space looking for a combination of parameters that are able to fulfill our basic requirement, i.e., sustained oscillations.

As a proof of concept, we derived a minimal working model of the *C. crescentus* cell cycle core regulatory network. As it is very simple, we explicitly define the function corresponding to each reaction. The model contains three variables for CtrA, DivK, and DivL, respectively, and it is, therefore, a crude approximation with respect to the real system. Besides representing an approximation, the requirement for simplicity allows studying the fundamental structure of a circuit, as the simplest arrangement able to reproduce the known properties of the system. In our toy model, no distinction is made between phosphorylated and non-phosphorylated DivK and CtrA proteins; both proteins have to be considered in the active form. This simplification follows is usually applied in approaching the implementation of models where there are processes proceeding on very different time scales. In practice, one can here consider phosphorylation as immediate with respect to the time-scale of the model (changes in protein concentrations happen in a time scale of minutes, while phosphorylation is an enzymatic reaction for which we can assume relaxation toward a steady state in seconds). By using a more rigorous terminology, we apply the *quasi-steady state* approximation to phosphorylation, with respect to the relevant time-scale of the model. In the model, CtrA degradation is constant instead of being regulated by phosphorelay. The phosphorylation cascade controlled by DivL (the CckA/ChpT phosphorelay) is lumped in a single reaction affecting CtrA: when DivL is high it feeds the phosphorelay; when DivL is low the phosphorelay is off and it cannot activate CtrA. The positive term in Eq. (2) corresponds to lumping together both the phosphorelay and CtrA transcriptional regulation (which is here independent of CtrA itself, another simplification of the model).

We first identified a set of parameters allowing sustained oscillations and without considering biological plausibility. Once found a parameterization able to originate sustained oscillations, we explored how this property is affected by different parameters by performing parameter scans and checking the model for its ability to oscillate. In a real case study, the analysis should be performed much more exhaustively and parameters sampled in a biologically plausible range, but for this example, we decided to avoid complications and show a very simple and naïve toy model able to illustrate the concept. In summary, parameter scanning consists in changing the value of one parameter at a time and then checking if oscillations are still possible. This can be checked by first calculating the *Jacobian* of the model and then its eigenvalues, thus we can immediately know if the system will undergo sustained oscillations at a steady state. In this toy example, we find that the exponent of the Hill function used to model the positive effect of the phosphorelay over CtrA, is critical for oscillations. Multiple phosphotransfer steps make the

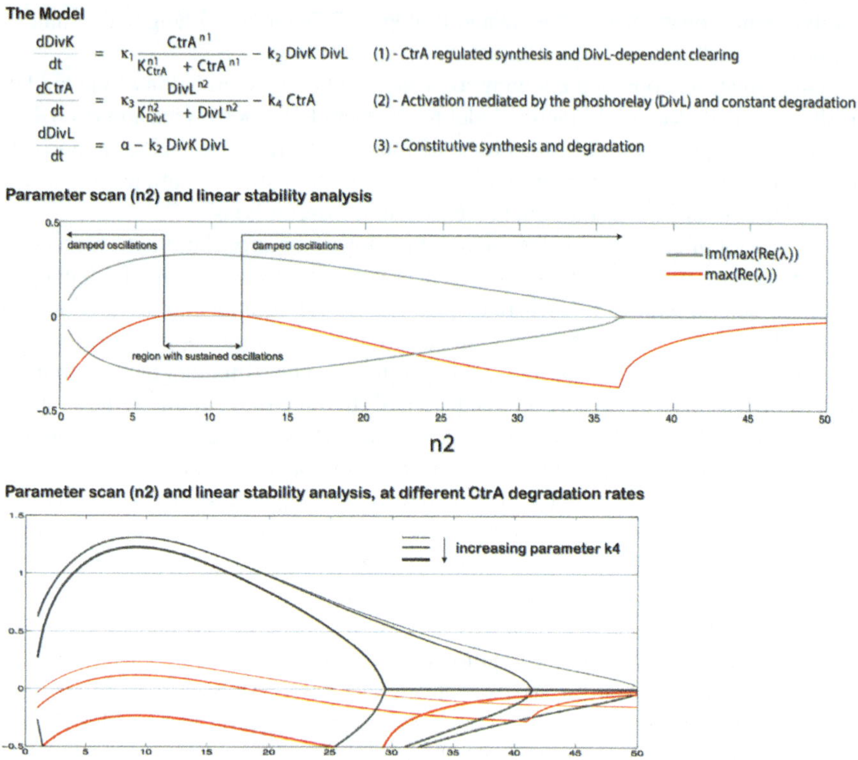

Fig. 3 Linear stability analysis of a toy model representing a minimal cell cycle regulatory network of *C. crescentus*. Given this model architecture, oscillations strongly depend on the activation speed of CtrA. In the second plot, we show how the n_2 range where the model oscillates changes when changing the degradation rate of CtrA (k_4): with a reduced CtrA degradation rate the oscillations are allowed also for smaller n_2 values

system ultrasensitive (Csikász-nagy et al. 2011), we may therefore conclude that with these parameters, the system would oscillate only with an ultrasensitive phosphorelay. Starting from this model one might additionally explore how the regulatory feedbacks on CtrA might change this situation for the effect of a single parameter over the behavior of the model can drastically change after modifying other parameter values. As an example, we repeated the n_2 parameter scan by first changing the degradation rate of CtrA (bottom chart in Fig. 3). Interestingly, when the degradation rate is smaller, the range of n_2 that allows for oscillations, is larger, i.e., the model is more robust to changes in the Hill exponent of the phosphorelay.

This kind of approach can therefore inform on the importance of certain parameters for the ability of the system to achieve a certain task; moreover, when several models describing the same system in different organisms are available, one could understand what are the differences that affect the system dynamics the most and how.

5 Conclusions

Possible perspectives follow from a better comprehension of the cell cycle regulatory machinery in the class *Alphaproteobacteria*; this taxonomic group comprises some important pathogens (such as *Bartonella* and *Brucella* spp.) for which there are indications that the cell cycle be interlaced with virulence (De Bolle et al. 2015); a deeper knowledge of the cell cycle in these organisms would imply additional information concerning their virulence, eventually disclosing new antibiotic targets or life stages where the bacterium is more/less sensitive to treatments; this can happen for instance if some of the antibiotic targets are cell cycle dependent such that only when they are expressed a pathogen is resistant/sensitive to a certain treatment. If this really happens, we can expect that in heterogeneous populations, where individual cells are distributed over the entire cell cycle, drug antibiotic treatment might have reduced utility.

Similarly, the cell cycle is affected during bacteroid development in symbiotic *S. meliloti*, and understanding how the plant interacts with the cell cycle of the symbiont might enable the production of more efficient crops, therefore requiring much less ammonia in the form of chemical fertilizers for growth. Synthetic Biology is moreover gaining more and more importance in the Biological Sciences, and this idea opens the possibility of manipulating such a fundamental process as the cell cycle to modulate its dynamical properties.

In this chapter, we grasped at an ambitious task: putting together Systems and Evolutionary Biology to suggest an approach that might provide a deep understanding of the evolution and dynamical properties of a fundamental system in Bacteria. This question is only approachable with a multi-disciplinary strategy where comparative genomics and mathematical modeling work in synergy to provide a global view of the properties of the regulatory circuits governing the system in alphaproteobacteria from different regions of the phylogenetic tree as to provide the most information.

References

Beaufay F, Coppine J, Mayard A et al (2015) A NAD-dependent glutamate dehydrogenase coordinates metabolism with cell division in *Caulobacter crescentus*. EMBO J 34:1786–1800. https://doi.org/10.15252/embj.201490730

Belas R, Horikawa E, Aizawa S-I, Suvanasuthi R (2009) Genetic determinants of *Silicibacter* sp. TM1040 motility. J Bacteriol 191:4502–4512. https://doi.org/10.1128/JB.00429-09

Bellefontaine A-F, Pierreux CE, Mertens P et al (2002) Plasticity of a transcriptional regulation network among alpha-proteobacteria is supported by the identification of CtrA targets in *Brucella abortus*. Mol Microbiol 43:945–960

Beroual W, Brilli M, Biondi EG (2018) Non-coding RNAs potentially controlling cell cycle in the model *Caulobacter crescentus*: a bioinformatic approach. Front Genet 9:1–9. https://doi.org/10.3389/fgene.2018.00164

Biondi EG, Skerker JM, Arif M et al (2006) A phosphorelay system controls stalk biogenesis during cell cycle progression in *Caulobacter crescentus*. Mol Microbiol 59:386–401. https://doi.org/10.1111/j.1365-2958.2005.04970.x

Bird TH, MacKrell A (2011) A CtrA homolog affects swarming motility and encystment in *Rhodospirillum centenum*. Arch Microbiol 193:451–459. https://doi.org/10.1007/s00203-011-0676-y

Borgmann J, Schäkermann S, Bandow JE, Narberhaus F (2018) A small regulatory RNA controls cell wall biosynthesis and antibiotic resistance. MBio 9:1–17. https://doi.org/10.1128/mBio.02100-18

Braun M, Silhavy TJ (2002) Imp/OstA is required for cell envelope biogenesis in *Escherichia coli*. Mol Microbiol 45:1289–1302

Brilli M, Fondi M, Fani R et al (2010) The diversity and evolution of cell cycle regulation in alpha-proteobacteria: a comparative genomic analysis. BMC Syst Biol 4. https://doi.org/10.1186/1752-0509-4-52

Cheng Z, Miura K, Popov VL et al (2011) Insights into the CtrA regulon in development of stress resistance in obligatory intracellular pathogen *Ehrlichia chaffeensis*. Mol Microbiol 82:1217–1234. https://doi.org/10.1111/j.1365-2958.2011.07885.x

Csikász-nagy A, Cardelli L, Soyer OS et al (2011) Response dynamics of phosphorelays suggest their potential utility in cell signalling. J R Soc Interface 8:480–488. https://doi.org/10.1098/rsif.2010.0336

De Bolle X, Crosson S, Matroule J-YY, Letesson J-JJ (2015) *Brucella abortus* cell cycle and infection are coordinated. Trends Microbiol 23:812–821. doi: https://doi.org/10.1016/j.tim.2015.09.007

De Nisco NJ, Abo RP, Wu CM et al (2014) Global analysis of cell cycle gene expression of the legume symbiont Sinorhizobium meliloti. Proc Natl Acad Sci U S A 111:3217–3224. https://doi.org/10.1073/pnas.1400421111

Deghelt M, Mullier C, Sternon J-F et al (2014) G1-arrested newborn cells are the predominant infectious form of the pathogen *Brucella abortus*. Nat Commun 5:4366. https://doi.org/10.1038/ncomms5366

diCenzo GC, Benedict AB, Fondi M, et al. (2018) Robustness encoded across essential and accessory replicons of the ecologically versatile bacterium *Sinorhizobium meliloti*

Domian IJ, Quon KC, Shapiro L (1996) The control of temporal and spatial organization during the *Caulobacter* cell cycle. Curr Opin Genet Dev 6:538–544. https://doi.org/10.1016/S0959-437X(96)80081-5

Fioravanti A, Fumeaux C, Mohapatra SS et al (2013) DNA binding of the cell cycle transcriptional regulator GcrA depends on N6-adenosine methylation in *Caulobacter crescentus* and other alphaproteobacteria. PLoS Genet 9:e1003541. https://doi.org/10.1371/journal.pgen.1003541

Francez-Charlot A, Kaczmarczyk A, Vorholt J a. (2015) The branched CcsA/CckA-ChpT-CtrA phosphorelay of *Sphingomonas melonis* controls motility and biofilm formation. Mol Microbiol 97:47–63. https://doi.org/10.1111/mmi.13011

Francis N, Poncin K, Fioravanti A et al (2017) CtrA controls cell division and outer membrane composition of the pathogen *Brucella abortus*. Mol Microbiol 103:780–797. https://doi.org/10.1111/mmi.13589

Gamba P, Rietkötter E, Daniel RA, Hamoen LW (2015) Tetracycline hypersensitivity of an *ezrA* mutant links GalE and TseB (YpmB) to cell division. Front Microbiol 6. https://doi.org/10.3389/fmicb.2015.00346

Gora KG, Cantin A, Wohlever M et al (2013) Regulated proteolysis of a transcription factor complex is critical to cell cycle progression in *Caulobacter crescentus*. Mol Microbiol 87:1277–1289. https://doi.org/10.1111/mmi.12166

Gottesman S, Storz G (2011) Bacterial small RNA regulators: versatile roles and rapidly evolving variations. Cold Spring Harb Perspect Biol 3:1–16. https://doi.org/10.1101/cshperspect.a003798

Greene SESE, Brilli M, Biondi EGEG, Komeili A (2012) Analysis of the CtrA pathway in *Magnetospirillum* reveals an ancestral role in motility in alphaproteobacteria. J Bacteriol 194:2973–2986. https://doi.org/10.1128/JB.00170-12

Hallez R, Bellefontaine A, Letesson J et al (2004) Morphological and functional asymmetry in a-proteobacteria. Trends Microbiol 12:361–365. https://doi.org/10.1016/j.tim.2004.06.002

Joshi KK, Chien P (2016) Regulated proteolysis in bacteria: *Caulobacter*. Annu Rev Genet 50:423–445. https://doi.org/10.1146/annurev-genet-120215-035235

Kim HS, Caswell CC, Foreman R et al (2013a) The *Brucella abortus* general stress response system regulates chronic mammalian infection and is controlled by phosphorylation and proteolysis. J Biol Chem 288:13906–13916. https://doi.org/10.1074/jbc.M113.459305

Kim J, Heindl JE, Fuqua C (2013b) Coordination of division and development influences complex multicellular behavior in *Agrobacterium tumefaciens*. PLoS One 8:e56682. https://doi.org/10.1371/journal.pone.0056682

Kobayashi H, De Nisco NJ, Chien P, et al (2009) *Sinorhizobium meliloti* CpdR1 is critical for co-ordinating cell cycle progression and the symbiotic chronic infection. Mol Microbiol 73:586–600. doi: https://doi.org/10.1111/j.1365-2958.2009.06794.x

Komatsu H, Yamamoto J, Suzuki H et al (2018) Involvement of the response regulator CtrA in the extracellular DNA production of the marine bacterium *Rhodovulum sulfidophilum*. J Gen Appl Microbiol 64:103–107. https://doi.org/10.2323/jgam.2017.08.003

Landt SG, Abeliuk E, McGrath PT et al (2008) Small non-coding RNAs in *Caulobacter crescentus*. Mol Microbiol 68:600–614. https://doi.org/10.1111/j.1365-2958.2008.06172.x

Landt SG, Lesley JA, Britos L, Shapiro L (2010) CrfA, a small noncoding RNA regulator of adaptation to carbon starvation in *Caulobacter crescentus*. J Bacteriol 192:4763–4775. https://doi.org/10.1128/JB.00343-10

Lang AS, Beatty JT (2001) The gene transfer agent of *Rhodobacter capsulatus* and "constitutive transduction" in prokaryotes. Arch Microbiol 175:241–249. https://doi.org/10.1007/s002030100260

Lang AS, Beatty JT (2006) Importance of wide spread gene transfer agent genes in alphaproteobacteria. Trends Microbiol 15:54–62

Laub MT, Chen SL, Shapiro L, Mcadams HH (2002) Genes directly controlled by CtrA, a master regulator of the *Caulobacter* cell cycle. Proc Natl Acad Sci U S A 99:4632–4637. https://doi.org/10.1073/pnas.062065699

Li X, Nicklas RB (1995) Mitotic forces control a cell-cycle checkpoint. Nature 373:630–632

Lin Y, Crosson S, Scherer NF (2010) Single-gene tuning of *Caulobacter* cell cycle period and noise, swarming motility, and surface adhesion. Mol Syst Biol 6:445. https://doi.org/10.1038/msb.2010.95

Liu D, Chang X, Liu Z et al (2011) Bistability and oscillations in gene regulation mediated by small noncoding RNAs. PLoS One 6:e17029. https://doi.org/10.1371/journal.pone.0017029

Mercer RG, Callister SJ, Lipton MS et al (2010) Loss of the response regulator CtrA causes pleiotropic effects on gene expression but does not affect growth phase regulation in *Rhodobacter capsulatus*. J Bacteriol 192:2701–2710. https://doi.org/10.1128/JB.00160-10

Mitarai N, Andersson AMC, Krishna S, et al. (2008) Efficient degradation and expression prioritization with small RNAs. Arxiv.

Mitarai N, Benjamin J-AM, Krishna S et al (2009) Dynamic features of gene expression control by small regulatory RNAs. Proc Natl Acad Sci U S A 106:10655–10659. https://doi.org/10.1073/pnas.0901466106

Nagao N, Hirose Y, Misawa N et al (2015) Complete genome sequence of *Rhodovulum sulfidophilum* DSM 2351, an extracellular nucleic acid-producing bacterium. Genome Announc 3:3–4. https://doi.org/10.1128/genomea.00388-15

Overlöper A, Kraus A, Gurski R et al (2014) Two separate modules of the conserved regulatory RNA AbcR1 address multiple target mRNAs in and outside of the translation initiation region. RNA Biol 11:624–640. https://doi.org/10.4161/rna.29145

Penterman J, Abo RP, De Nisco NJ et al (2014) Host plant peptides elicit a transcriptional response to control the *Sinorhizobium meliloti* cell cycle during symbiosis. Proc Natl Acad Sci U S A 111:3561–3566. https://doi.org/10.1073/pnas.1400450111

Pérez-Gil J, Rodríguez-Concepción M (2015) Metabolic plasticity for isoprenoid biosynthesis in bacteria. Biochem J 452:19–25. https://doi.org/10.1042/bj20121899

Pini F, Frage B, Ferri L et al (2013) The DivJ, CbrA and PleC system controls DivK phosphorylation and symbiosis in *Sinorhizobium meliloti*. Mol Microbiol 90:54–71. https://doi.org/10.1111/mmi.12347
Pini F, De Nisco NJNJ, Ferri L et al (2015) Cell cycle control by the master regulator CtrA in *Sinorhizobium meliloti*. PLoS Genet 11:e1005232. https://doi.org/10.1371/journal.pgen.1005232
Rivera-Osorio A, Osorio A, Poggio S et al (2018) Architecture of divergent flagellar promoters controlled by CtrA in *Rhodobacter sphaeroides*. BMC Microbiol 18:1–14. https://doi.org/10.1186/s12866-018-1264-y
Robledo M, Frage B, Wright PR, Becker A (2015) A stress-induced small RNA modulates alpha-rhizobial cell cycle progression. PLoS Genet 11:e1005153. https://doi.org/10.1371/journal.pgen.1005153
Robledo M, Peregrina A, Millán V et al (2017) A conserved α-proteobacterial small RNA contributes to osmoadaptation and symbiotic efficiency of rhizobia on legume roots. Environ Microbiol 19:2661–2680. https://doi.org/10.1111/1462-2920.13757
Robledo M, Schlüter JP, Loehr LO et al (2018) An sRNA and cold shock protein homolog-based feedforward loop post-transcriptionally controls cell cycle master regulator ctrA. Front Microbiol 9:1–14. https://doi.org/10.3389/fmicb.2018.00763
Rodríguez-Díaz J, Rubio-del-Campo A, Yebra MJ (2012) Regulatory insights into the production of UDP-N-acetylglucosamine by *Lactobacillus casei*. Bioengineered 3:339–342
Schrader JM, Zhou B, Li G-W et al (2014) The coding and noncoding architecture of the *Caulobacter crescentus* genome. PLoS Genet 10:e1004463. https://doi.org/10.1371/journal.pgen.1004463
Semsey S, Benjamin J-AM, Mitarai N et al (2009) Dynamics of the recovery from sRNA-mediated gene silencing. Cell Cycle 8:2863–2864
Shapland EB, Reisinger SJ, Bajwa AK, Ryan KR (2011) An essential tyrosine phosphatase homolog regulates cell separation, outer membrane integrity, and morphology in *Caulobacter crescentus*. J Bacteriol 193:4361–4370. https://doi.org/10.1128/JB.00185-11
Sheehan LM, Caswell CC (2017) A 6-nucleotide regulatory motif within the AbcR small RNAs of *Brucella abortus* mediates host-pathogen interactions. MBio 8:1–15. https://doi.org/10.1128/mbio.00473-17
Slingsby C, Wistow GJ, Clark AR (2013) Evolution of crystallins for a role in the vertebrate eye lens. Protein Sci 22:367–380. https://doi.org/10.1002/pro.2229
Sperber AM, Herman JK (2017) Metabolism shapes the cell. J Bacteriol 199:1–14. https://doi.org/10.1128/jb.00039-17
Steuer R, Gross T, Selbig J, Blasius B (2006) Structural kinetic modeling of metabolic networks. Proc Natl Acad Sci U S A 103:11868–11873. https://doi.org/10.1073/pnas.0600013103
Storz G, Vogel J, Wassarman KMM, Storz G (2011) Regulation by small RNAs in bacteria: expanding frontiers. Mol Cell 43:880–891. https://doi.org/10.1016/j.molcel.2011.08.022
Subramanian K, Paul MR, Tyson JJ (2015) Dynamical localization of DivL and PleC in the asymmetric division cycle of *Caulobacter crescentus*: a theoretical investigation of alternative models. PLoS Comput Biol 11:e1004348. https://doi.org/10.1371/journal.pcbi.1004348
Taguchi A, Welsh MA, Marmont LS et al (2018) FtsW is a peptidoglycan polymerase that is activated by its cognate penicillin-binding proteins. bioRxiv. https://doi.org/10.1101/358663
Tiricz H, Szücs A, Farkas A et al (2013) Antimicrobial nodule-specific cysteine-rich peptides induce membrane depolarization-associated changes in the transcriptome of *Sinorhizobium meliloti*. Appl Environ Microbiol 79:6737–6746. https://doi.org/10.1128/AEM.01791-13
Torres-Quesada O, Reinkensmeier J, Schluẗer JP et al (2014) Genome-wide profiling of Hfq-binding RNAs uncovers extensive post-transcriptional rewiring of major stress response and symbiotic regulons in *Sinorhizobium meliloti*. RNA Biol 11:563–579. https://doi.org/10.4161/rna.28239
Typas A, Banzhaf M, Gross CA, Vollmer W (2012) From the regulation of peptidoglycan synthesis to bacterial growth and morphology. Nat Rev Microbiol 10:123–136. https://doi.org/10.1038/nrmicro2677

Ulvé VM, Sevin EW, Chéron A, Barloy-Hubler F (2007) Identification of chromosomal alpha-proteobacterial small RNAs by comparative genome analysis and detection in *Sinorhizobium meliloti* strain 1021. BMC Genomics 8:467. https://doi.org/10.1186/1471-2164-8-467

Wang H, Ziesche L, Frank O et al (2014) The CtrA phosphorelay integrates differentiation and communication in the marine alphaproteobacterium *Dinoroseobacter shibae*. BMC Genomics 15:130. https://doi.org/10.1186/1471-2164-15-130

Wheeler RT, Shapiro L (1999) Differential localization of two histidine kinases controlling bacterial cell differentiation. Mol Cell 4:683–694

Wilms I, Voss B, Hess WR et al (2011) Small RNA-mediated control of the *Agrobacterium tumefaciens* GABA binding protein. Mol Microbiol 80:492–506. https://doi.org/10.1111/j.1365-2958.2011.07589.x

Wu T, McCandlish AC, Gronenberg LS et al (2006) Identification of a protein complex that assembles lipopolysaccharide in the outer membrane of *Escherichia coli*. Proc Natl Acad Sci U S A 103:11754–11759. https://doi.org/10.1073/pnas.0604744103

Zan J, Heindl JE, Liu Y et al (2013) The CckA-ChpT-CtrA phosphorelay system is regulated by quorum sensing and controls flagellar motility in the marine sponge symbiont *Ruegeria sp.* KLH11. PLoS One 8:e66346. https://doi.org/10.1371/journal.pone.0066346

Temporal Control of Promoter Activity During the *Caulobacter* Cell Cycle

Marie Delaby and Patrick H. Viollier

Abstract

A cascade of cellular events must occur to allow cells to complete one round of cell division. Such a successful cell division cycle relies on the predetermined and sequential production of specific proteins that execute dedicated functions. Protein production is typically governed by transcriptional control occurring at the promoter of the genes encoding the proteins whose function are needed at a specific time in the cell cycle. Here we review the basis for the cell-cycle-controlled promoter activation in the synchronizable model bacterium *Caulobacter crescentus*, a Gram-negative alpha-proteobacterium. We detail which promoters fire at the same time and we reason why this is the case.

1 Introduction

In this chapter we review the regulatory path controlling the differential activity of selected promoters during the cell cycle of *Caulobacter crescentus*. A precondition to studying gene activation/repression during the cell cycle is the ability to synchronize populations of cells. *C. crescentus* has been used as a prominent model to study cell cycle as it is genetically tractable and can be easily synchronized by density gradient centrifugation on the basis of capsulation properties on different cell cycle stages (Ardissone et al. 2014).

C. crescentus is a Gram-negative (diderm) bacterium living in oligotrophic aquatic environments. Like many alpha-proteobacteria, *C. crescentus* undergoes an asymmetric cell division (Hallez et al. 2004), giving two progeny cells with

M. Delaby · P. H. Viollier (✉)
Faculty of Medicine, Department of Microbiology and Molecular Medicine/Centre Médical Universitaire, University of Geneva, Genève 4, Switzerland
e-mail: Patrick.Viollier@unige.ch

E. Biondi (ed.), *Cell Cycle Regulation and Development in Alphaproteobacteria*,
https://doi.org/10.1007/978-3-030-90621-4_2

distinct morphologies and developmental fates. The two daughter cells feature different transcriptional programs and regulatory fluxes controlling them (Laub et al. 2007). The replicative stalked cell is sessile and encapsulated. The adventurous swarmer cell is piliated, flagellated and possesses a chemotaxis apparatus at one pole. It resides in a non-replicative G1-like state and must differentiate into the stalked cell before proceeding to division (Fig. 1). During this morphological transition, the pili are retracted (Ellison et al. 2017), the chemosensory machines are lost, and the flagellum is shed from the old pole. A stalk elaborates from the vacated old cell pole while the cell starts chromosome replication. It grows into an asymmetric predivisional cell by building a new flagellum, chemosensory and pilus secretion complexes at the pole opposite the stalk. Once chromosome segregation is completed, cytokinesis can ensue with the release of two daughter cells, a stalked daughter cell and a swarmer daughter cell having differences in key proteins and mRNAs and this is also reflected in the differential transcriptional regulation during the cell cycle observed in synchronized populations.

Control of protein levels imposed through transcript abundance is a key element underpinning these developmental and morphological transitions during the cell cycle, although additional layers of regulation acting at the level of translation, protein stability, protein phosphorylation, and/or protein localization are often superimposed. Due to the short half-life of most transcripts in bacteria, determining changes in transcripts (mRNA) levels during the *C. crescentus* cell cycle has emerged as a straightforward and reliable proxy to describe developmentally regulated mRNA synthesis using systematic approaches. These approaches reveal between a third to a fifth of total mRNA of *C. crescentus* to fluctuate in abundance during cell cycle progression. Here, we focus on those mRNAs that are regulated at the level of synthesis, specifically discuss those that are transcribed from cell-cycle-controlled promoters in *C. crescentus*.

The first global view of transcriptional control during the cell cycle of synchronized *C. crescentus* cells was given by an analysis conducted on DNA microarrays probing steady-state mRNA from which they identified 553 cell-cycle-regulated genes (Laub et al. 2000) (Fig. 6). This study was the first one to capture the transcriptional regulatory network that controls the *C. crescentus* cell cycle. Further global transcriptomic analysis (Laub et al. 2000; Fang et al. 2013; McGrath et al. 2007) identified around 1500 cell-cycle-regulated transcripts via RNA deep sequencing (RNA-seq). These studies determine the *steady-state* levels of mRNAs and are generally assumed to reflect the rate of synthesis (transcription), even though regulation of transcript stability can also contribute to limited mRNA abundance during the cell cycle. Further confounding the use of RNA-seq data for inferences on transcription (initiation) and promoter activity is that it is not always possible to assign transcription start points (TSPs) from such data, as transcripts may be cleaved at the 5′ end. Finally, one must bear in mind that cell-cycle-controlled mRNA abundance does not imply change in protein concentration or activity during the cell cycle, as proteins are subject to variable synthesis and degradation, while activities may be regulated at the post-translational level such as folding, covalent maturation/modification, and by co-factor availabilities.

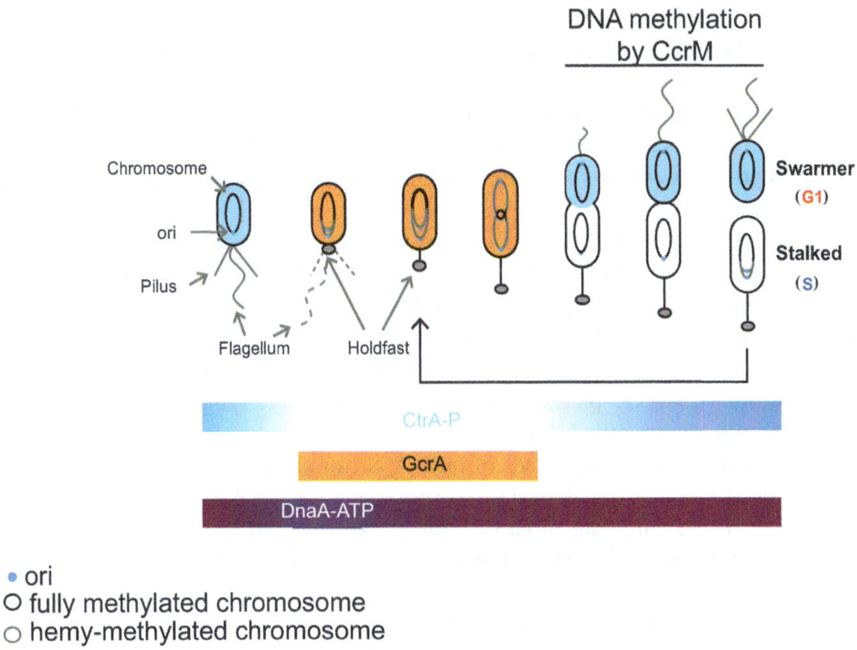

Fig. 1 Cell cycle progression in *C. crescentus* and the conserved cell cycle regulators. *C. crescentus* undergoes an asymmetric cell division giving at the end of each cell cycle two progeny cells with distinct morphologies and fates: the sessile replicating stalked cell that is encapsulated and the adventurous quiescent swarmer cell that resides in G1-like non-replicative state and must differentiate into the stalked cell before proceeding to division. The two daughter cells feature different transcriptional and developmental programs orchestrated by three main transcriptional regulators. The swarmer cell is piliated, flagellated and possesses a chemotaxis apparatus at one pole. These apparatuses are synthetized in predivisional cells and the expression of flagellar, pili, and chemotaxis genes requires the essential DNA-binding response regulator CtrA (blue) activity. In the swarmer stage, CtrA binds to the origin of replication (blue dots, *Cori*) and suppresses the initiation of DNA activated by DnaA (purple) during the swarmer to stalked transition when CtrA is proteolyzed. During this developmental change, the pili are retracted, the chemotaxis receptors and the flagellum present at the old pole are lost and replaced by the stalk. In early S-phase, the cell starts the chromosome replication and 5′-GAnTC-3′ sites methylated by the CcrM methyltransferase are converted from fully-methylated state to hemi-methylated state. By contrast, DnaA will activate the conserved cell-cycle-regulated transcription factor GcrA (orange) that directly controls the expression of CtrA. As a result, the stalked cell grows and becomes an asymmetric predivisional cell by building a new flagellum and pilus secretion system at the pole opposite the stalk. Once chromosome segregation is completed, cell division occurs, the two daughter cells are liberated and can start a new round of cell cycle

Some genes appear to be cell cycle regulated and their protein levels are constant (Brun 2001). In order to these forms of regulation, ribosome profiling was coupled to RNA-Seq to decipher the role of the translational control (Schrader et al. 2016) and to proteomics by high-throughput mass spectrometry. Moreover, variations of RNA-seq have been used to determine TSPs in *C. crescentus WT* and mutant strains on the basis of 5′ triphosphate end (5′-ppp) that is present of naïve (primary) transcripts, while processed transcripts feature a monophosphate (5′-p) or hydroxyl group (5′-OH) at their 5′ end (Ardissone et al. 2016). TSP-probing by RNA-seq and/or simply chromatin precipitation coupled to deep-sequencing experiments with antibodies to transcription factors or to components of the transcription initiation machinery are ideal proxies to study cell-cycle-regulated promoters and the factors controlling their firing.

Below we will summarize our current knowledge of such developmental promoter control during the *C. crescentus* cell cycle and the transcriptional reprogramming that occur in response to changing environmental conditions. We detail the functions, transcripts, and transcriptional factors that are known to be active in three consecutive cell cycle phases: early S-phase, late S-phase (predivisional), and (postdivisional) G1-phase, as well as the events underlying the developmental reprogramming during the G1➔S transition.

Three general temporal classes of mRNA accumulation (early S-, late S-phase, and G1-phase) are easily discernible. As a general rule, it is now clear that the transcripts that accumulate in early S-phase are required for the early events of cell division (septal growth), cell envelope extension cell and for setting up the polarity for polar structures, as well those that function in the later events of DNA replication/metabolism. Late S-phase is concerned with the completion of cell division (cell separation) and the implementation of polarity, i.e. the assembly of trans-envelope machines, specifically those that are then required in the subsequent G1 (swarmer) daughter cell, such as the flagellar and pilus assembly machines, chemosensory apparatus, holdfast secretion, and attachment proteins. Recently, it has been proposed that cell-cycle transcripts are controlled by three main regulatory modules, the CtrA/MucR, the GcrA/CcrM, and finally the CtrA/SciP modules that will successively control transcription of the G1-, S-, and late S-/G2-phase (Panis et al. 2015) (Fig. 2).

1.1 Early S-Phase Transcriptional Control

Under replete environmental conditions, the swarmer cell differentiates into the replication-competent stalked cell before proceeding to division (Fig. 1). The early S-phase is the period of DNA synthesis that is controlled by the DNA-binding protein DnaA. It is also tightly regulated at the transcriptional level by the GcrA/CcrM module that activates the S-phase promoters in order to ensure the presence of transcripts required for the early events of cell division (septal growth) and cell envelope extension cell and setting up the polarity for polar structures, as well those that function in the later events of DNA replication/metabolism (Fig. 3).

1.1.1 DnaA and Its Targets

Among bacteria, the highly conserved DnaA protein is the central player implicated in DNA replication (Katayama et al. 2010). An important feature of *C. crescentus* is that it only replicates once per cell cycle and only the stalk cell is replicative,

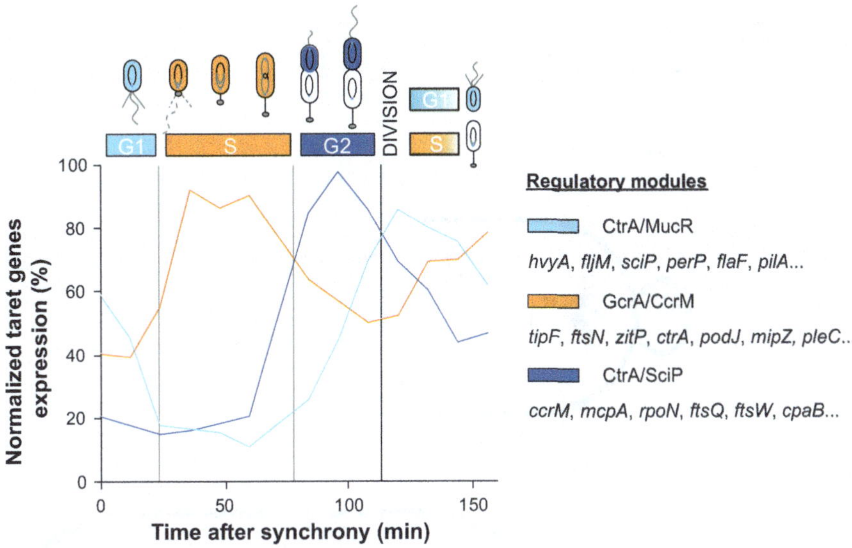

Fig. 2 Fluctuation and fine-tuning of the transcripts through the regulatory modules during cell cycle. CtrA is implicated in two different transcriptional modules, the CtrA/MucR (light blue) and the CtrA/SciP (dark blue) that will sequentially control transcription in G1- and late S-/G2-phase, respectively. GcrA (orange) controls early S-phase transcripts. For each module, a list of validated target genes was generated based on the ChIP-seq, RNA-seq, and microarrays data from Fiebig et al. (2014), Fioravanti et al. (2013), Fumeaux et al. (2014), Gora et al. (2010), Holtzendorff et al. (2004), Murray et al. (2013), Tan et al. (2010). Adapted from Panis et al. (2015)

harboring a single polarly localized origin of replication (referred to *Cori* hereafter) and implicating that replication is silenced within the swarmer cell. Replication has to be temporally and specially regulated.

These roles are dedicated to DnaA and the CtrA, respectively (Jonas et al. 2011). Besides its role as a transcription factor (see below), CtrA is able to bind to the *Cori* in order to silence it (Quon et al. 1998) (see part b)i.). DnaA is an AAA+ (ATPases associated with diverse cellular activities) protein and both DnaA-ATP and DnaA-ADP can bind to the *Cori*. However, studies in *E. coli* showed that only the DnaA-ATP is active for replication initiation as it forms stable oligomerized filaments (Duderstadt et al. 2011) (Fig. 3). Binding of DnaA to *Cori* is mediated through seven DnaA boxes found in *Cori* that are essential for the replication activity (Taylor et al. 2011; Shaheen et al. 2009), two have moderate affinity for DnaA and the five others have low affinity. In addition to these DnaA-boxes the *Cori* has several strong CtrA-binding sites overlapping partially with the latter ones. The presence of both DnaA- and CtrA-binding sites explains the dual control of the replication in *C. crescentus* as well as the competition that happens between the two regulators during the G1-phase (Quon et al. 1998; Taylor et al. 2011).

Besides its major role in replication initiation, DnaA is also a transcriptional regulator (Hottes et al. 2005). In vitro studies showed that in *C. crescentus* DnaA

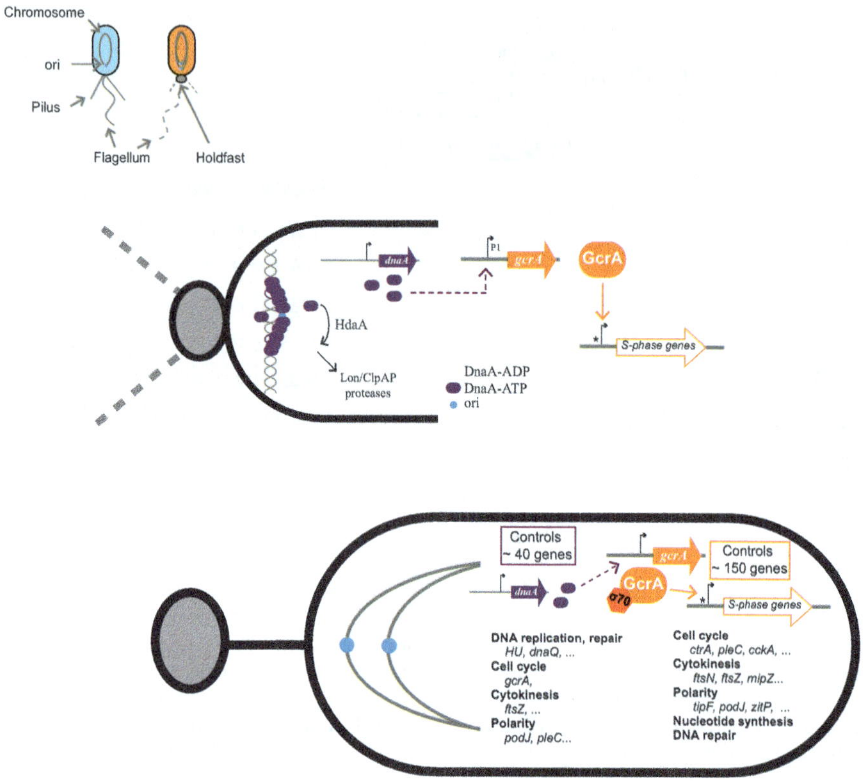

Fig. 3 Schematic of the *C. crescentus* early S-phase transcriptional control. During the G1- to S-phase transition, CtrA-P will be degraded by the ClpXP protease to allow the initiation of DNA replication, while the flagellum will be ejected, and pili will be retracted. The AAA+ ATPase DnaA (purple) promotes DNA replication by binding to the DnaA boxes at the *Cori* (blue dot) and promotes the recruitment of the replisome. Besides its activity of replication initiator, DnaA appears to be a transcriptional regulator and controls transcription directly or indirectly of around 40 genes implicated in DNA replication and repair, cytokinesis, polarity, and cell cycle regulation with the activation of the cell cycle regulator GcrA (orange), in which 13 harbor DnaA boxes. GcrA accumulates in early S-phase and activates transcription from 5′GANTC3′ methylated (*) S-phase promoters of *ctrA* (P1), *ftsN*, *ftsZ*, *tipF*, and *zitP* along with σ^{70} RNA polymerase holoenzyme. For each transcriptional regulator, a list of validated target genes was generated from ChIP-seq datasets for GcrA (Murray et al. 2013; Haakonsen et al. 2015) and microarrays for DnaA (Hottes et al. 2005)

binds to at least 13 promoters regions (Hottes et al. 2005; Collier et al. 2006; Fernandez-Fernandez et al. 2011), including the *gcrA* (Fig. 3, upper part) and *ftsZ* promoters, two genes encoding proteins require to the activation of S-phase promoter and cell division, respectively. Moreover, study on one hyper-replicative mutant of DnaA, the $DnaA^{R357A}$ mutated in its AAA+ domain, shows that this domain and thus the nucleotide-bound state of DnaA may not be implicated in its activity as a transcriptional factor (Fernandez-Fernandez et al. 2011). However, it is difficult to really define the DnaA (mutant) regulon, as ChIP-seq data are missing.

1.1.2 Regulation of DnaA

Despite being an unstable protein, degraded by the Lon protease, DnaA levels are relatively constant during *C. crescentus* cell cycle (Jonas et al. 2013). This suggests that the transcriptional control and steady-state level of DnaA are less likely to be solely responsible for the single round of replication even if it can be responsible for regeneration of DnaA-ATP and, therefore, the regulation of the DnaA activity is more likely responsible for the replication periodicity. Indeed, similar to what was shown in *E. coli* (Skarstad and Katayama 2013; Camara et al. 2005), the HdaA (homolog of the *E. coli* Hda) is involved in the conversion of DnaA-ATP to DnaA-ADP (Fernandez-Fernandez et al. 2011; Felletti et al. 2019) (Fig. 3, upper part). This protein, part of the regulatory inactivation of DnaA or RIDA, together with the DnaN protein (β-clamp of the DNA polymerase), blocks extra round of chromosome replication. Consequently, the level of DnaA-ATP drops thus preventing new round of replication during the S-phase. Interestingly, *hdaA* is also one of the genes proposed to be under the control of DnaA, providing a negative feedback for its activity (Collier and Shapiro 2009). Moreover, it has been shown that the protease ClpAP has an accessory role in DnaA degradation (Liu et al. 2016) (Fig. 3, upper part).

1.1.3 GcrA and Its Targets

In *C. crescentus* after DNA replication has started and activation by the DnaA transcription factor, the atypical transcription factor GcrA accumulates (Holtzendorff et al. 2004) (Figs. 2 and 3). GcrA is well conserved in the alpha-proteobacteria class (except for the rickettsial branch) and has interestingly co-evolved with the CcrM adenine methyltransferase, that methylates 5′-GANTC-3′ (hereafter GANTC) sites on double strand DNA (Fig. 4).

GcrA is a small protein of 173 residues comprising two different domains connected by an unstructured linker (residues 46–107). The N-terminal domain (residues 1–45) contains the DNA-binding domain, whereas the C-terminal domain (residues 108–173) mediates the interaction with the RNAP through interaction with σ2 (Fioravanti et al. 2013; Haakonsen et al. 2015; Wu et al. 2018) that differs from canonical transcription factor that usually interacts either with σ4 or the α-CTD subunit of the RNAP (see below). GcrA is not essential (Murray et al. 2013) but loss of GcrA leads to cells with defect in cell division resulting in elongated cell with extra-chromosomes (Murray et al. 2013; Haakonsen et al. 2015). It has been proposed that GcrA and CcrM work as a regulatory module as most of the GcrA targets harbor m6A GANTC marks previously introduced by CcrM methyltransferase during the G2-phase (Fioravanti et al. 2013; Panis et al. 2015; Zweiger et al. 1994). In addition, Haakonsen et al. have proposed that GcrA forms a stable complex with the RNAP through its interaction with σ2 and binds to nearly all the σ^{70} active promoters without necessarily activating transcription to all the promoters to which it binds (Haakonsen et al. 2015). Rather, and in agreement with

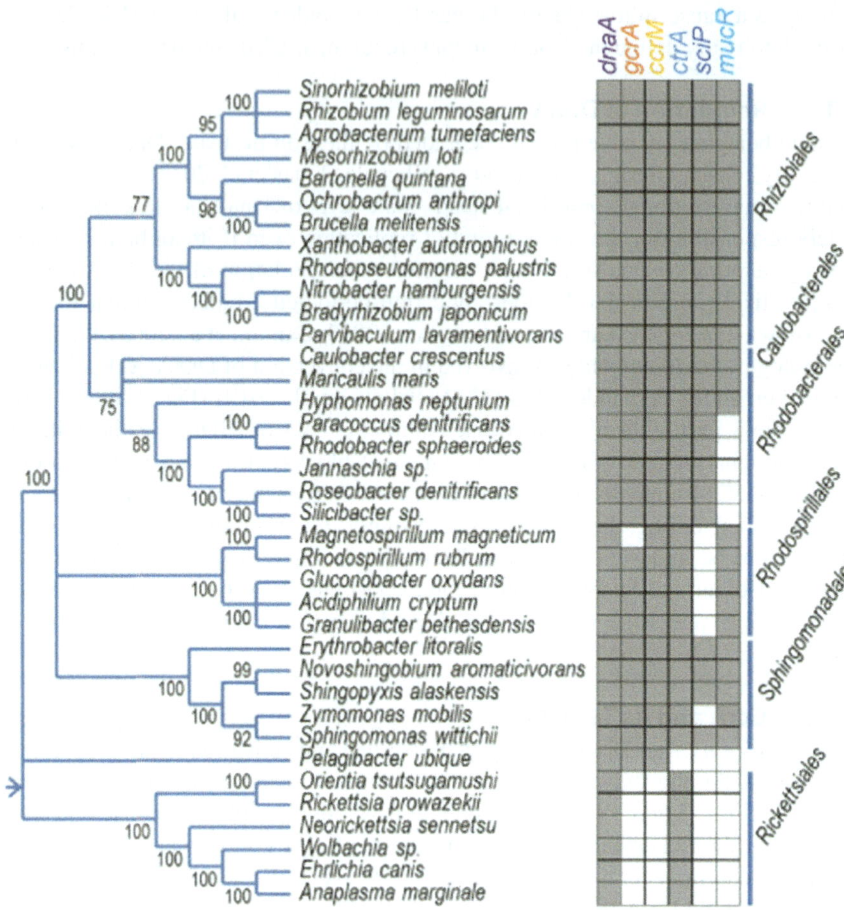

Fig. 4 Conservation of global transcriptional regulators in alpha-proteobacteria. The color used for the transcriptional regulators is the same used as in Fig. 7. Adapted from Panis et al. (2015)

the previous model proposed in which GcrA and CcrM act as pair (Fioravanti et al. 2013; Murray et al. 2013), GcrA will preferentially activate transcription at those promoters harboring GANTC methylation sites (Fioravanti et al. 2013; Murray et al. 2013). Moreover, they suggested that GcrA provides a powerful mechanism to control transcription initiation by promoting the isomerization of the RNAP as well as combinatory mechanisms of control by other transcription factors as it lets free the access to the σ4 domain (Haakonsen et al. 2015).

ChIP-seq analysis and DNA microarray analysis have been used to identify the binding profile of GcrA and showed that GcrA directly activates expression of around 150 genes implicated in cell cycle, cytokinesis, polar morphogenesis but also DNA repair and nucleotide synthesis (Fioravanti et al. 2013; Holtzendorff et al. 2004; Haakonsen et al. 2015) (Fig. 3, lower part). As previously mentioned,

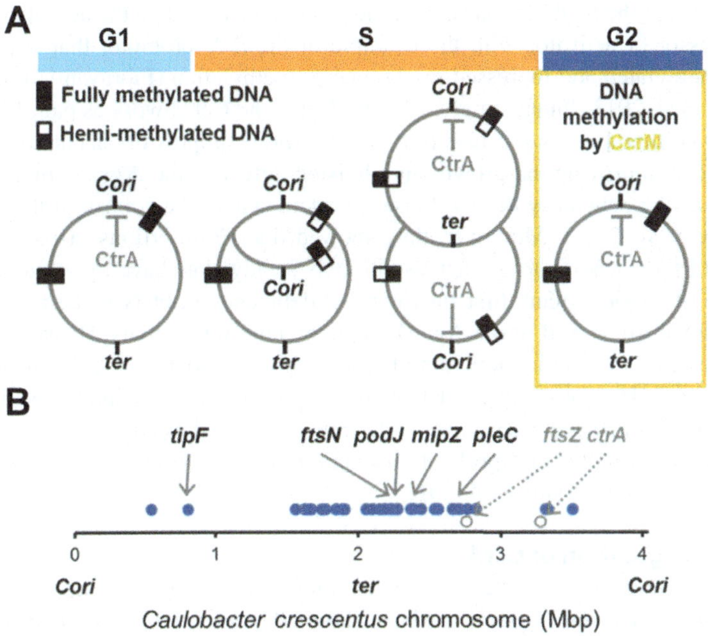

Fig. 5 Methylation state of the *C. crescentus* chromosome during replication and positions of some GcrA/CcrM targets along the chromosome. (**a**) Schematics of the *C. crescentus* chromosome and methylation state of the GANTC sites with progression of the replication fork. The GANTC sites are converted from the fully-methylated state to the hemi-methylated state until CcrM accumulates in the G2-phase under CtrA control. *Cori* represents the *C. crescentus* origin of replication and *ter* the terminus. (**b**) Positions of putative GcrA/CcrM targets along the *C. crescentus* chromosome and many targets are located close to the terminus (*ter*). Adapted from Panis et al. (2015)

GANTC methylation enhances GcrA binding to its target promoter. However, the methylation state of those sites changes with the DNA replication machinery that progresses and converts GANTC fully methylated to a hemi-methylated form (Fig. 5a).

Indeed, CcrM only accumulates in G2-phase under the dependency of CtrA and the hemi-methylated GANTC sites will be re-methylated only during this phase. Interestingly, many GcrA targets reside near the *ter* region of the *C. crescentus* chromosome (Nierman et al. 2001) and remain fully methylated for most of the cell cycle, whereas the *ctrA* gene, under the control of GcrA through its P1, is located at proximal position to the *Cori* (Fig. 5b).

Thus, although this variation of methylation state can be seen as a molecular clock and provides an easy way to control transcription during cell cycle (Collier et al. 2007), it seems that GcrA transcriptional activity is not sensitive to the change in promoter methylation state after replication (Fioravanti et al. 2013; Murray et al. 2013) with the exception of the P1 *ctrA* promoter. Besides, it has also been proposed that the methylation state of the promoter can dictate the activity of GcrA as repressor or activator with the example of the *ftsN* promoter that is activated when methylated and repressed in absence of methylation (Fioravanti et al. 2013; Murray et al. 2013). Interestingly, whereas GcrA and CcrM work as pair, Murray et al. showed that Transposon-insertion (Tn-insertion) suppressors accumulate in the *ccrM* gene improving the growth and division defect of the Δ*gcrA* single mutant as well as the deficiency of key division proteins under the control of this module (Murray et al. 2013). Moreover, they showed that Tn-insertions in the promoter region of *ftsN*, leading to an increase in FtsN steady state have the same effect in the Δ*gcrA* single mutant, first showing that deficiency in FtsN is the major cause of the defects observed in the Δ*gcrA* single mutant. Intriguingly, Haakonsen et al. also showed that ectopic induction of (p)ppGpp can also alleviate the difficulty to disrupt *gcrA* (Haakonsen et al. 2015), and proposed that (p)ppGpp helps to restore the balance of cell cycle activities while slowing down the cell growth and the rate of initiation of replication probably through an impact on DnaA stability (Lesley and Shapiro 2008; Boutte et al. 2012).

1.1.4 Regulation of GcrA

During the swarmer to stalked cell transition GcrA accumulation depends on two different mechanisms. First, it has been shown that a CtrA-binding motif overlaps the −10 region of the *gcrA* transcript and that CtrA acts as a repressor for the *gcrA* transcription (Holtzendorff et al. 2004). Indeed, CtrA is degraded during the swarmer to stalked cell transition by the ClpXP protease (Jenal and Fuchs 1998) (see below) relieving both repression at the *Cori* and the *gcrA* promoter. Second, as mentioned before with the transcriptional activity of DnaA, *gcrA* is under positive control of this transcriptional factor (Collier et al. 2006). Moreover, as CtrA starts to accumulate again at the late S-phase stage following its transcription activation by GcrA, the master cell cycle regulator will then in turn repress *gcrA* transcription. Until now, no evidence of specific cell cycle degradation have been proposed for this non-canonical transcription factor.

1.2 Late S-Phase Transcriptional Program

During the early S-phase, DnaA and GcrA will activate successively genes required for the late S-phase transcriptional program that is concerned with the completion of cell division and the implementation of polarity, specifically those whose function are then required in the subsequent G1 (swarmer) daughter cell, such as the early flagellar and pilus assembly proteins, chemosensory apparatus, holdfast secretion and attachment proteins. GcrA will activate the master cell cycle regulator CtrA that is responsible for the main implementation of this late S-phase transcriptional program together with the MucR1/2 repressors (Fig. 6) for example on the promoters that will drive the synthesis of the late pilus and flagellar assembly proteins.

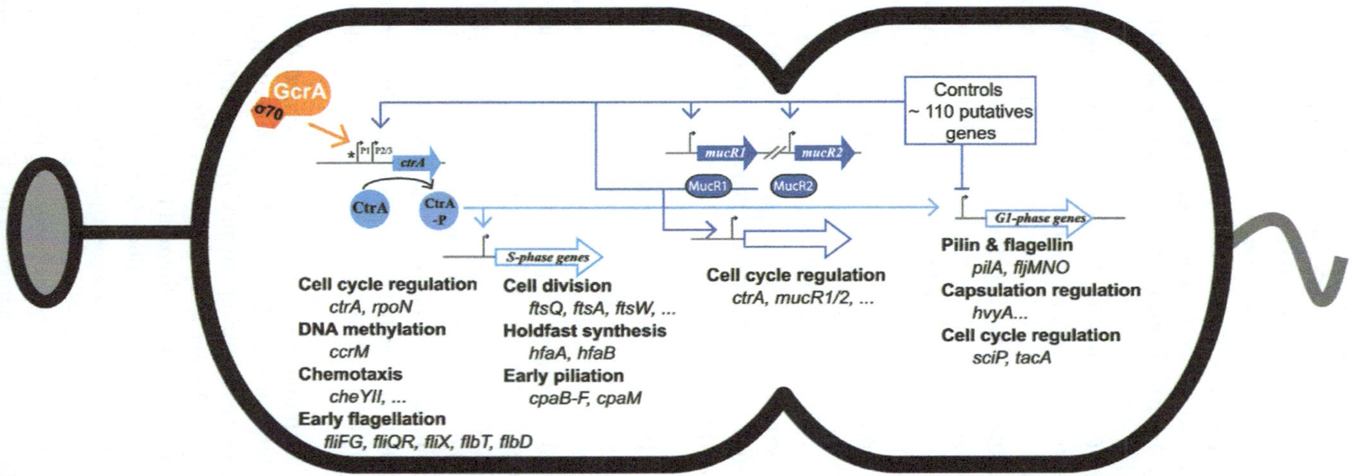

Fig. 6 Schematic of the *C. crescentus* late S-phase transcriptional control. During early S-phase, CtrA (light blue) re-accumulates upon transcription activation by GcrA (orange) through P1 promoter and is then phosphorylated in predivisional cells. CtrA-P activates late S-phase promoters such as those controlling early flagellar, chemotaxis, cell division, early piliation, and DNA methylation genes with the methyltransferase CcrM. CcrM accumulates and introduces methylation marks on the 5′GANTC3′ sequences in order to promote GcrA's affinity in the next round of cell cycle. Although CtrA-P binds to around 183 genes, the genes encoded do not fire at the same time. During the late S-phase, the ancestral zinc-finger transcription factor paralogs MucR1/2 (dark blue) acts negatively on G1-phase promoters activated by CtrA-P such as *tacA* and *sciP*. MucR1/2 also positively controls genes including *ctrA*, reinforcing the S to G1 transition. For each transcriptional regulator, a list of validated target genes was generated from the ChIP-seq, microarray, and RNA-seq datasets (Laub et al. 2000; Fiebig et al. 2014; Fumeaux et al. 2014)

1.2.1 CtrA and Its Targets

CtrA is an essential OmpR-like response regulator with an N-terminal receiver/dimerization domain and a C-terminal DNA-binding domain (DBD). CtrA oscillates during cell cycle: it is present in G1-phase, degraded at the G1 → S transition, and is resynthesized later in the S-phase. CtrA plays a global role in establishing the transcriptional program in the late S-phase: it directly binds at least 110 promoters many of which induce or repress genes that act in cell cycle regulation, cell division, morphogenesis, or other functions (Laub et al. 2007; Fiebig et al. 2014; Quon et al. 1996) (Fig. 6).

All the functions conferred by CtrA seem to be through its ability to bind DNA, as mutations in the DBD have profound effects on target promoter activity. Mutations in CtrA were isolated first in order to identify transcriptional factor or proteins implicated in transcriptional regulation of the flagellum and events critical for cell cycle progression and viability. Quon and colleagues were the first ones to identify in such screen CtrA. A thermo-sensitive (ts) mutant in the DNA-binding domain, *ctrA(ts)* (also known as *ctrA401*), in which the threonine codon 170 is mutated to isoleucine (T170I) impairs motility at the permissive temperature (30 °C) and viability at the non-permissive temperature (37 °C) (Quon et al. 1996). Interestingly, when the T170 codon is mutated to alanine (T170A) instead of isoleucine, the mutant *ctrA* allele becomes hypermorphic and it was proposed that this increases CtrA's capacity to compete against negative regulators such as MucR (Delaby et al. 2019; see below part iii.) that normally act antagonistically to CtrA at many target promoters (Fumeaux et al. 2014).

Global analyses of total mRNA in the *ctrA(ts)* mutant implicated CtrA in controlling, directly or indirectly, more than 25% of the 553 cell-cycle-regulated mRNAs (Laub et al. 2000) (Fig. 6). However, mapping the in vivo binding sites of CtrA by ChIP-chip (chromatin immunoprecipitation followed by microarray-based hybridization experiments) suggested that CtrA directly regulates 95 genes in 55 operons (Laub et al. 2002). Recent ChIP-Seq and TSS profiling experiments estimate at least 183 high confidence target sites in the *C. crescentus* genome, most of which reside in or near putative promoters (Zhou et al. 2015). The target promoters control genes implicated in wide range of functions such as regulation, flagellation, piliation, capsulation, DNA methylation, cell division and cell wall and other biosynthetic pathways.

The CtrA target sequence in promoters is the 15-mer inverted dyad repeat sequence 5′-TTAA-N7-TTAA-3 (Quon et al. 1996). Modifications within the TTAA upstream or downstream sequences lead to variability in CtrA recognition and therefore modulate transcription levels (Ouimet and Marczynski 2000). Interestingly, Zhou et al proposed that CtrA possesses three different binding motifs: (i) the full motif previously identified, in which the 5′ nucleotide is positioned near the −35 promoter element, in agreement with CtrA as a transcriptional activator for the 52 targets identified, (ii) the half motif 5′-TTAA-3′ repressor, and (iii) the half motif activator. In the case of the 24 CtrA half repressor containing promoters identified, the binding site was positioned near the −10 element of the promoter regions, again consistent with its activity as repressor. Finally, 107 promoters were identified with half activator sites also located near the −35 region. Moreover, Zhou and colleagues also showed that expression of the 52 full motif target peaks at the same time as CtrA in late S-phase whereas the 107-half binding show similar pattern to CtrA throughout the cell cycle. Conversely, the 24 repressed ones showed opposite pattern to CtrA (Zhou et al. 2015).

Phosphorylated CtrA is present in G1-phase (Fig. 2 and further discussed in c)i.) then degraded during the swarmer to stalked transition and resynthesized later in predivisional cell within the future motile compartment where it activates late S-phase promoter, including class II flagellar genes encoding the MS-ring, flagellar switch and the export apparatus, a σ54-transcriptional activator *flbD* and its regulator *fliX* (Ardissone and Viollier 2015). Flagellar genes expression starts early in predivisional cells and it is maintained in G1-phase where after cytokinesis, CtrA is maintained in the swarmer daughter cell where it induces firing of G1 promoters. CtrA also regulates genes for polar morphogenesis. DNA microarray showed that genes required for new flagellum and piliation system are transcribed in late S-phase when CtrA re-accumulates (Laub et al. 2000). By contrast, transcription of *pilA*, the structural subunit of the pilus filament, is also under the control of CtrA bit its transcription occurs last and is limited to the swarmer compartment once the division septum has fomed (Laub et al. 2000, 2002; Skerker and Shapiro 2000). Constitutive expression of *pilA* leads to premature assembly of the pilus in predivisional cells (Skerker and Shapiro 2000; Radhakrishnan et al. 2010), attesting to the tight control of the switch between S-phase transcriptional control and G1-phase transcriptional program (Fumeaux et al. 2014). Several mechanisms can be implicated in this differential promoter activation by CtrA process. Modification within the CtrA boxes can explain part modulation in the CtrA-P affinity for the promoter (Ouimet and Marczynski 2000) but also competition with different negative regulators (Fumeaux et al. 2014; Gora et al. 2010; Tan et al. 2010; Reisenauer et al. 1999). The "just in time" transcription of several cell-cycle-regulated genes thus depends on a tight control of the key cell cycle regulator CtrA and this is accomplished through a conserved regulatory module MucR/SciP/CtrA further discussed below, in which MucR1/2 represses the G1-phase promoters in late S-phase including the promoter of the gene encoding SciP, the third transcriptional regulator that will turn OFF late S-phase genes in the swarmer cell (Fumeaux et al. 2014).

Additionally phosphorylated CtrA also controls essential cell cycle processes such as DNA methylation and cell division (Laub et al. 2002). CtrA directly regulates six cell division genes: it represses *ftsZ*, a tubulin-like GTPase essential for cell division in bacteria (Kelly et al. 1998) and activates *ftsA*, *ftsQ*, *ftsO*, *ftsW*, and *ftsI* necessary for initiation and progression of cell division septum during late S-phase. CtrA also targets the promoters of two genes linked to DNA methylation: *ccrM* which encodes for an adenine DNA methyltransferase, late in the cell cycle and *CC_0050* encoding for a S-adenosylmethionine (SAM) responsible for synthesis of the CcrM substrate that is used to methylate DNA (Laub et al. 2002). Finally, as mentioned above, GcrA and CtrA are functionally linked since CtrA will repress transcription of GcrA whereas GcrA first activates CtrA transcription in early S-phase.

As mentioned above, CtrA activates the transcription of *ccrM* in late S-phase and the CcrM gene product will then methylate the GANTC sites preferentially bound by GcrA during the early S-phase of the esnsuing cell cycle. After the replication fork passage, the GANTC sites previously fully methylated remain hemi-methylated until late S-phase until CcrM starts to accumulate. Thus the period during which a locus remains fully- or hemi-methylated depends on its position on the chromosome. Indeed, a gene close to the *ter* will mainly stay as fully-methylated during the cell cycle conversely to a gene close to the Cori (Collier and Shapiro 2009).

1.2.2 Regulation of CtrA

Synthesis of the *ctrA* mRNA is governed by three promoters (Zhou et al. 2015), P1–P3. Of these, *P1* is a methylation-sensitive promoter that is active in S-phase and located 122 basepairs (bp) of the translational start site of CtrA. As the *ctrA* gene and its promoters are located at a proximal position to the origin of replication (*Cori*) on the circular chromosome of *C. crescentus*, it is duplicated early during S-phase. Prior to DNA replication (i.e., in G1 cells), the chromosome is adenine-methylated (at the N6 position) on both strands in the context of 5′-GANTC-3′ sequences by the CcrM DNA methyltransferase that was active in the previous cell cycle after completion of replication (Fig. 5). Once DNA replication starts and proceeds bidirectionally from *Cori*, *ctrA* is duplicated and hemi-methylated GANTC sequences emerge. They remain hemi-methylated until CcrM is again expressed (by CtrA) in late S-phase.

Firing of the *ctrA* P1 promoter in S-phase requires binding of GcrA. The GANTC site in P1 is hemi-methylated when it fires, while the stronger *P2* promoter that is located 65 bp from the ATG is activated by CtrA itself later in S-phase that in turn inhibits *P1* (Domian et al. 1999). Recently, *P3*, an additional cell-cycle-regulated promoter located between P1 and P2 was identified and it shows a similar pattern of regulation to P1 (Zhou et al. 2015). This leads to an accumulation of CtrA in late S-phase, but as CtrA activity is regulated by phosphorylation on its conserved aspartate (D51) as for other response regulators, the activation of CtrA via the phosphorelay is needed for transcriptional activity. The phosphorylation of CtrA is directly controlled by the phosphorelay system CckA/ChpT (Biondi et al. 2006a). The kinase domain of the membrane anchored hybrid histidine kinase CckA autophosphorylates and then transfers the phosphoryl group to its receiver domain. The phosphoryl group ultimately ends up on CtrA via the soluble histidine phosphotransferase intermediate ChpT. Phosphorylated and dimerized CtrA is abundant in G1/swarmer cells where it binds to DNA and therefore modulates transcription and prevents replication initiation. At times in the cell cycle when CtrA activity is not desired, the phosphate flow of the CckA-ChpT system is reversed and this switch is controlled by several accessory factors (Fig. 7), including the single-domain response regulator DivK.

Briefly, DivK will act on the CckA/ChpT phosphorelay through its action on the histidine kinase DivL: DivK will be phosphorylated by the DivJ histidine kinase

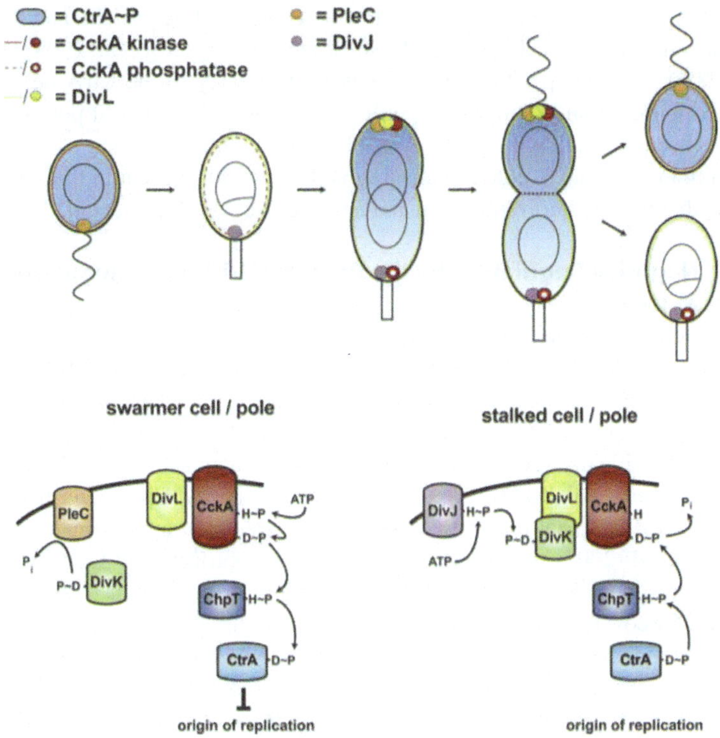

Fig. 7 Regulation of CtrA activity and cell cycle transitions through phosphorelay during cell cycle in *C. crescentus*. Image from Tsokos et al. (2011)

that will in turn downregulates the phosphorylation of the DivL tyrosine kinase through direct binding and finally DivL leads to the inhibition of CckA and to the inactivation of CtrA (Biondi et al. 2006a; Tsokos et al. 2011; Iniesta et al. 2006; Wu et al. 1998, 1999) (Fig. 7).

Remarkably, temporal control of CtrA stability is also governed by the CckA-ChpT system. CtrA is degraded during the G1→S transition by the essential ClpXP ATP-dependent protease (Jenal and Fuchs 1998) and proper activation of this degradation process requires several components including the single-domain response regulator CpdR. While CpdR functions as proteolytic adaptor that interacts with the ATP-dependent unfoldase, it is also a phosphorylation substrate by the CckA-ChpT system. However, unlike CtrA, CpdR is activated as an adaptor in the non-phosphorylated state. Thus, activation and proteolysis of CtrA are inversely linked through the phosphorelay CckA/ChpT, resulting in a boost of CtrA activity in G1- and late S-phase when the phosphoflux is toward CtrA and no CtrA activity in early S-phase when the phosphoryl groups are drained from CtrA (Iniesta et al. 2006). The degradation of CtrA is a complex mechanism that requires several adaptors to the ClpXP protease as CpdR alone cannot substantially enhance CtrA proteolysis in vitro (Smith et al. 2014). PopA, a cdG-binding protein, and RcdA are also actors that allow localization of the substrate CtrA (Duerig et al. 2009; McGrath et al. 2006) for the ClpXP machinery and are necessary to the CtrA degradation. In late S-phase, unphosphorylated CpdR is sequestrated to the old pole where the stalk has been built allowing proteolysis of CtrA and thus enforcing its asymmetric distribution in the late S-phase to ensure that CtrA is not present in the stalked cell progeny that will re-start replication and a new cell cycle (Iniesta et al. 2006).

1.2.3 Negative Regulation by the Ancestral Virulence Regulators, MucR1/2

CtrA is active in both G1- and late S-phase to ensure promoter firing to the correct cell cycle phase (Laub et al. 2000, 2002; Fiebig et al. 2014; Fumeaux et al. 2014). Fumeaux et al. proposed that this role is encompassed by the two conserved ancestral virulence regulators MucR1 and MucR2 and that these regulators specifically repress the G1 specific-CtrA-activated genes during late S-phase (Fumeaux et al. 2014) (Fig. 6). Knowing that the *pilA* promoter is under the control of CtrA and that *pilA* mRNA peaks in G1-phase at approximately 120 min (McGrath et al. 2007) they used forward genetic approaches to identify negative regulators responsible for the premature activation of the CtrA G1-regulon in late S-phase. Moreover, ChIP-seq analysis (Fumeaux et al. 2014) and global 5′ RACE (RNA ligase-mediated rapid amplification of 5′ cDNA ends), allowing identification of the TSS coupled to identification of regulatory factor binding sites (Zhou et al. 2015), showed that more than 75% of the MucR1/2 cell-cycle-regulated regulon harbors CtrA-binding site such as *pilA* or the late S-phase target repressor gene *sciP* (see section c)ii.).

Interestingly, MucR1/2 can also act as a transcriptional activator, indeed whereas it will repress the G1-phase target promoters of CtrA, it also activates the *ctrA* promoter as well as its own promoter (Fumeaux et al. 2014) being part of regulatory module that will control the G1-phase transcriptional module during the late S-phase transcriptional program.

A study combining restriction-enzyme-cleavage deep sequencing (REC-Seq) with single-molecule real-time (SMRT) sequencing showed that MucR leads to specific site of hypomethylation by preventing the DNA adenine methyltransferase CcrM to access these sites through competition in *C. crescentus* but also in *Sinorhizobium meliloti*, suggesting that hypomethylation control by MucR during cell cycle is conserved in alpha-proteobacteria (Ardissone et al. 2016). Besides, by using RNA-Seq-based strategy for exact mapping of transcriptome 5′-ends (EMOTE), they identify several TSS activated in absence of MucR1/2, showing that this hypomethylation can also control sense and anti-sense transcription with potential regulatory roles. Interestingly, they also showed that environmental cues such as phosphate depletion could also influence local hypomethylation of DNA (Ardissone et al. 2016).

Intriguingly, newly identified NAPs in *C. crescentus* (Ricci et al. 2016; Taylor et al. 2017; Arias-Cartin et al. 2017; Guo et al. 2018) called GapR have been shown to bind at active promoters controlled by master regulators of the cell cycle progression with an overlap of more than 90% with the MucR targets. GapR has been shown to specifically recognize and encircle overtwisted DNA and further allow DNA replication through stimulation of the gyrase and the topoisomerase IV in order to relax positive supercoils (Guo et al. 2018). Despite its major role in DNA replication, the fact that GapR binding profiles determined by ChIP-seq highlights an overlap in occupancy with major cell cycle regulators shows the interplay of all the cell cycle players as well as the cell type compartmentalization. Indeed, as GapR is expressed constitutively but accumulates preferentially in the swarmer daughter cell of late predivisional cells (Ricci et al. 2016), it is suggested that GapR influences the nucleoid conformation within the swarmer cell and that it could interplay with the program of gene expression that occurs in swarmer cell and during the S→G1 transition (concomitantly with MucR1).

During the late S-phase, predivisional cells are concerned with transcriptional program that will allow completion of cell division but also with the implementation of the asymmetry. It is also the time when CtrA accumulates and becomes finally asymmetrically distributed in the future swarmer or G1-phase cells.

1.3 The G1-Phase Transcriptional Program

Upon cell division, a new transcriptional program is setting up in the SW/G1-phase cells. CtrA plays a global role in establishing this transcriptional program in the SW (G1) daughter cell together with the small CtrA inhibitory protein SciP (see above) (Fig. 8) that will restrict firing of G1-phase promoters.

1.3.1 CtrA and Its Target

During the G1-phase CtrA controls key processes such as capsulation through activation of the *hvyA* promoter and follows up the implementation of the asymmetry by activation of some flagellins promoters and the *pilA* promoter that only fire in the SW compartment (Laub et al. 2000, 2002; Skerker and Shapiro 2000) (Fig. 8). Besides the flagellum and pili that are needed during the G1-phase, the holdfast, found at the tip of the stalk, is synthetized early during the swarmer cell development at the flagellar pole (Levi and Jenal 2006). Indeed, transcription of the *hfa* genes, implicated in the holdfast biogenesis, is cell cycle regulated (Janakiraman and Brun 1999) and under the control of CtrA (Laub et al. 2002). Despite being implicated in the biogenesis of the holdfast, CtrA does not directly control stalk biogenesis, instead it activates other signaling pathways that fulfill this role in SW cells. Dissection of stalk less mutants lead to the identification of some factors required for stalk synthesis, two proteins under the direct control of CtrA are necessary: the alternative sigma factor encoded by *rpoN* and its activator encoded by *tacA* (see section c)iii.) (Fumeaux et al. 2014; Laub et al. 2002; Janakiraman et al. 2016).

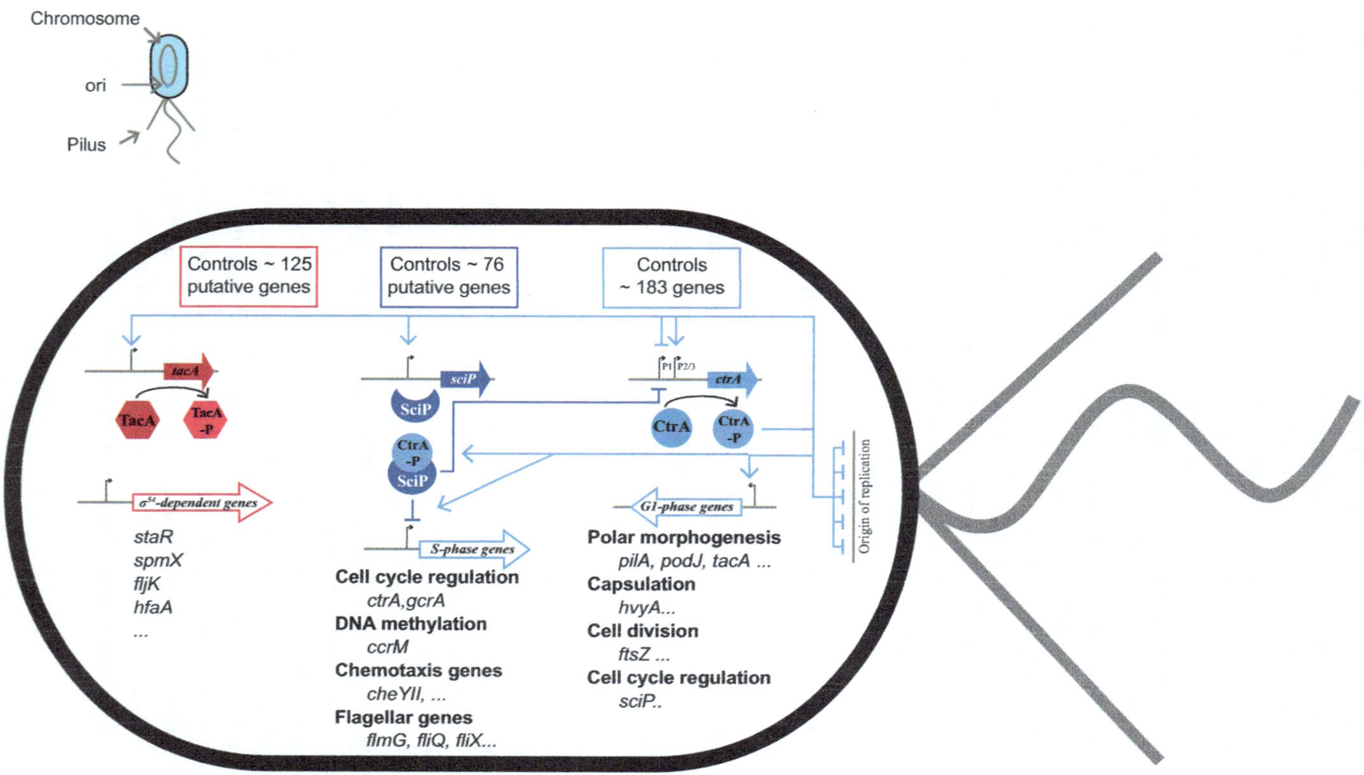

Fig. 8 Schematic of the *C. crescentus* swarmer cell/G1-phase transcriptional control. G1-phase is mainly controlled by the cell cycle transcriptional regulator A (CtrA, light blue), activated by phosphorylation through the CckA→ChpT→CtrA phospho-cascade and acts with different negative regulators to restrict

1.3.2 Negative Regulation of the S-Phase Promoters by the Small CtrA Inhibitory Protein SciP

As mentioned previously, CtrA binds and activates different promoter regions that do not all fire at the same time. Through bioinformatics screen Gora and colleagues identify the essential small CtrA inhibitory protein or SciP as a 93 amino acid protein that possesses helix–turn–helix domain and only present in G1-phase (Gora et al. 2010; Tan et al. 2010). SciP appears to be necessary for swarmer cell development and a key regulator of cell cycle progression. They also demonstrate that SciP binds to CtrA and impairs its binding to RNA polymerase (Gora et al. 2010). Besides its crucial role in cell cycle regulation, they showed that SciP is activated by CtrA and that the mRNA is accumulating in predivisional cell and swarmer cell; meanwhile, the protein only accumulates in G1 cells and feeds back the transcriptional activity of CtrA for the S-phase promoters (Gora et al. 2010; Tan et al. 2010). Tan and colleagues propose that SciP can also bind to specific DNA motif on genes targeted by CtrA at the $5'$-TGTCGCG-$3'$ sites and demonstrated by ChIP followed by real-time PCR that SciP binds directly to the promoter regions of genes encoding CtrA, such as *ccrM* encoding DNA methyltransferase, as well as to the promoter regions of a subset of flagellar and chemotaxis genes and bioinformatics analysis reveals that SciP has 76 potential binding sites in vivo (Fumeaux et al. 2014; Tan et al. 2010) (Fig. 8). Among the 50 top targets, a predicted MEME-based motif was proposed, having a half CtrA box (underline) $5'$-(G/A)TTAACCAT (A/G)-$3'$ (Fumeaux et al. 2014). In sum, even if both previous models, proposed by Gora et al. and Tan et al., agree in the fact that SciP acts as a repressor to provide negative feedback and control regulation of cell-cycle-dependent transcription, it is more likely that either SciP and CtrA compete for these sites or that repression involves a binding cooperation between these two factors (Fumeaux et al. 2014). Indeed, ChIP-seq analysis performed by

Fig. 8 (continued) promoter firing to the correct cell cycle phase. CtrA possesses a complex transcriptional control. As soon as CtrA-P accumulates, the weaker promoter P1 is repressed by CtrA itself and the negative regulator SciP and fully methylation state, the stronger P2 promoter is activated by CtrA-P. CtrA-P binds to and activates G1-phase promoter implicated in different functions such as polar morphogenesis, capsulation, cell division but also cell cycle regulation by activated *sciP* (dark blue) in order to fine-tune transcriptional regulation and binds to the origin of replication to suppress the initiation of DNA replication and stops cells in G1. SciP will work as a negative transcriptional regulator and associates preferentially with the late S-phase promoter in order to shut off genes encoding functions such as DNA methylation, flagellation, chemotaxis but also cell cycle regulation and allows G1-phase genes to fire. Similar to *sciP*, the mRNA of the highly conserved σ54-dependent transcriptional regulator *tacA* (pink) surges in G1 under the control of CtrA-P and TacA is activated through phosphorylation by the His-Asp ShkA-ShpA phosphorelay. TacA-P targets transcription of several developmental and cell cycle genes along with σ54-containing RNA polymerase holoenzyme such as *spmX*, a gene encoding for an important polar organizer and *staR* that encodes for a transcriptional regulator of stalk biogenesis. For each transcriptional regulator, a list of validated target genes was generated from the same ChIP-seq, microarray, and RNA-seq datasets (Fiebig et al. 2014; Fumeaux et al. 2014; Gora et al. 2010; Tan et al. 2010; Janakiraman et al. 2016; Laub et al. 2002)

Fumeaux and colleagues also demonstrates that SciP preferentially binds to the S-phase promoters targeted by CtrA and not to the G1-phase promoters (Fumeaux et al. 2014).

Conversely, the fact that SciP levels drops during the swarmer to S-phase transition, suggests that SciP may be also subject to proteolysis as CtrA (Gora et al. 2013). However, unlike CtrA, SciP is degraded by the Lon protease even if the two proteins stabilize each other (Gora et al. 2013). In conclusion, transcription and proteolysis work together ensuring that SciP peaks in G1-phase to negatively regulate late S-phase promoters and it is degraded before CtrA has to be active to regulate genes implicated in polar morphogenesis and cell division. Proteolysis of CtrA during G1- to S-phase transition will allow initiation of DNA replication and the removal of SciP allows transcription of late S-phase genes.

1.3.3 TacA and Its Targets

TacA is a transcriptional regulator that directs stalk elongation during the G1- to S-phase transition (Biondi et al. 2006b). Similar to CtrA, TacA is controlled at the level of transcription, phosphorylation, and proteolysis and oscillates in phase with CtrA (Biondi et al. 2006b; Joshi et al. 2015). The *tacA* mRNA surges in G1-phase and its transcription is directly activated by CtrA-P (Fumeaux et al. 2014; Biondi et al. 2006b) (Fig. 8). TacA is also regulated at the level of phosphorylation by the histidine kinase/response regulator hybrid ShkA and the phosphotransfer protein ShpA (Biondi et al. 2006b; Kaczmarczyk et al. 2020) and surprisingly, TacA is degraded by the ClpXP protease in a cell cycle dependent. Similar to CtrA, proteolysis of TacA relies on the response regulator CpdR and the RcdA adaptor but it does not require PopA. Microarray analysis identified 30 genes under the control of TacA (Biondi et al. 2006b) and more recently ChIP-seq experiments defined more than 125 putative targets for the TacA regulon, confirming that the active and phosphorylated form of TacA (TacA~P) controls stalk biogenesis, by acting on a subclass of σ^{54}-dependent genes (Janakiraman et al. 2016) (Fig. 8). Once activated through the ShkA-ShpA phosphorelay on the aspartate 54 (D54) (Biondi et al. 2006a), TacA-P interacts with the RNA polymerase-σ^{54} holoenzyme to activate transcription by binding on a consensus motif of a (inverted) dyad symmetry [5′-tTCgCct-(N)3-agGcGAa-3′] (Janakiraman et al. 2016). Among the TacA regulon, SpmX and StaR, a regulator of stalk length, were confirmed (Janakiraman et al. 2016; Biondi et al. 2006b; Radhakrishnan et al. 2008).

Finally, once the adventurous G1-phase cell finds appropriate nutrient conditions it will differentiate in the stalked cell that will start again a new round of replication and cell cycle. We have seen that *C. crescentus* cell cycle is regulated by several waves of transcriptional regulators that specifically fine-tune cell cycle functions through activation or repression of targeted promoters and therefore that regulation at the level of transcription initiation is a way for *C. crescentus* to ensure correct cell cycle together with translational and post-translational control. First, DnaA and the epigenetic module GcrA/CcrM will control the establishment of the early S-phase transcriptional program followed by the accumulation of CtrA that will together with MucR control the late S-phase prior to division and finally the CtrA/SciP module after derepression by MucR activates the G1-phase genes. In the next part we will describe how these transcriptional cascades specifically regulate the establishment of the different cell morphologies committed to the specific stages of the cell cycle.

1.4 Transcriptional Cascades in Cell-Cycle-Regulated Functional Modules

The *C. crescentus* late S-phase transcriptional program will implement polarity, i.e. the flagellar biogenesis and the chemotaxis machinery as well as the completion of cell division (Fig. 9). The cell cycle regulators previously described orchestrate this process through transcription activation of genes encoding structural proteins but also through the activation of polarization factors. Whereas the main polar morphogenesis transcriptional control occurs during the late S-phase, the early S-phase is concerned with the transcription activation of the polarity factors such as TipF or PodJ (see below) activated by GcrA, that respectively promote flagellum assembly and pilus assembly. Conversely, the G1-phase is concerned with the regulation of *C. crescentus* capsulation as well as the completion of the flagellum and pili assembly.

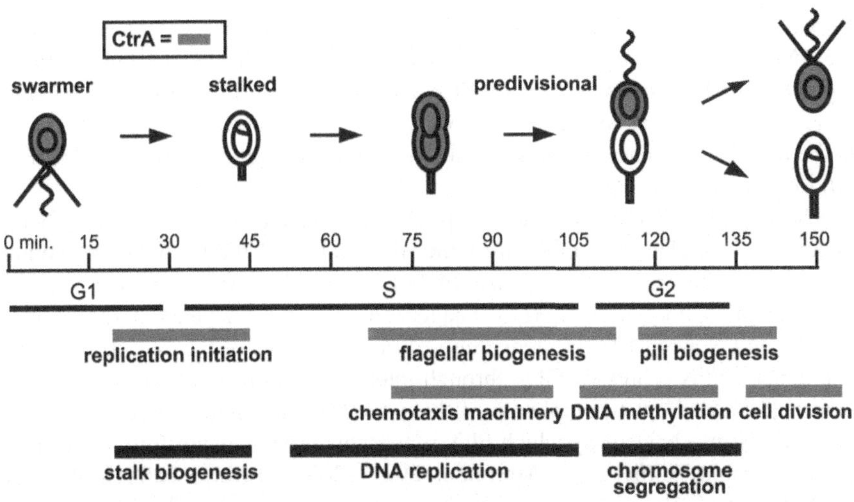

Fig. 9 Temporal coordination of cell cycle events. During cell cycle *C. crescentus* undergoes an asymmetric cell division that gives rise to two different daughter cells with different morphologies and fates. During the swarmer to stalked cell transition, the flagellum is lost and the pili are retracted. During the late S-phage, the new flagellum is built and the chemotaxis machinery is assembled. Timing of key cell-cycle-regulated events is indicating in black and gray bars indicated the ones regulated by CtrA. CtrA presence is denoted in gray. Schematic from Laub et al. (2002)

1.4.1 Transcriptional Control of Flagellar Biogenesis

Flagellar assembly is tightly coupled with cell cycle progression. It involves approximately 50 genes organized in several transcriptional cascades (Ely and Ely 1989). *ctrA* is a class I flagellar (regulatory) genes whose gene product will bind and activate the promoters of (early) class II flagellar genes that encode early structural components (assembling the secretion system) of the flagellum but also regulatory genes for the following steps (*flgBC*, *fliE*, *fliLM*, *flbT*, *fliX*, and *rpoN*) during late S-phase. The (middle) class III genes are then activated by FlbD/RpoN which also activates as well as several (late) class IV flagellin clusters (*fljKL* encoding two α-flagellins), except for *fljMNO* (encoding the three β-flagellins) that are activated by CtrA exclusively in G1-phase (Skerker and Laub 2004) (Fig. 10), while the fljJ flagellin gene is directly activated by CtrA in late S-phase.

The transcriptional cascade is set in motion in early S-phase by the GcrA regulator which induces transcription of CtrA and the flagellar assembly factor TipF. Once TipF is synthesized, it is recruited by the TipN landmark protein to the site where the new flagellum will be built (Huitema et al. 2006; Lam et al. 2006) and TipF will further recruit early flagellar assembly proteins such as FliG (class II genes) or the flagellar positioning factor PflI (Davis et al. 2013) that are expressed from operons activated by CtrA. Upon accumulation of CtrA in late S-phase and activation through phosphorylation, the class II structural components genes (Quon et al. 1996; Skerker and Laub 2004) and three transcriptional regulators implicated in the regulation of the late flagellar promoters (Ramakrishnan and Newton 1990; Brown et al. 2009) will be transcribed (Fig. 10). These transcriptional regulators are the RNA polymerase sigma subunit σ^{54} (RpoN), the σ^{54}-dependent transcriptional regulator FlbD and its own regulator FliX and will activate transcription of the class III and IVa genes that encode, respectively, proteins that form the basal body/hook in the periplasmic space and in the outer membrane and the three α-flagellins (Fig. 10) forming the flagellar filament.

The flagellar filament is also composed of β-flagellins that instead of being under the dependency of FlbD are under the control of the CtrA/MucR module and will be transcribed in G1-phase. FlbD is a NtrC-like transcriptional activator that upon phosphorylation binds to enhancer and drives σ^{54} dependent transcription (Mullin and Newton 1989). Interestingly, during assembly of the class II structural component, FliX represses FlbD through interaction (Muir et al. 2001; Muir and Gober 2002) and upon completion of the assembly, the complex becomes active, however the mechanism by which FliX senses this process is not fully understood (Ardissone and Viollier 2015; Muir et al. 2001; Muir and Gober 2002; Mohr et al. 1998; Wu et al. 1995) (Fig. 10), but likely involves a recently identified flagellar factor that is itself expressed from a CtrA-dependent promoter and that interacts with FliX and the FlhA component of the flagellar sceretion machine (Siwach et al. 2021).

During the assembly of the flagellar P(peptidoglycan)-ring, the L(LPS/OM)-ring, the rod and the hook, encoded by the class III genes (Fig. 10), the mRNA of the α-flagellins is not translated. Indeed, the mRNA binding protein FlbT, a class

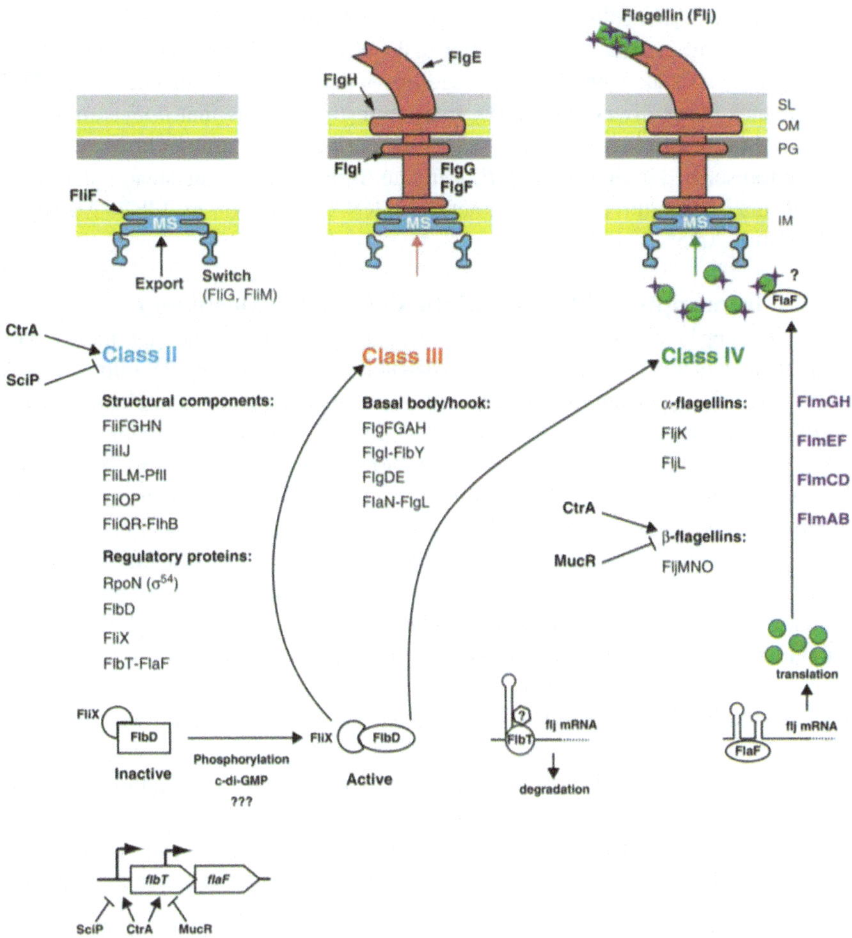

Fig. 10 Schematic of the regulation of flagellum assembly in *C. crescentus*. Class II promoter genes encoding structural components and regulatory proteins are expressed during the late S-phase under the control of CtrA and the SciP repressor. The regulatory genes encoded in this class II promoter are RpoN or σ^{54}, the σ^{54}-dependent activator FlbD, its partner FliX and the FlbT/FlaF proteins that control translation of the flagellin mRNAs. FliX is thought to form an inactive pair with FlbD until the class II structural components are assembled, in order to prevent FlbD activation of the class III and IV promoters. Once the inner membrane flagellar components (class II) are assembled, the FlbD/FliX complex is activated by an unknown mechanism and in turn activates transcription of the class III and IV components. Similar to the inactivation of the FlbD/FliX complex until the complete assembly of class II component, FlbT promotes degradation of the flagellins mRNA until full assembly of the basal body and hook structures. Flagellins are under the dependency of two different transcriptional controls. The α-flagellins (class IVa) are FlbD dependent whereas transcription of beta-flagellins (class IVb) is under control of the CtrA/MucR module and is activated in during the switch between late S- and G1-phase. As mentioned previously, *flbT* is expressed from CtrA/SciP promoter whereas *flaF*, that encodes a protein that counteracts FlbT repression and allows flagellins accumulation, can be either expressed in operon with *flbT* or from a CtrA/MucR dependent promoter. Adapted from Ardissone and Viollier (2015). The question mark corresponds to FljJ that has been recently implicated as co-regulator of FlbT (Ardissone et al. 2020)

II flagellar gene controlled by CtrA, sequestrates the mRNA through interaction with the 5′ untranslated region of the α-flagellin mRNA (Anderson and Gober 2000; Mangan et al. 1999). The α-flagellins will finally be translated once FlaF accumulates (Llewellyn et al. 2005). It is interesting to note that *flaF* is under the control of CtrA and MucR as the β-flagellins during the G1-phase; however, it can also be transcribed in operon with *flbT* in late S-phase. Thus, the steady-state level of FlaF may be required to counteract the mRNA repression by FlbT (Ardissone and Viollier 2015).

1.4.2 Transcriptional Control of the ChemotaxisMachinery

As fresh water bacterium, *C. crescentus* uses its flagellum to find new niches with optimal nutrient availabilities for growth and survival. Modulating flagellum activity allows bacteria to change their swimming trajectories in response to changes in the environment. Flagellar motors rotate either clockwise (CW) or counter clockwise (CCW) and *C. crescentus* swarmer cells show a three steps swimming pattern: Forward, reverse, and flick. The reversion from CW to CCW is controlled by the chemotaxis apparatus (Ely et al. 1986; Skerker et al. 2005). *C. crescentus* possesses 19 chemoreceptors; among them 2 CheAs and 12 CheYs. Briefly, membrane-bound chemoreceptors relayed the information to the CheA chemoreceptor coupled histidine kinase, which in turn phosphorylates a soluble response regulator, CheYII (functional homolog of *E. coli* CheY). Once phosphorylated, CheY-P interacts with the flagellar switch protein FliM. A new class of CheY-like proteins, which tunes flagellar activity in response to c-di-GMP but not phosphorylation was identified (Nesper et al. 2017).

Chemotaxis is spatially and temporally associated with the flagellum, with the chemoreceptor clusters localizing to the flagellated pole. Indeed, chemotaxis genes are induced in late S-phase at the same time as the early flagellar genes (Schrader et al. 2016; Zhou et al. 2015; Laub et al. 2002) and through CtrA activation (Laub et al. 2002).

1.4.3 Transcriptional Control of Pilus Biogenesis

During the late S-phase and prior to division, flagellum and the *C. crescentus* pilus assembly (Cpa) or tight adherence (Tad) pili are synthetized to the new cell pole. Bacterial type IV pilus systems are ubiquitous among bacteria and involved in several key processes such as surface sensing, motility, DNA uptake, and biofilm formation (Costa et al. 2015; Melville and Craig 2013; Craig et al. 2004). Type IV pili are cell surface structures and the machine that assembles and retract them is anchored in the cell enveloped (Fig. 11a). The genes encoding the core component of the Cpa system are encoded in the same loci with *cpaBCDEFGH* as an operon whereas additional assembly factors required for the pilus biogenesis such as ZitP and PodJ are trans-encoded (Fig. 11b). Like the synthesis of the flagellum, biogenesis of the pili is spatially and temporally regulated and subjects to transcriptional control by the cell cycle regulatory modules that will both activate pilus genes and genes encoding pilus factors (Fig. 11b).

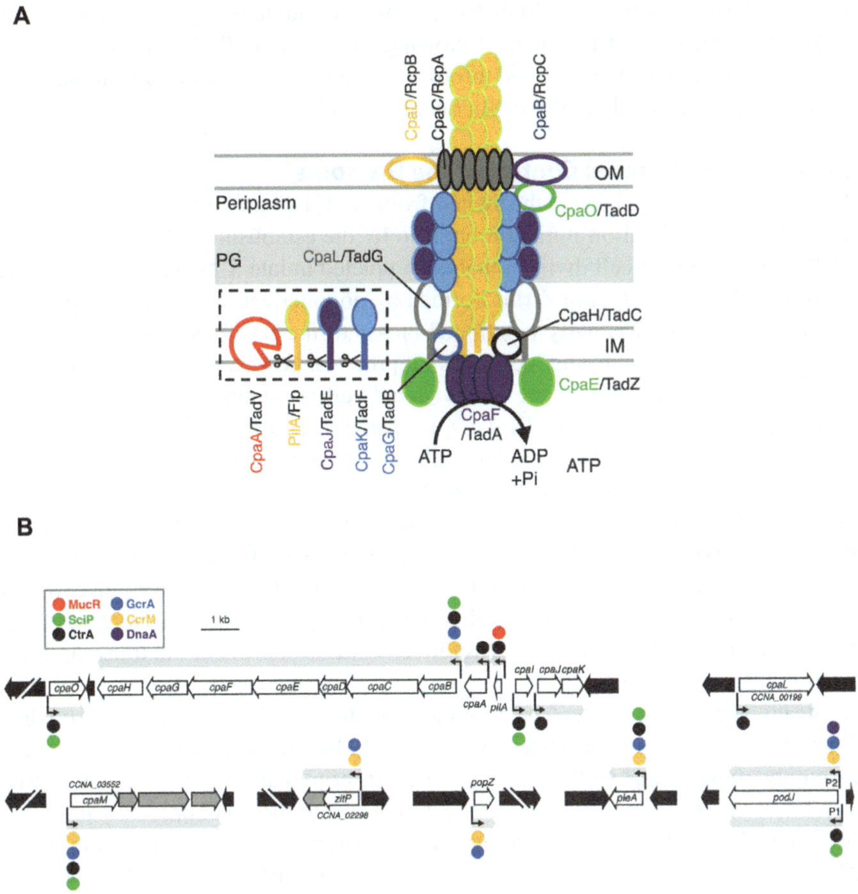

Fig. 11 Schematic of the Cpa/Tad secretion system structure. (**a**) Predicted structure of the *C. crescentus* Cpa secretion system. (**b**) Schematics of the genes implicated in the assembly or polar localization of the Cpa secretion system. Bent arrow shows the identified promoters and gray arrows indicate the operon structures. A color circle based on published data shows promoter occupancy by the cell cycle regulators. The blue circle refers to GrcA, the yellow to CcrM, the purple to DnaA, the red to MucR, the green to SciP, and finally the black circle refers to CtrA. Images adapted from Mignolet et al. (2018)

During the early S-phase GcrA accumulates and activates transcription of ZitP and PodJ, polar landmark proteins necessary for the later pili biogenesis. ZitP coordinates the insertion of the secretin CpaC at the new cell pole (Mignolet et al. 2016), PodJ is necessary for the localization of pilus assembly protein CpaE and the secretion proteins at the newborn pole (Viollier et al. 2002a, b). Following activation of CtrA by GcrA, during the late S-phase, the *cpa* genes under the control of the CtrA/SciP regulatory module will fire followed by the prepilin peptidase CpaA controlled by CtrA solely and lately *pilA* will be transcribed in the G1-phase as it is under the control the MucR/CtrA regulatory module (Fig. 11b) (Laub et al. 2000; Mignolet et al. 2018). Thus, following cell division, PilA is polymerized to form the filament which is secreted through the CpaC pore to assemble into the pilus filament (Mignolet et al. 2018; Viollier et al. 2002b).

1.4.4 Transcriptional Control of the Divisome

As already mentioned before, the switch from the G1-phase to the early S-phase is marked by the replication initiation but also by the establishment in early S-phase of the early events of cell division that is completed in late S-phase. Indeed, during *C. crescentus* cell cycle, the Z-ring formed by polymerization of the FtsZ tubulin-like GTPase proteins is early positioned closer to the flagellated pole and not to mid-cell and will recruit other cell division proteins, however the cytokinesis ends later once the asymmetry of the predivisional cell is established. The Z-ring that acts as a scaffold to build the divisome is spatially regulated by the MipZ protein coordinating the replication initiation with cell division through its interaction with centromere-binding protein ParB (Thanbichler and Shapiro 2006; Kiekebusch et al. 2012). Transcription of the *ftsZ* gene is under the control of several transcriptional regulators. First DnaA (Hottes et al. 2005) and the GcrA/CcrM module during the early S-phase activate *ftsZ* transcription (Panis et al. 2015; Gonzalez and Collier 2013) and finally CtrA, after accumulation in late S-phase will act as a repressor of *ftsZ* expression (Kelly et al. 1998). Moreover, *mipZ* is also under the transcriptional control of GcrA (Fig. 12).

Steady-state levels of FtsZ peak in early predivisional cells and decrease when CtrA-P starts to accumulate (Sackett et al. 1998). Following activation of CtrA and assembly of the Z-ring, *ftsA* and *ftsQ* will be transcriptonally activated by CtrA (Fig. 12). Interestingly, *ftsQ*, a regulator of the Z-ring constriction will only be transcribed in late S-phase when the DNA replication is completed (Wortinger et al. 2000). Moreover, the last essential protein recruited to the divisome is FtsN (Möll and Thanbichler 2009), which contributes to the progress of constriction under positive GcrA control when the promoter is methylated but negatively regulated by GcrA in absence of methylation (Murray et al. 2013). Finally, compartmentalization of the cells during cytokinesis allows the clearing of CtrA-P in the stalk cell (see IV.ii.2) leading to new round of replication (Jonas et al. 2011).

1.4.5 Transcriptional Control of Capsulation

C. crescentus cells undergo a buoyancy switch conferred by a polysaccharidic capsule that is cell cycle regulated. Indeed, in the SW daughter cell the capsulation is prevented (Ardissone et al. 2014) through expression of the *hvyA* gene encoding a bacterial transglutaminase-like cysteine protease that inhibits capsulation (Ardissone et al. 2014) (Fig. 13a, b).

This buoyancy switch has been used to study cell cycle as it allows synchronization of the *C. crescentus* population by density gradient centrifugation. ST and PD cells, that are capsulated (Fig. 13b), sediment in the upper part of the gradient allowing isolation of the non-capsulated SW population that sediments in the lower part (Fig. 13a). *hvyA* is expressed from a promoter that is controlled by CtrA

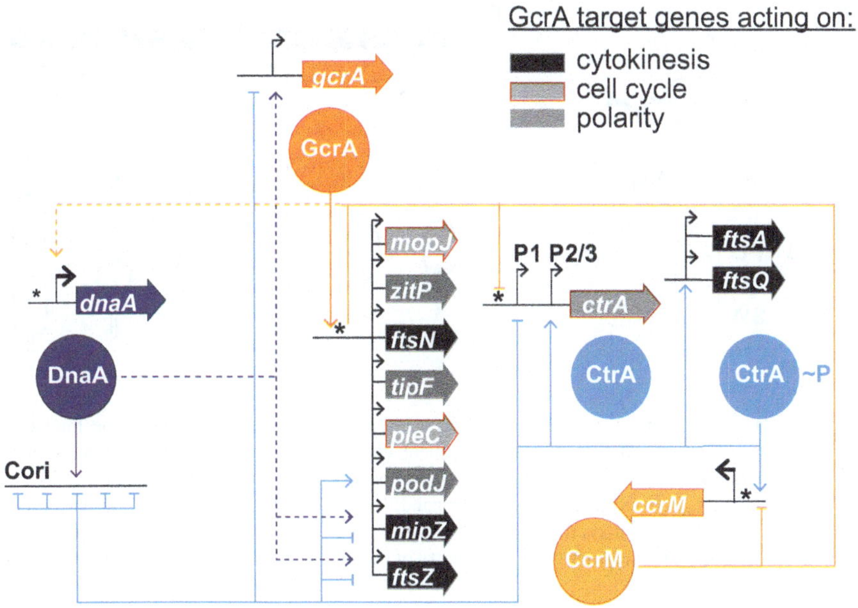

Fig. 12 Schematic of *C. crescentus* targeted genes by the GcrA/CcrM module. Schematic of the regulatory interaction acting on cell-cycle-regulated targeted promoters. In S-phase, the module GcrA/CcrM activates different promoters allowing transcription of genes implicated in cytokinesis, polarity, and cell cycle regulation. The orange line represents transcriptional control by GcrA. Yellow lines represent methylation control by CcrM on 5′-GAnTC-3′ sites (represented with *) on promoters. The blue lines represent positive or negative transcriptional control by CtrA. DnaA control at the promoter of the *gcrA*, *podJ*, and *ftsZ* genes is represented as dotted lines because the occupancy of this promoter by DnaA has only been demonstrated in vitro, but not yet in vivo

activation in G1-phase and MucR repression in late S-phase. Interestingly, in a CtrA thermo-sensitive mutant, transcription of *hvyA* is strongly affected (Ardissone et al. 2014) in contrast to the class II flagellar genes also under the control of CtrA (Quon et al. 1996). Capsulation also confers resistance against the generalized transducing *C. crescentus* bacteriophage ϕCr30, meaning that only the swarmer population is sensitive to ϕCr30 infection.

1.4.6 Transcriptional Control of Holdfast Formation and Stalk Biogenesis

During the G1-phase, implementation of the future cellular structures present in stalk cell starts. The transition from swarmer- to stalk cell is marked by the ejection of the flagellum and retraction of the pili followed by the synthesis of the stalk at the same pole. However, while synthesis of the stalk only starts during the transition, the holdfast formation, that is structure found at the tip of stalk and responsible for permanent surface attachment (Levi and Jenal 2006; Bodenmiller et al. 2004;

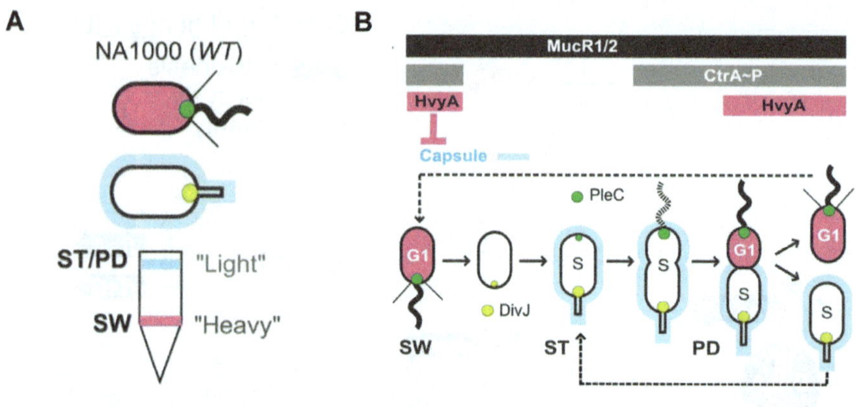

Fig. 13 Capsulation of *C. crescentus* is cell cycle regulated. (**a**) Schematics of the buoyancy switch of *C. crescentus* after cell density gradient centrifugation. Swarmer cells (SW) show a "heavy" phenotype as they sediment in the lower band whereas the "light" stalked (ST) and predivisional cells (PD) sediment in the upper band. (**b**) Schematic of the capsulation during cell cycle and the regulatory interactions that drive capsule (light blue) pattern. Pink cells denote the presence of HvyA under the transcriptional control of the Mucr/CtrA module. Images adapted from Ardissone et al. (2014)

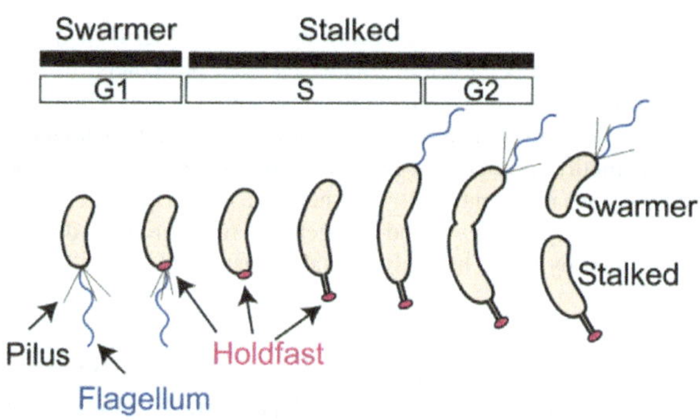

Fig. 14 Development of polar structures during *C. crescentus* cell cycle. The holdfast is shown in pink. Image adapted from Fiebig et al. (2014)

Entcheva-Dimitrov and Spormann 2004), begins at the late swarmer stage (Levi and Jenal 2006) (Fig. 14).

Several loci in *C. crescentus* chromosome encode the holdfast genes implicated in biosynthesis (*hfsE*, *hfsH*, *hfsG*, and *hfsH*), attachment (*hfaA*, *hfaB*, *hfaD*, and *hfaD*) and transport (*hfsD*, *hfsA*, *hfsB*, and *hfsC*). Expression of these genes peaks in late S-phase by activation of CtrA (Fig. 6). Fiebig and colleagues showed that a newly identified holdfast inhibitor gene, *hfiA*, encoding a protein implicated in the negative regulation of HfsJ, a predicted glycosyltransferase required for the holdfast development machinery, is under positive control of the two cell cycle regulators CtrA and GcrA as well as the negative control of StaR (Fiebig et al.

2014). Indeed, *hfsJ* is also under the control the CtrA/SciP module and then produced during late S-phase. Thus, this complex transcriptional control of *hfiA*, i.e. activation by GcrA and CtrA in early and late S-phase, respectively, allows repression of the holdfast synthesis during those stages, while repression of *hfiA* in swarmer cell by StaR correlates with the developmental timing of holdfast synthesis (Fiebig et al. 2014). Moreover, they also showed that environmental changes could override the cell cycle developmental control. Indeed, cells grown in minimal medium accumulate more *hfiA* transcripts leading to a decrease in the population that possesses holdfast. Similarly, overexpression of the LovK/LovR two-component system, implicated in general stress response (Foreman et al. 2012) and modulating cell adhesion (Purcell et al. 2008) leads to an increase in holdfast occurrence (Fiebig et al. 2014). Interestingly, a recent study showed that a fully assembled flagellum promotes synthesis of HfiA while defective mutant in flagellum assembly impairs *hfiA* expression through a mechanism that requires c-di-GMP and promotes holdfast synthesis (Berne et al. 2018) suggesting a role of c-di-GMP in the transcriptional control of *hfiA*. Intriguingly, in *Pseudomonas aeruginosa* a σ^{54}-dependent motility master regulator, FleQ, is known to bind c-di-GMP in order to regulate flagellar gene expression (Hickman and Harwood 2008). It is thus conceivable that a similar mechanism occurs in *C. crescentus* showing how signaling molecule could interfere with cell cycle transcriptional control such as (p)ppGpp in response to environmental cues (Hallez et al. 2017).

Once the SW to ST cell transition started, the stalk organelle is produced at the place formerly occupied by the flagellum. TacA, which accumulates during the G1-phase under the dependency of the CtrA, will control stalk biogenesis through activation of the stalk biogenesis regulator StaR (McGrath et al. 2007) and the polar organizer protein SpmX (Janakiraman et al. 2016). Interestingly, a *tacA* deletion generates stalkless cells while a *staR* deletion does not, suggesting that additional downstream targets of TacA-σ^{54} contribute to the control of stalk biosynthesis. However, similarly to the holdfast synthesis, environmental cues can override stalk synthesis. Indeed, while a *tacA* deletion strain appears stalkless in rich medium, this mutant harbors stalks in medium depleted for phosphate (Biondi et al. 2006b), suggesting that phosphate starvation induces somehow the stalk synthesis and that *tacA* mutant is impaired in the regulation of the stalk biogenesis but not in its synthesis (Biondi et al. 2006b).

1.5 Transcriptional Reprogramming in Response to Changing Environmental Conditions

Besides the transcriptional changes that occur throughout the cell cycle, global transcriptional responses also occur in response to changing environmental conditions, particularly those involving stress responses. When at least one source of nutrients becomes limiting for growth, when toxic compounds accumulate or in case of environmental stress, the cells enter in stationary phase (Bergkessel et al. 2016). A common feature within bacteria to react with those situations and maintain viability is to reduce ribosomal activity but also adapt metabolism, reprogram transcription and translation. In the *E. coli*, the genetic switch occurs mainly in response to the alternative sigma factor *rpoS* and the stress alarmone (p)ppGpp (Bergkessel et al. 2016; Potrykus and Cashel 2008). In *C. crescentus*, adaptation to stationary phase leads to drastic morphological changes (elongated and helicoidal cell morphologies) and increase to stress resistance (Wortinger et al. 1998), however the underlying gene regulation and transcriptional control are not well understood. When *C. crescentus* enters into stationary phase, DNA replication is suppressed and the oscillation of the cell cycle transcriptional regulator CtrA during the cell cycle (degradation in S-phase) is abrogated (Hallez et al. 2017; Leslie et al. 2015; Sanselicio and Viollier 2015). This maintenance of CtrA arrests cells either in G1 or G2 state. CtrA is conserved and plays an important role in alpha-proteobacterial cell cycle and developmental control, even in obligate intracellular pathogens from the order *Rickettsiales* such as *Ehrlichia chaffeensis* (Cheng et al. 2011). Upon infection in human monocytes, CtrA is induced at the late growth stage during the differentiation from large reticulate cells (RCs) to small dense-cored cells (DCs) where it induces stress-resistance traits as *surE*, a stationary phase survival protein and factors to initiate new round of infection (Cheng et al. 2011; Rikihisa 2015) showing that CtrA or other conserved master regulators may also have an important role of transcriptional factor in stationary phase which can also be affected by the alarmone (p)ppGpp that can also affect cell.

1.5.1 (p)ppGpp

Following stress, starvation, or entry in stationary phase, bacteria accumulate guanosine tetraphosphate and guanosine pentaphosphate (referred to as (p)ppGpp) in order to modulate several key cellular processes such as transcription, translation or ribosome biogenesis, growth and cell cycle (Hallez et al. 2017; Potrykus and Cashel 2008). Cell cycle control and mechanism of action of (p)ppGpp in response to starvation in *C. crescentus* is not fully understood, however it has been demonstrated that in response to nitrogen starvation, the activation is based on the nitrogen-related phosphoenolpyruvate (PEP)-phosphotransferase system (PTS^{Ntr}) (Hallez et al. 2017; Ronneau et al. 2016). In *C. crescentus*, SpoT is a bifunctional protein, responsible for both synthesis and hydrolysis of the (p)ppGpp. Moreover, recent study showed that this PTS^{Ntr} system and particularly phosphorylated $EIIA^{Ntr}$ inhibits the hydrolase activity of SpoT in response to nitrogen starvation (Ronneau et al. 2019). Even if a Δ*spoT* mutant does not have a viability defect in stationary phase, (p)ppGpp seems to be necessary to cause the G1 arrest in stationary phase or upon carbon starvation as some cells accumulate in G2-phase (Leslie et al. 2015). Moreover, upon entry in stationary phase or carbon starvation, levels of DnaA decrease leading to inhibition of replication initiation and growth arrest in G1-phase (Felletti et al. 2019; Lesley and Shapiro 2008). In a Δ*spoT* mutant, the DnaA levels are no longer reduced in contrast to the stabilization of CtrA suggesting that (p)ppGpp might be implicated in CtrA activity and/or stability (Hallez et al. 2017; Leslie et al. 2015; Boutte and Crosson 2011). As a main effector of G1-phase and central cell cycle regulator, CtrA might play a role in regulating cell cycle in stationary phase.

While deciphering the role of the single PAS domain protein MopJ that accumulates in stationary phase in a (p)ppGpp-dependent way in order to enhance CtrA levels, it was shown that transcription of GcrA-targeted promoters is increased in stationary phase and this may also link (p)ppGpp and reprogramming of cell cycle transcription (Hallez et al. 2017; Sanselicio and Viollier 2015; Sanselicio et al. 2015).

1.6 Conclusions

Several transcriptional factors that affect cell cycle transcription in *C. crescentus* have been characterized over the last 40 years and the cell cycle regulated transcripts and promoters have been identified. Many transcription factors and thus likely the mechanisms that they are part of are conserved in alpha-proteobacteria. Efforts so far have focused on transcription factors that bind DNA independently of RNA polymerase, factors that bind or affect RNA polymerase directly such as (p)ppGpp are well positioned to affect cell cycle transcription. Future research should also investigate these avenues further as well.

References

Anderson PE, Gober JW (2000) FlbT, the post-transcriptional regulator of flagellin synthesis in Caulobacter crescentus, interacts with the 5′ untranslated region of flagellin mRNA. Mol Microbiol 38:41–52

Ardissone S, Viollier PH (2015) Interplay between flagellation and cell cycle control in Caulobacter. Curr Opin Microbiol 28:83–92

Ardissone S et al (2014) Cell cycle constraints on capsulation and bacteriophage susceptibility. elife 3:e03587

Ardissone S et al (2016) Cell cycle constraints and environmental control of local DNA hypomethylation in α-proteobacteria. PLoS Genet 12:e1006499

Ardissone S, Kint N, Petrignani B, Panis G, Viollier PH (2020) Secretion relieves translational co-repression by a specialized flagellin paralog. Dev Cell 55(4):500–513.e4. https://doi.org/10.1016/j.devcel.2020.10.005

Arias-Cartin R et al (2017) Replication fork passage drives asymmetric dynamics of a critical nucleoid-associated protein in Caulobacter. EMBO J 36:301–318

Beaufay F et al (2015) A NAD-dependent glutamate dehydrogenase coordinates metabolism with cell division in Caulobacter crescentus. EMBO J 34:1786–1800

Bergkessel M, Basta DW, Newman DK (2016) The physiology of growth arrest: uniting molecular and environmental microbiology. Nat Rev Microbiol 14:549–562

Berne C et al (2018) Feedback regulation of Caulobacter crescentus holdfast synthesis by flagellum assembly via the holdfast inhibitor HfiA. Mol Microbiol 110:219–238

Biondi EG et al (2006a) Regulation of the bacterial cell cycle by an integrated genetic circuit. Nature 444:899–904

Biondi EG et al (2006b) A phosphorelay system controls stalk biogenesis during cell cycle progression in Caulobacter crescentus. Mol Microbiol 59:386–401

Bodenmiller D, Toh E, Brun YV (2004) Development of surface adhesion in Caulobacter crescentus. J Bacteriol 186:1438–1447
Boutte CC, Crosson S (2011) The complex logic of stringent response regulation in Caulobacter crescentus: starvation signaling in an oligotrophic environment. Mol Microbiol 80:695–714
Boutte CC, Henry JT, Crosson S (2012) ppGpp and Polyphosphate Modulate Cell Cycle Progression in Caulobacter crescentus. J Bacteriol 194:28–35
Brown PJB, Hardy GG, Trimble MJ, Brun YV (2009) Complex regulatory pathways coordinate cell-cycle progression and development in Caulobacter crescentus. Adv Microb Physiol 54:1–101
Brun YV (2001) Global analysis of a bacterial cell cycle: tracking down necessary functions and their regulators. Trends Microbiol 9:405–407
Camara JE et al (2005) Hda inactivation of DnaA is the predominant mechanism preventing hyperinitiation of Escherichia coli DNA replication. EMBO Rep 6:736–741
Cheng Z, Miura K, Popov VL, Kumagai Y, Rikihisa Y (2011) Insights into the CtrA regulon in development of stress resistance in obligatory intracellular pathogen Ehrlichia chaffeensis. Mol Microbiol 82:1217–1234
Collier J, Shapiro L (2009) Feedback control of DnaA-mediated replication initiation by replisome-associated HdaA protein in Caulobacter. J Bacteriol 191:5706–5716
Collier J, Murray SR, Shapiro L (2006) DnaA couples DNA replication and the expression of two cell cycle master regulators. EMBO J 25:346–356
Collier J, McAdams HH, Shapiro L (2007) A DNA methylation ratchet governs progression through a bacterial cell cycle. Proc Natl Acad Sci U S A 104:17111–17116
Costa TRD et al (2015) Secretion systems in Gram-negative bacteria: structural and mechanistic insights. Nat Rev Microbiol 13:343–359
Craig L, Pique ME, Tainer JA (2004) Type IV pilus structure and bacterial pathogenicity. Nat Rev Microbiol 2:363–378
Davis NJ et al (2013) De- and repolarization mechanism of flagellar morphogenesis during a bacterial cell cycle. Genes Dev 27:2049–2062
Delaby M, Panis G, Viollier PH (2019) Bacterial cell cycle and growth phase switch by the essential transcriptional regulator CtrA. Nucleic Acids Res 47(20):10628–10644. https://doi.org/10.1093/nar/gkz846
Domian IJ, Reisenauer A, Shapiro L (1999) Feedback control of a master bacterial cell-cycle regulator. Proc Natl Acad Sci U S A 96:6648–6653
Duderstadt KE, Chuang K, Berger JM (2011) DNA stretching by bacterial initiators promotes replication origin opening. Nature 478:209–213
Duerig A et al (2009) Second messenger-mediated spatiotemporal control of protein degradation regulates bacterial cell cycle progression. Genes Dev 23:93–104
Ellison CK et al (2017) Obstruction of pilus retraction stimulates bacterial surface sensing. Science 358:535–538
Ely B, Ely TW (1989) Use of pulsed field gel electrophoresis and transposon mutagenesis to estimate the minimal number of genes required for motility in Caulobacter crescentus. Genetics 123:649–654
Ely B et al (1986) General nonchemotactic mutants of Caulobacter crescentus. Genetics 114:717–730
Entcheva-Dimitrov P, Spormann AM (2004) Dynamics and control of biofilms of the oligotrophic bacterium Caulobacter crescentus. J Bacteriol 186:8254–8266
Fang G et al (2013) Transcriptomic and phylogenetic analysis of a bacterial cell cycle reveals strong associations between gene co-expression and evolution. BMC Genomics 14:450
Felletti M, Omnus DJ, Jonas K (2019) Regulation of the replication initiator DnaA in Caulobacter crescentus. Biochim Biophys Acta Gene Regul Mech 1862(7):697–705. https://doi.org/10.1016/j.bbagrm.2018.01.004
Fernandez-Fernandez C, Gonzalez D, Collier J (2011) Regulation of the Activity of the Dual-Function DnaA Protein in Caulobacter crescentus. PLoS One 6:e26028

Fiebig A et al (2014) A cell cycle and nutritional checkpoint controlling bacterial surface adhesion. PLoS Genet 10:e1004101
Fioravanti A et al (2013) DNA binding of the cell cycle transcriptional regulator GcrA depends on N6-adenosine methylation in Caulobacter crescentus and other alphaproteobacteria. PLoS Genet 9:e1003541
Foreman R, Fiebig A, Crosson S (2012) The LovK-LovR two-component system is a regulator of the general stress pathway in Caulobacter crescentus. J Bacteriol 194:3038–3049
Fumeaux C et al (2014) Cell cycle transition from S-phase to G1 in Caulobacter is mediated by ancestral virulence regulators. Nat Commun 5
Gonzalez D, Collier J (2013) DNA methylation by CcrM activates the transcription of two genes required for the division of Caulobacter crescentus. Mol Microbiol 88:203–218
Gora KG et al (2010) A cell-type-specific protein-protein interaction modulates transcriptional activity of a master regulator in Caulobacter crescentus. Mol Cell 39:455–467
Gora KG et al (2013) Regulated proteolysis of a transcription factor complex is critical to cell cycle progression in Caulobacter crescentus. Mol Microbiol 87:1277–1289
Guo MS, Haakonsen DL, Zeng W, Schumacher MA, Laub MT (2018) A bacterial chromosome structuring protein binds overtwisted DNA to stimulate type II topoisomerases and enable DNA replication. Cell 175:583–597.e23
Haakonsen DL, Yuan AH, Laub MT (2015) The bacterial cell cycle regulator GcrA is a σ70 cofactor that drives gene expression from a subset of methylated promoters. Genes Dev 29:2272–2286
Hallez R, Bellefontaine A-F, Letesson J-J, De Bolle X (2004) Morphological and functional asymmetry in alpha-proteobacteria. Trends Microbiol 12:361–365
Hallez R, Delaby M, Sanselicio S, Viollier PH (2017) Hit the right spots: cell cycle control by phosphorylated guanosines in alphaproteobacteria. Nat Rev Microbiol 15:137–148
Hickman JW, Harwood CS (2008) Identification of FleQ from Pseudomonas aeruginosa as a c-di-GMP-responsive transcription factor. Mol Microbiol 69:376–389
Holtzendorff J et al (2004) Oscillating global regulators control the genetic circuit driving a bacterial cell cycle. Science 304:983–987
Hottes AK, Shapiro L, McAdams HH (2005) DnaA coordinates replication initiation and cell cycle transcription in Caulobacter crescentus. Mol Microbiol 58:1340–1353
Huitema E, Pritchard S, Matteson D, Radhakrishnan SK, Viollier PH (2006) Bacterial birth scar proteins mark future flagellum assembly site. Cell 124:1025–1037
Iniesta AA, McGrath PT, Reisenauer A, McAdams HH, Shapiro L (2006) A phospho-signaling pathway controls the localization and activity of a protease complex critical for bacterial cell cycle progression. Proc Natl Acad Sci U S A 103:10935–10940
Janakiraman RS, Brun YV (1999) Cell cycle control of a holdfast attachment gene in Caulobacter crescentus. J Bacteriol 181:1118–1125
Janakiraman B, Mignolet J, Narayanan S, Viollier PH, Radhakrishnan SK (2016) In-phase oscillation of global regulons is orchestrated by a pole-specific organizer. Proc Natl Acad Sci U S A 113:12550–12555
Jenal U, Fuchs T (1998) An essential protease involved in bacterial cell-cycle control. EMBO J 17:5658–5669
Jonas K, Chen YE, Laub MT (2011) Modularity of the bacterial cell cycle enables independent spatial and temporal control of DNA replication. Curr Biol 21:1092–1101
Jonas K, Liu J, Chien P, Laub MT (2013) Proteotoxic stress induces a cell-cycle arrest by stimulating Lon to degrade the replication initiator DnaA. Cell 154:623–636
Joshi KK, Bergé M, Radhakrishnan SK, Viollier PH, Chien P (2015) An adaptor hierarchy regulates proteolysis during a bacterial cell cycle. Cell 163:419–431
Kaczmarczyk A, Hempel AM, von Arx C, Böhm R, Dubey BN, Nesper J, Schirmer T, Hiller S, Jenal U (2020) Precise timing of transcription by c-di-GMP coordinates cell cycle and morphogenesis in Caulobacter. Nat Commun 11(1):816. https://doi.org/10.1038/s41467-020-14585-6

Katayama T, Ozaki S, Keyamura K, Fujimitsu K (2010) Regulation of the replication cycle: conserved and diverse regulatory systems for DnaA and *oriC*. Nat Rev Microbiol 8:163–170

Kelly AJ, Sackett MJ, Din N, Quardokus E, Brun YV (1998) Cell cycle-dependent transcriptional and proteolytic regulation of FtsZ in Caulobacter. Genes Dev 12:880–893

Kiekebusch D, Michie KA, Essen L-O, Löwe J, Thanbichler M (2012) Localized dimerization and nucleoid binding drive gradient formation by the bacterial cell division inhibitor MipZ. Mol Cell 46:245–259

Lam H, Schofield WB, Jacobs-Wagner C (2006) A landmark protein essential for establishing and perpetuating the polarity of a bacterial cell. Cell 124:1011–1023

Laub MT, McAdams HH, Feldblyum T, Fraser CM, Shapiro L (2000) Global analysis of the genetic network controlling a bacterial cell cycle. Science 290:2144–2148

Laub MT, Chen SL, Shapiro L, McAdams HH (2002) Genes directly controlled by CtrA, a master regulator of the Caulobacter cell cycle. Proc Natl Acad Sci U S A 99:4632–4637

Laub MT, Shapiro L, McAdams HH (2007) Systems biology of Caulobacter. Annu Rev Genet 41:429–441

Lesley JA, Shapiro L (2008) SpoT regulates DnaA stability and initiation of DNA replication in carbon-starved Caulobacter crescentus. J Bacteriol 190:6867–6880

Leslie DJ et al (2015) Nutritional control of DNA replication initiation through the proteolysis and regulated translation of DnaA. PLoS Genet 11:e1005342

Levi A, Jenal U (2006) Holdfast formation in motile swarmer cells optimizes surface attachment during Caulobacter crescentus development. J Bacteriol 188:5315–5318

Li S, Brazhnik P, Sobral B, Tyson JJ (2009) Temporal controls of the asymmetric cell division cycle in Caulobacter crescentus. PLoS Comput Biol 5

Liu J, Francis LI, Jonas K, Laub MT, Chien P (2016) ClpAP is an auxiliary protease for DnaA degradation in Caulobacter crescentus. Mol Microbiol 102:1075–1085

Llewellyn M, Dutton RJ, Easter J, O'Donnol D, Gober JW (2005) The conserved flaF gene has a critical role in coupling flagellin translation and assembly in Caulobacter crescentus. Mol Microbiol 57:1127–1142

Mangan EK et al (1999) FlbT couples flagellum assembly to gene expression in Caulobacter crescentus. J Bacteriol 181:6160–6170

McGrath PT, Iniesta AA, Ryan KR, Shapiro L, McAdams HH (2006) A dynamically localized protease complex and a polar specificity factor control a cell cycle master regulator. Cell 124:535–547

McGrath PT et al (2007) High-throughput identification of transcription start sites, conserved promoter motifs and predicted regulons. Nat Biotechnol 25:584–592

Melville S, Craig L (2013) Type IV pili in Gram-positive bacteria. Microbiol Mol Biol Rev 77:323–341

Mignolet J et al (2016) Functional dichotomy and distinct nanoscale assemblies of a cell cycle-controlled bipolar zinc-finger regulator. elife 5

Mignolet J, Panis G, Viollier PH (2018) More than a Tad: spatiotemporal control of Caulobacter pili. Curr Opin Microbiol 42:79–86

Mohr CD, MacKichan JK, Shapiro L (1998) A membrane-associated protein, FliX, is required for an early step in Caulobacter flagellar assembly. J Bacteriol 180:2175–2185

Möll A, Thanbichler M (2009) FtsN-like proteins are conserved components of the cell division machinery in proteobacteria. Mol Microbiol 72:1037–1053

Muir RE, Gober JW (2002) Mutations in FlbD that relieve the dependency on flagellum assembly alter the temporal and spatial pattern of developmental transcription in Caulobacter crescentus. Mol Microbiol 43:597–615

Muir RE, O'Brien TM, Gober JW (2001) The Caulobacter crescentus flagellar gene, fliX, encodes a novel trans-acting factor that couples flagellar assembly to transcription. Mol Microbiol 39:1623–1637

Mullin DA, Newton A (1989) Ntr-like promoters and upstream regulatory sequence ftr are required for transcription of a developmentally regulated Caulobacter crescentus flagellar gene. J Bacteriol 171:3218–3227

Murray SM, Panis G, Fumeaux C, Viollier PH, Howard M (2013) Computational and genetic reduction of a cell cycle to its simplest, primordial components. PLoS Biol 11:e1001749

Nesper J et al (2017) Cyclic di-GMP differentially tunes a bacterial flagellar motor through a novel class of CheY-like regulators. elife 6:e28842

Nierman WC et al (2001) Complete genome sequence of Caulobacter crescentus. Proc Natl Acad Sci U S A 98:4136–4141

Ouimet M-C, Marczynski GT (2000) Analysis of a cell-cycle promoter bound by a response regulator11Edited by M. Yaniv. J Mol Biol 302:761–775

Panis G, Murray SR, Viollier PH (2015) Versatility of global transcriptional regulators in alpha-Proteobacteria: from essential cell cycle control to ancillary functions. FEMS Microbiol Rev 39:120–133

Potrykus K, Cashel M (2008) (p)ppGpp: still magical? Annu Rev Microbiol 62:35–51

Purcell EB, Boutte CC, Crosson S (2008) Two-component signaling systems and cell cycle control in Caulobacter crescentus. Adv Exp Med Biol 631:122–130

Quon KC, Marczynski GT, Shapiro L (1996) Cell cycle control by an essential bacterial two-component signal transduction protein. Cell 84:83–93

Quon KC, Yang B, Domian IJ, Shapiro L, Marczynski GT (1998) Negative control of bacterial DNA replication by a cell cycle regulatory protein that binds at the chromosome origin. Proc Natl Acad Sci U S A 95:120–125

Radhakrishnan SK, Thanbichler M, Viollier PH (2008) The dynamic interplay between a cell fate determinant and a lysozyme homolog drives the asymmetric division cycle of Caulobacter crescentus. Genes Dev 22:212–225

Radhakrishnan SK, Pritchard S, Viollier PH (2010) Coupling prokaryotic cell fate and division control with a bifunctional and oscillating oxidoreductase homolog. Dev Cell 18:90–101

Ramakrishnan G, Newton A (1990) FlbD of Caulobacter crescentus is a homologue of the NtrC (NRI) protein and activates sigma 54-dependent flagellar gene promoters. Proc Natl Acad Sci U S A 87:2369–2373

Reisenauer A, Quon K, Shapiro L (1999) The CtrA response regulator mediates temporal control of gene expression during the caulobacter cell cycle. J Bacteriol 181:2430–2439

Ricci DP et al (2016) Cell cycle progression in Caulobacter requires a nucleoid-associated protein with high AT sequence recognition. Proc Natl Acad Sci U S A 113:E5952–E5961

Rikihisa Y (2015) Molecular pathogenesis of Ehrlichia chaffeensis infection. Annu Rev Microbiol 69:283–304

Ronneau S, Petit K, De Bolle X, Hallez R (2016) Phosphotransferase-dependent accumulation of (p)ppGpp in response to glutamine deprivation in Caulobacter crescentus. Nat Commun 7:11423

Ronneau S et al (2019) Regulation of (p)ppGpp hydrolysis by a conserved archetypal regulatory domain. Nucleic Acids Res 47:843–854

Sackett MJ, Kelly AJ, Brun YV (1998) Ordered expression of ftsQA and ftsZ during the Caulobacter crescentus cell cycle. Mol Microbiol 28:421–434

Sanselicio S, Viollier PH (2015) Convergence of alarmone and cell cycle signaling from trans-encoded sensory domains. mBio 6:e01415-15

Sanselicio S, Bergé M, Théraulaz L, Radhakrishnan SK, Viollier PH (2015) Topological control of the Caulobacter cell cycle circuitry by a polarized single-domain PAS protein. Nat Commun 6:7005

Schrader JM et al (2016) Dynamic translation regulation in Caulobacter cell cycle control. Proc Natl Acad Sci U S A 113:E6859–E6867

Shaheen SM, Ouimet M-C, Marczynski GT (2009) Comparative analysis of Caulobacter chromosome replication origins. Microbiol Read Engl 155:1215–1225

Siwach M, Kumar L, Palani S, Muraleedharan S, Panis G, Fumeaux C, Mony BM, Sanyal S, Viollier PH, Radhakrishnan SK (2021) An organelle-tethering mechanism couples flagellation to cell division in bacteria. Dev Cell 56(5):657–670.e4. https://doi.org/10.1016/j.devcel.2021.01.013

Skarstad K, Katayama T (2013) Regulating DNA replication in bacteria. Cold Spring Harb Perspect Biol 5:a012922

Skerker JM, Laub MT (2004) Cell-cycle progression and the generation of asymmetry in *Caulobacter crescentus*. Nat Rev Microbiol 2:325–337

Skerker JM, Shapiro L (2000) Identification and cell cycle control of a novel pilus system in Caulobacter crescentus. EMBO J 19:3223–3234

Skerker JM, Prasol MS, Perchuk BS, Biondi EG, Laub MT (2005) Two-component signal transduction pathways regulating growth and cell cycle progression in a bacterium: a system-level analysis. PLoS Biol 3:e334

Smith SC et al (2014) Cell cycle-dependent adaptor complex for ClpXP-mediated proteolysis directly integrates phosphorylation and second messenger signals. Proc Natl Acad Sci U S A 111:14229–14234

Tan MH, Kozdon JB, Shen X, Shapiro L, McAdams HH (2010) An essential transcription factor, SciP, enhances robustness of Caulobacter cell cycle regulation. Proc Natl Acad Sci U S A 107:18985–18990

Taylor JA, Ouimet M-C, Wargachuk R, Marczynski GT (2011) The Caulobacter crescentus chromosome replication origin evolved two classes of weak DnaA binding sites. Mol Microbiol 82:312–326

Taylor JA, Panis G, Viollier PH, Marczynski GT (2017) A novel nucleoid-associated protein coordinates chromosome replication and chromosome partition. Nucleic Acids Res 45:8916–8929

Thanbichler M, Shapiro L (2006) MipZ, a spatial regulator coordinating chromosome segregation with cell division in Caulobacter. Cell 126:147–162

Tsokos CG, Perchuk BS, Laub MT (2011) A dynamic complex of signaling proteins uses polar localization to regulate cell-fate asymmetry in Caulobacter crescentus. Dev Cell 20:329–341

Viollier PH, Sternheim N, Shapiro L (2002a) Identification of a localization factor for the polar positioning of bacterial structural and regulatory proteins. Proc Natl Acad Sci U S A 99:13831–13836

Viollier PH, Sternheim N, Shapiro L (2002b) A dynamically localized histidine kinase controls the asymmetric distribution of polar pili proteins. EMBO J 21:4420–4428

Wortinger MA, Quardokus EM, Brun YV (1998) Morphological adaptation and inhibition of cell division during stationary phase in Caulobacter crescentus. Mol Microbiol 29:963–973

Wortinger M, Sackett MJ, Brun YV (2000) CtrA mediates a DNA replication checkpoint that prevents cell division in Caulobacter crescentus. EMBO J 19:4503–4512

Wu J, Benson AK, Newton A (1995) Global regulation of a sigma 54-dependent flagellar gene family in Caulobacter crescentus by the transcriptional activator FlbD. J Bacteriol 177:3241–3250

Wu J, Ohta N, Newton A (1998) An essential, multicomponent signal transduction pathway required for cell cycle regulation in Caulobacter. Proc Natl Acad Sci U S A 95:1443–1448

Wu J, Ohta N, Zhao J-L, Newton A (1999) A novel bacterial tyrosine kinase essential for cell division and differentiation. Proc Natl Acad Sci U S A 96:13068–13073

Wu X et al (2018) Structural insights into the unique mechanism of transcription activation by Caulobacter crescentus GcrA. Nucleic Acids Res 46:3245–3256

Zhou B et al (2015) The global regulatory architecture of transcription during the Caulobacter cell cycle. PLoS Genet 11:e1004831

Zweiger G, Marczynski G, Shapiro L (1994) A Caulobacter DNA methyltransferase that functions only in the predivisional cell. J Mol Biol 235:472–485

Cell Cycle Signal Transduction and Proteolysis in *Caulobacter*

Justin J. Zik and Kathleen R. Ryan

Abstract

Caulobacter crescentus is an extremely informative model system for studies of the bacterial cell division cycle. *Caulobacter* resembles eukaryotes in that it performs chromosome replication once and only once per cell division. The ability to isolate G1-phase cells and examine them during synchronous passage through the division cycle has enabled studies that are difficult or impossible in other bacteria. Groundbreaking studies in *Caulobacter* have revealed that subcellular localization and non-canonical interactions are important for the function of two-component signaling proteins. In addition, several functions and mechanisms of action of the small signaling molecule cyclic-di-GMP have been uncovered in studies of *Caulobacter* cell division and the development of its polar organelles. Finally, the temporally controlled degradation of key *Caulobacter* proteins is mediated by ATP-dependent proteases in concert with novel, hierarchical adaptor complexes. This review describes the signal transduction systems which orchestrate the *Caulobacter* cell cycle and cell fate decisions. We pay particular attention to non-canonical mechanisms in two-component signaling, functions of cyclic-di-GMP, roles and mechanisms of regulated proteolysis, and environmental effects upon the cell cycle regulatory network.

J. J. Zik · K. R. Ryan (✉)
Department of Plant and Microbial Biology, University of California, Berkeley, Berkeley, CA, USA
e-mail: krr@berkeley.edu

E. Biondi (ed.), *Cell Cycle Regulation and Development in Alphaproteobacteria*,
https://doi.org/10.1007/978-3-030-90621-4_3

1 Introduction

Caulobacter cell division is asymmetric, producing one motile swarmer cell and one sessile stalked cell (Fig. 1a) (Kirkpatrick and Viollier 2012). The stalked progeny immediately begins a new round of chromosome replication (S-phase) and cell division, while the swarmer progeny is temporarily suspended in a nonreplicating state (G1). When the appropriate environmental conditions are present, the swarmer cell differentiates into a stalked cell by remodeling its polar organelles. The polar flagellum is ejected and is replaced by the stalk, a narrow extension of the cell envelope bearing adhesive holdfast material at the tip. At the same time, the cell gains the ability to initiate chromosome replication, so that entry into the cell cycle is associated with a motile-to-sessile lifestyle decision. The cell division cycle and its associated morphological changes are orchestrated by a core network of two-component signaling proteins and proteins that make, degrade, and respond to the

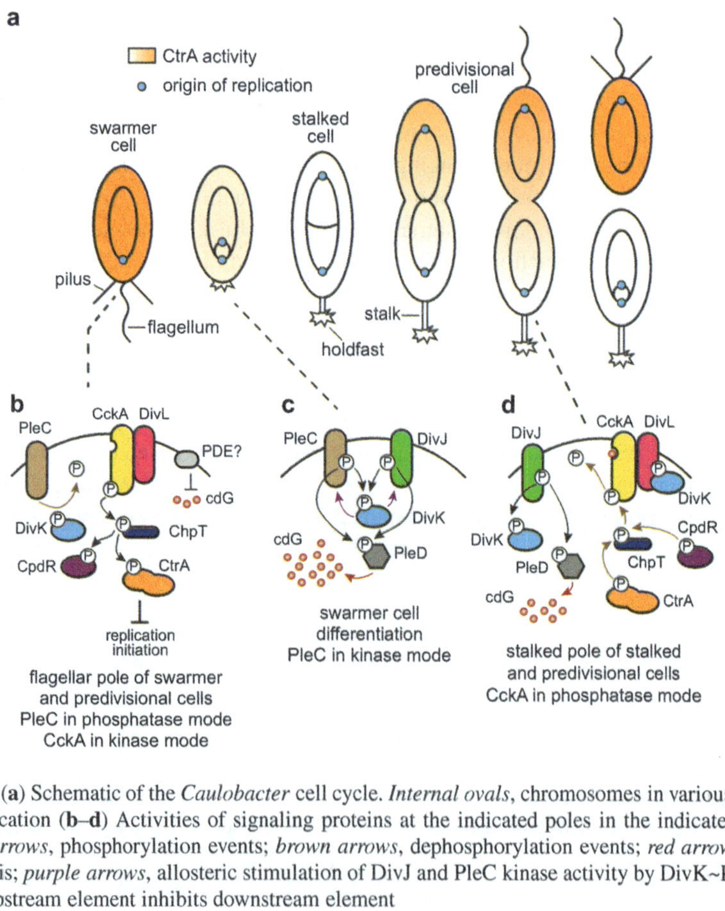

Fig. 1 (**a**) Schematic of the *Caulobacter* cell cycle. *Internal ovals*, chromosomes in various stages of replication (**b–d**) Activities of signaling proteins at the indicated poles in the indicated cells. *Black arrows*, phosphorylation events; *brown arrows*, dephosphorylation events; *red arrows*, cdG synthesis; *purple arrows*, allosteric stimulation of DivJ and PleC kinase activity by DivK~P; *black bars*, upstream element inhibits downstream element

signaling molecule cyclic-di-GMP. Here we provide brief descriptions of these systems to serve as background for their specific functions in *Caulobacter*.

1.1 The Two-Component Signaling Paradigm

Two-component signaling systems are ubiquitous in bacteria, where they mediate responses to both intracellular and extracellular cues (Zschiedrich et al. 2016). Histidine kinases possess a variety of sensory domains, but all share the conserved dimerization and histidine phosphotransfer (DHp) domain and the ATP-binding catalytic domain (CA). In a canonical two-component signaling pathway, the histidine kinase senses a specific signal and autophosphorylates on a conserved histidine residue within the DHp domain, using the terminal phosphoryl group of a bound ATP molecule. The histidine kinase then serves as a phosphodonor for its cognate response regulator, the output component of the system. Response regulators contain a conserved receiver domain (RD) which catalyzes transfer of the phosphoryl group to a conserved aspartate residue within the RD. Phosphorylation of the conserved aspartate triggers conformational changes in the RD, which lead to downstream responses. Most response regulators exert their effects through additional domains whose activity is regulated by receiver domain phosphorylation (Gao and Stock 2010). For example, response regulators with sequence-specific DNA-binding domains alter gene expression in response to upstream signals, and those with attached diguanylate cyclase or phosphodiesterase domains function by synthesizing or breaking down cyclic-di-GMP, respectively. Response regulators consisting of an isolated RD work via phosphorylation-induced changes in protein–protein interactions.

Signal shut-off in two-component systems occurs via several mechanisms. Purified response regulator proteins have different degrees of auto-phosphatase activity, with some phosphorylated species persisting only for seconds, while others are stable for minutes to hours (Bourret 2010). Some signaling pathways include dedicated phosphatases that dephosphorylate specific response regulators (Silversmith 2010). Finally, histidine kinases of the HisKA family are often bifunctional, mediating either phosphorylation or dephosphorylation of the cognate response regulator (Huynh and Stewart 2011). The output of a HisKA protein therefore depends upon factors that modulate its signaling state.

1.2 Cyclic-di-GMP-Dependent Signaling

Cyclic-di-GMP (cdG) is a second messenger found throughout the *Bacteria* that functions as a key regulator of lifestyle decisions (Jenal et al. 2017). Low intracellular levels of cdG typically favor a motile, planktonic lifestyle, while higher concentrations trigger surface attachment and biofilm formation. cdG is synthesized by diguanylate cyclases (DGCs) and degraded by phosphodiesterases (PDEs). These enzymes often contain additional signaling domains such as RD and Per-Arnt-Sim

(PAS), which regulate DGC or PDE activity in response to upstream signals. cdG exerts its effects by binding to proteins and allosterically regulating their activity (Chou and Galperin 2016). PilZ domain-containing proteins are common cdG effectors, as are catalytically inactive DGC domains repurposed to serve as cdG sensors. Because some DGC and PDE enzymes are themselves response regulators, and because cdG can directly or indirectly affect the activity of histidine kinases (Dubey et al. 2016; Mann et al. 2016), cdG-based signaling systems and two-component systems can be intimately interconnected.

2 Two Core Phosphorylation Pathways Regulate the *Caulobacter* Cell Cycle and Development

DivJ-PleC-DivK *Caulobacter* development and cell cycle progression are orchestrated by a pair of two-component signaling pathways that are interconnected by the small molecule c-di-GMP and by non-canonical interactions among histidine kinases and response regulators. One pathway consists of the membrane-bound histidine kinase DivJ, the bifunctional, membrane-bound kinase PleC, and the single-domain response regulator DivK. In predivisional cells, DivJ is located at the stalked pole and functions as a DivK kinase (Fig. 1d) (Wheeler and Shapiro 1999). PleC is located at the flagellar pole opposite the stalk (Wheeler and Shapiro 1999), where it dephosphorylates DivK (Fig. 1b) (Hecht et al. 1995; Wheeler and Shapiro 1999). When cell division is under way, and the cytoplasm of the predivisional cell becomes separated in to stalked and swarmer compartments, phosphorylated DivK (DivK~P) accumulates in the stalked compartment, while unphosphorylated DivK accumulates in the swarmer compartment (Matroule et al. 2004). In *Caulobacter* progeny, high levels of DivK~P are associated with the replicative stalked cell fate, whereas unphosphorylated DivK is associated with the motile, non-replicating swarmer state.

CckA-ChpT-CtrA The second core two-component system is composed of the bifunctional histidine kinase CckA, the histidine phosphotransferase ChpT, and the two response regulators CtrA and CpdR (Quon et al. 1996; Jacobs et al. 1999; Biondi et al. 2006b; Iniesta et al. 2006). All are essential for viability except CpdR, which is dispensable. When CckA is in kinase mode, the phosphoryl group from ATP is transferred from the conserved histidine residue in the DHp domain to a conserved aspartate residue within a contiguous receiver domain. From there, it is passed to the histidine phosphotransferase protein ChpT, which in turn phosphorylates either CtrA or CpdR (Biondi et al. 2006b). CckA resides at both poles of the *Caulobacter* predivisional cell, but it acts primarily as a phosphatase at the stalked pole (Fig 1d) and as a kinase at the pole opposite the stalk (Fig. 1a) (Chen et al. 2009, 2011; Iniesta et al. 2010). Thus, when cell division occurs, the stalked progeny inherits unphosphorylated CtrA and CpdR, while the swarmer progeny inherits phosphorylated CtrA and CpdR.

CtrA contains an N-terminal receiver domain and a C-terminal DNA-binding domain. When phosphorylated, CtrA directly promotes or represses the expression of ~100 *Caulobacter* genes (Laub et al. 2002), and its indirect regulon includes one third of the transcripts whose levels vary during the *Caulobacter* cell cycle (Laub et al. 2000). CtrA modulates the expression of genes for flagellar motility, pilus production, DNA methylation, and cell division, among other processes. Although CtrA is essential for cell cycle progression, it also represses the initiation of DNA replication by binding to sites within the chromosomal replication origin (*C*ori) (Quon et al. 1998). To satisfy these conflicting requirements, levels of CtrA protein and CtrA phosphorylation oscillate during the cell cycle, with high levels of CtrA~P in swarmer and predivisional cells, and low levels in stalked cells that are initiating chromosome replication (Fig. 1a) (Domian et al. 1997; Jacobs et al. 2003).

CpdR is a single-domain response regulator that is active when unphosphorylated (see Sect. 6). In this state, CpdR promotes the degradation of some substrates by the ATP-dependent protease ClpXP (Iniesta et al. 2006; Rood et al. 2012; Lau et al. 2015). Importantly, CpdR is required for the regulated degradation of CtrA that occurs during swarmer cell differentiation and in the stalked compartment of the predivisional cell (Iniesta et al. 2006; Smith et al. 2014). Thus, when CckA is in kinase mode, both CtrA and CpdR are phosphorylated, generating active, stable CtrA (Fig. 1b). When CckA functions as a phosphatase, CtrA is both dephosphorylated (Fig. 1d) and subject to proteolysis by ClpXP (Figs. 2d, e), assisted by CpdR and other factors discussed in Sect. 6.2.

Caulobacter swarmer cells are characterized by high levels of CtrA~P and low levels of DivK~P, while the reverse is true of stalked cells (Domian et al. 1997; Jacobs et al. 2001, 2003; Matroule et al. 2004). The absence of CtrA~P allows the newly born stalked cell to initiate chromosome replication immediately (Fig. 1a). In contrast, the swarmer cell is incapable of beginning chromosome replication until CtrA~P has been eliminated. The CtrA protein is rapidly degraded by ClpXP during swarmer cell differentiation (Domian et al. 1997; Jenal and Fuchs 1998). This reaction requires unphosphorylated CpdR, as well as two additional adaptor proteins that deliver CtrA to the protease (Iniesta et al. 2006; McGrath et al. 2006; Duerig et al. 2009). However, even when amino acid substitutions are made which render CtrA immune to proteolysis, the existing CtrA protein becomes dephosphorylated during the SW-ST transition (Domian et al. 1997). Thus, cell cycle-regulated deactivation of CtrA is initiated by CckA acting in phosphatase mode.

3 Non-canonical Interactions Between the DivJ-PleC-DivK and CckA-ChpT-CtrA Pathways Control Polarity and Development

The differentiation of a non-replicating, motile swarmer cell into a sessile, replicating stalked cell requires a complete change in the signaling status of the regulatory network. As described above, CckA must be converted from a kinase,

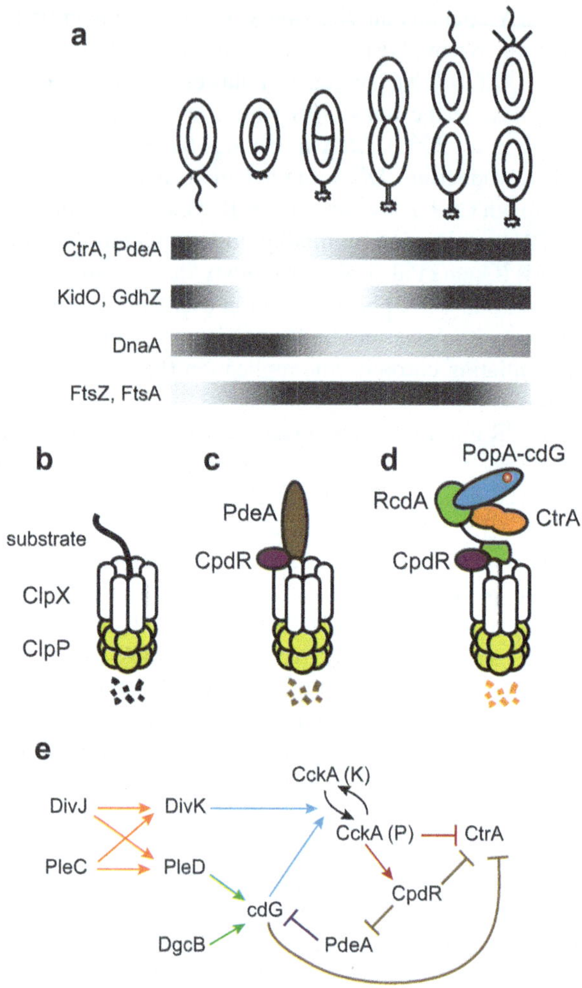

Fig. 2 (**a**) Relative abundances of proteins whose levels and/or stability change during the cell cycle. (**b**) ClpXP proteolysis unassisted by adaptors. (**c**) ClpXP degrades PdeA with the assistance of the priming factor CpdR. (**d**) ClpXP degrades CtrA with the assistance of a multi-component adaptor comprising CpdR, RcdA, and cdG-bound PopA. (**e**) Signaling events leading to the degradation of PdeA and CtrA during swarmer cell differentiation. *Orange arrows*, phosphorylation events; *CckA (K)*, CckA in kinase mode; *CckA (P)*, CckA in phosphatase mode; *blue arrows*, DivK~P and cdG promote the phosphatase activity of CckA; *red arrow*, CckA activates CpdR via dephosphorylation; *red bar*, CckA deactivates CtrA via dephosphorylation; *green arrows*, cdG synthesis; *purple bar*, cdG hydrolysis; *brown bars*, upstream components stimulate ClpXP-mediated proteolysis of downstream components

phosphorylating CtrA and CpdR, to a phosphatase, dephosphorylating both targets. Changes in the activity of the DivJ-PleC-DivK pathway drive this developmental transition.

DivK is an essential protein, and mutants harboring conditional alleles of *divK* arrest in the G1 phase of the cell cycle, unable to initiate chromosome replication (Hecht et al. 1995; Hung and Shapiro 2002). This phenotype provided the first clue that DivK is responsible for deactivating CtrA, and DivK~P is now known to switch CckA from its kinase to its phosphatase signaling mode (Tsokos et al. 2011). During SW-ST differentiation, DivJ becomes localized at the flagellated pole, and PleC is subsequently released (Wheeler and Shapiro 1999; Paul et al. 2008). DivJ localization requires the polar organizing protein PopZ and the localization factor SpmX, which also stimulates DivJ's kinase activity (Ebersbach et al. 2008;

Radhakrishnan et al. 2008; Perez et al. 2017). Although PleC usually functions as a DivK~P phosphatase, the transient colocalization of DivJ and PleC initiates a positive feedback loop, in which DivK~P phosphorylated by DivJ allosterically stimulates the kinase activities of both DivJ and PleC (Fig. 1c) (Paul et al. 2008). The resulting increase in DivK~P levels toggles CckA from the kinase to the phosphatase mode, which leads to the elimination of CtrA~P and the commencement of S-phase (Tsokos et al. 2011; Childers et al. 2014).

The effects of DivK upon CckA are mediated by another two-component signaling protein, the atypical histidine kinase homolog DivL. Also essential for *Caulobacter* viability, DivL resembles histidine kinase proteins, but has a tyrosine residue in place of the conserved, phosphorylated histidine (Wu et al. 1999). Cells with conditional mutations in *divL* have the opposite phenotype of those lacking DivK; they fail to divide, contain many chromosomes per cell, and have less CtrA~P than wild-type cells (Pierce et al. 2006; Reisinger et al. 2007), indicating that DivL is important for the activation of CtrA. However, rather than functioning as a kinase itself, DivL is required for CckA localization at the pole opposite the stalk and for the kinase activity of CckA (Fig. 1b) (Reisinger et al. 2007; Iniesta et al. 2010; Tsokos et al. 2011; Childers et al. 2014). DivL can be co-immunoprecipitated with CckA from *Caulobacter* lysates, but it is unknown whether DivL interacts directly with CckA, or if it interacts through intermediary proteins (Iniesta et al. 2010). In either case, the activation of a bona fide histidine kinase by a pseudokinase is a novel, non-canonical interaction between two-component proteins, and it may point to a function for pseudokinases encoded in other bacterial genomes.

DivK~P toggles CckA from kinase to phosphatase mode by interacting directly with DivL (Fig. 1d) (Tsokos et al. 2011). The binding site for DivK~P on DivL includes the region where a cognate response regulator would normally dock during phosphotransfer, but specific binding to the phosphorylated form of DivK also requires three PAS domains that lie N-terminal to the DHp domain in DivL (Tsokos et al. 2011, Childers et al. 2014). The PAS domains are not thought to interact directly with DivK~P, but to influence the positioning of the catalytic CA domain, thereby impacting the DivK~P binding site. In the current model (Fig. 1d), an interaction between DivL and CckA promotes CckA kinase activity, whereas a ternary complex including DivK~P favors CckA phosphatase activity (Tsokos et al. 2011, Childers et al. 2014). However, it is also formally possible that DivK~P binding causes DivL to release CckA, causing it to revert to phosphatase mode. Thus, DivL functions as sensor for a cytoplasmic response regulator, DivK~P, using the DHp and CA domains that are traditionally involved in histidine kinase output. In an interaction that awaits further dissection, DivL transmits the information of its interaction with DivK~P to CckA, linking the activity of the DivJ-PleC-DivK pathway to the CckA-ChpT-CtrA pathway.

4 cdG Signaling Regulates the *Caulobacter* Cell Cycle and Polar Development

4.1 The Cell Cycle Signaling Network Generates and Is Modulated by cdG Oscillations

Measurement of cdG levels in individual *Caulobacter* cells using a fluorescent biosensor demonstrated that the swarmer progeny contains <100 nM cdG, while the stalked progeny contains ~500 nM cdG (Christen et al. 2010). Measurement of cdG by LC-MS in *Caulobacter* cultures undergoing synchronous passage through the cell cycle also revealed a transient increase in cdG levels from <100 nM to ~275 nM during swarmer cell differentiation, followed by a slow decay back to ~100 nM in predivisional cells (Abel et al. 2013). Here we describe the signaling pathways that produce oscillations in the level of cdG, as well as mechanisms by which cdG impacts cell cycle progression and polar morphogenesis.

DivJ and PleC each interact with a second response regulator, PleD, which possesses two tandem receiver domains followed by a DGC domain (Paul et al. 2004). Phosphorylation of the first receiver domain causes PleD to dimerize and activates the production of cdG (Paul et al. 2007). During swarmer cell differentiation, when DivJ and PleC are temporarily colocalized at the developing flagellar pole, phosphorylation of DivK by DivJ initiates a positive feedback loop, in which DivK~P stimulates the kinase activity of DivJ and also causes PleC to enter kinase mode. Both DivJ and PleC then act as kinases for PleD, leading to a surge in cdG production by PleD~P (Fig. 1c) (Paul et al. 2008; Abel et al. 2011). Switching PleC from phosphatase to kinase mode is an important step in polar morphogenesis, as cells expressing a variant of PleC which lacks kinase activity (Matroule et al. 2004) do not experience an increase in cdG levels are impaired in holdfast and stalk biogenesis (Paul et al. 2008).

Increased production of cdG by PleD contributes to, but is not sufficient for, the increase in [cdG] during swarmer cell development. A second diguanylate cyclase, DgcB, produces cdG throughout the cell cycle, and its effects are counteracted specifically in swarmer cells by the phosphodiesterase PdeA (Abel et al. 2011). Importantly, PdeA is proteolyzed by ClpXP during swarmer cell differentiation (Fig. 2e), so that cdG synthesis by PleD and DgcB is temporarily unopposed. In addition to ClpXP, PdeA degradation requires the unphosphorylated form of; thus toggling CckA into phosphatase mode is important for the surge in cdG that occurs during swarmer cell development (Abel et al. 2011; Rood et al. 2012).

cdG produced by the action of cell cycle-regulated DGC and PDE enzymes feeds back to modulate key steps in the cell cycle network. First, cdG binds directly to CckA and promotes its phosphatase activity (Figs. 1d and 2e) (Lori et al. 2015; Dubey et al. 2016; Mann et al. 2016), thereby playing an important role, along with DivK~P, in switching CckA to phosphatase mode. CckA is the second histidine kinase demonstrated to respond directly to cdG, after SgmT, which regulates the expression of extracellular matrix proteins in *Myxococcus xanthus*

(Petters et al. 2012). Reinforcing its effects on the CckA-ChpT-CtrA pathway, cdG is also necessary for regulated proteolysis of CtrA by ClpXP during swarmer cell differentiation (Fig. 2d, e). cdG binds directly to an adaptor protein, PopA, that functions in the temporally regulated degradation of CtrA and other substrates (Duerig et al. 2009). Together these effects promote the initiation of chromosome replication and entry into the cell division cycle.

4.2 cdG Signaling Is Required for Polar Morphogenesis

By generating a *Caulobacter* mutant lacking all DGC enzymes (cdG^0), Abel et al. (2013) demonstrated that cdG is necessary for the proper construction of all polar organelles; the mutant strain is stalkless and lacks the flagellum, pili, and the adhesive holdfast. Synthesis of the stalk and holdfast is initiated during swarmer cell differentiation, when [cdG] is at its peak, while synthesis of the flagellum and pili occurs in predivisional cells, when the level of cdG is lower (Poindexter 1964; Shapiro and Maizel Jr. 1973; Smit and Agabian 1982; Bodenmiller et al. 2004; Levi and Jenal 2006; Christen et al. 2010; Abel et al. 2013). The different steps in polar morphogenesis may occur at different threshold levels of cdG, an inference that was generally supported by examining polar structures in cdG^0 mutants containing various levels of cdG produced by a heterologous DGC enzyme (Abel et al. 2013). However, in *Caulobacter* mutants with cdG produced only by a constitutively active heterologous enzyme, each cellular process that was measured occurred at a higher [cdG] concentration than it did in wild-type cells (Abel et al. 2013). These findings suggest that measurements of [cdG] in wild-type *Caulobacter* underestimate the true concentrations, or possibly that cdG produced by native enzymes that are temporally regulated or spatially localized is more efficient at promoting downstream effects. Although cdG is a rapidly diffusing small molecule, and there is no direct evidence of anisotropy in its distribution within individual *Caulobacter* cells (Christen et al. 2010), we cannot entirely rule out the possibility that localized production, degradation, and sensing of cdG are involved in some aspects of *Caulobacter* cell polarity.

cdG has not yet been mechanistically linked to every polar structure, in part because it is still easier to identify proteins that synthesize and degrade cdG than to identify cdG-binding effectors. However, great progress in this area has been achieved using affinity binding techniques (Nesper et al. 2012), and the *Caulobacter* proteins revealed by this method include cdG effectors that modulate flagellar motor function and participate in holdfast biosynthesis (Hug et al. 2017; Nesper et al. 2017; Sprecher et al. 2017). Here we focus on cdG-regulated polar morphogenesis events that occur at distinct times in the *Caulobacter* cell cycle and development.

Predivisional cells of the cdG^0 mutant fail to synthesize even the earliest substructures in flagellar biogenesis, the MS-ring and switch complex (Abel et al. 2013). Flagellar construction is initiated when the assembly factor TipF binds cdG using an enzymatically inactive PDE domain (Davis et al. 2013). Upon cdG binding, TipF localizes to the pole opposite the stalk and there recruits the flagellar placement factor PflI and FliG of the switch complex, which attract additional flagellar proteins. cdG is also thought to stimulate the expression of flagellar genes later in the transcriptional hierarchy by a mechanism unrelated to TipF, but the effector(s) responsible are unknown (Davis et al. 2013).

The flagellum is built in predivisional cells and begins to rotate shortly before cell separation. The Δ*pleC* mutant is nonmotile because the polar flagellum is paralyzed (Sommer and Newton 1989). A possible linkage between PleC and flagellar rotation is the cdG sensor DgrA. DgrA binds cdG via a PilZ domain and in the cdG-bound state inhibits flagellar motility (Christen et al. 2007). One hypothesis is that PleC acting in phosphatase mode deactivates PleD or activates a PDE enzyme at the flagellar pole of the predivisional cell. This process would cause the observed drop in [cdG] in the swarmer compartment (Christen et al. 2010), preventing DgrA from inhibiting flagellar rotation and, in turn, yielding motile swarmer progeny.

When swarmer cells differentiate, they eject the flagellum and synthesize an adhesive holdfast at the same pole. The Δ*pleD* mutant fails to eject the flagellum because the MS-ring protein FliF is not proteolyzed by ClpAP (Aldridge and Jenal 1999; Grunenfelder et al. 2004). This relationship suggests that there may be a cdG-dependent adaptor for ClpAP-mediated proteolysis of FliF, analogous to PopA.

The swarmer cell begins to synthesize holdfast even before the flagellum is ejected, because optimal surface attachment requires both flagellar motility and the holdfast (Bodenmiller et al. 2004; Levi and Jenal 2006). After synchronization, when swarmer cells are released into dilute liquid culture, holdfast synthesis begins after about 20 min. In contrast, isolated swarmer cells begin expressing the holdfast only 1–2 min after exposure to a surface (Li et al. 2012). These results suggested that holdfast production is under cell cycle control but can be accelerated by surface contact. A recent study found that the inhibition of flagellar rotation by a nearby surface is sensed by the cdG synthase DgcB. DgcB interacts directly or indirectly with the flagellar stator component MotA, and a change in motor function or in some property of the cytoplasmic membrane is thought to stimulate its activity (Hug et al. 2017). The same study identified HfsJ, a glycosyltransferase essential for holdfast production, as a cdG-dependent effector protein. Therefore, surface sensing via the flagellum and DgcB is thought to generate a local increase in cdG, which stimulates HfsJ activity and triggers holdfast production in advance of when it would begin in a planktonic swarmer cell.

Together, these studies demonstrate the wide range of mechanisms by which cdG can modulate bacterial behavior. Although we do not yet know how cdG is connected to *Caulobacter* pilus or stalk synthesis, we are likely to encounter the same level of complexity in their relationship to cdG.

5 Spatial Regulation of CckA Activity

Caulobacter CckA was the first histidine kinase shown to be located at a distinct position within a bacterial cell (Jacobs et al. 1999), other than CheA within the chemoreceptor complex (Maddock and Shapiro 1993). Prior to these observations, it had been assumed that signaling proteins performed their functions while diffusing throughout the cytoplasm or cytoplasmic membrane. The importance of protein localization for the correct signaling output is clear for enzymes such as DivJ and PleC, whose activities need to be at opposite poles of the cell to generate progeny with asymmetric replicative fates and polar organelles (Matroule et al. 2004; Tropini and Huang 2012; Subramanian et al. 2015). PleC switches its activity temporally, from a phosphatase to a kinase, during swarmer cell differentiation (Fig. 1c). The positive feedback mechanism responsible for the switch is described in Sect. 3. Here we consider several mechanisms proposed to account for distinct CckA kinase and phosphatase activities at the two poles of the *Caulobacter* predivisional cell.

Swarmer cells contain high levels of CtrA~P (Fig. 1a) and low levels of DivK~P, while the reverse is true in stalked cells (Domian et al. 1997; Jacobs et al. 2001). The stimulation of CckA phosphatase activity by DivK~P would seem to enforce this inverse relationship, yet *Caulobacter* predivisional cells contain high levels of both CtrA~P and DivK~P (Domian et al. 1997; Jacobs et al. 2001). Therefore, some fraction of the CckA protein in predivisional cells must be protected from the effects of DivK~P. Studies using *Caulobacter* cells treated with the division inhibitor cephalexin showed that, when an elongated cell contains two chromosomes, DNA replication is five times more likely to commence at the chromosomal origin near the stalked pole than at the origin near the flagellar pole (Chen et al. 2011). Using mutants in which CckA kinase or phosphatase activity was selectively impaired, it was demonstrated that replicative asymmetry required both activities. The model that emerged is that CckA acts as a kinase at the flagellar pole (Fig. 1b) and as a phosphatase at the stalked pole (Fig. 1d), generating a gradient of CtrA activity along the length of the cell (Fig. 1a) (Chen et al. 2011). At the flagellar pole, where [CtrA~P] is highest, replication initiation is blocked, but at the stalked pole, where [CtrA~P] is lowest, replication can commence even in the absence of cell division.

Several processes have been proposed to explain how CckA performs different functions at the two poles. DivL accumulates at the flagellar pole, which suggests that it may specifically promote CckA kinase activity at this site (Iniesta et al. 2010). However, DivL is also distributed around the cell membrane, and the swarmer and stalked progeny inherit roughly equal amounts of DivL (Sciochetti et al. 2005). Therefore, it can't be assumed that DivL only interacts with CckA at the flagellar pole.

Features intrinsic to CckA could spatially regulate its activity. Reconstitution of CckA in liposomes showed that, in the absence of other factors, CckA is more likely to work as a kinase when it is present at higher densities in a membrane (Mann et al. 2016). This property could bias CckA that is diffusely located around the cytoplasmic membrane toward the phosphatase state, but it does not easily explain how CckA can have opposing activities at the two cell poles. Although foci of fluorescently labeled CckA proteins are often brighter at the flagellar pole than at the stalked pole (Jacobs et al. 1999; Angelastro et al. 2010), we do not know if this corresponds to a higher local density of CckA at the flagellar pole.

CckA may be biased toward the kinase state at the flagellar pole because there, it is protected from cdG and DivK~P, which promote the switch to phosphatase mode. With respect to cdG, this possibility was investigated using a truncated variant of

CckA lacking only its cytoplasmic membrane anchor, *cckAΔTM*, which diffuses throughout the cytoplasm. *cckAΔTM* cannot compensate for a deletion of the native *cckA* gene, and when expressed in a wild-type strain, it causes over-replication of chromosomal DNA, suggesting that it functions as a phosphatase to deactivate CtrA (Jacobs et al. 1999; Chen et al. 2009). In confirmation, a double *cckA* mutant lacking the transmembrane domain and also lacking phosphatase activity has no effect when overexpressed in wild-type cells (Chen et al. 2009). Importantly, a double *cckA* mutant lacking the membrane anchor and unable to bind cdG (or *cckAΔTM* expressed in the cdG^0 background) leads to a strong G1 arrest, indicating that the cytoplasmic pool of cdG prevents kinase activity in CckA molecules that are delocalized from the flagellar pole (Lori et al. 2015). These results suggest that the flagellar pole may be depleted of cdG by an unknown, localized phosphodiesterase (Fig. 1b).

The localized phosphatase activity of PleC is thought to protect CckA at the flagellar pole from DivK~P (Fig. 1b). A temperature-sensitive *pleC* mutant contained only 18% of the wild-type amount of CckA~P after a short incubation at the nonpermissive temperature, indicating that PleC inactivation rapidly and profoundly reduces CckA kinase activity (Tsokos et al. 2011). A single-domain PAS protein called MopJ is also partially responsible for maintaining CckA kinase activity in the face of DivK~P. The Δ*mopJ* strain has moderately reduced CtrA activity in exponentially growing cells, and it is far more sensitive than a wild-type strain to DivK overexpression, suggesting that MopJ counteracts DivK activity (Sanselicio et al. 2015). Overexpression of MopJ increases DivK localization to both poles but does not affect the cellular level of DivK~P. In contrast, DivL is delocalized from the flagellar pole in the Δ*mopJ* strain (Sanselicio et al. 2015). Together, these results suggest that MopJ promotes CckA kinase activity by promoting the flagellar pole localization of DivL and by reducing the impact of DivK~P through an uncharacterized mechanism (Fig. 3b).

6 Proteolytic Regulation of the *Caulobacter* Cell Cycle

Regulated degradation of select protein substrates is a crucial process in maintaining proper cell physiology and homeostasis in all organisms. In bacteria, regulated proteolysis is important for numerous processes, including the response to envelope stress (Barchinger and Ades 2013), spore formation (Moliere and Turgay 2013), and the clearance of misfolded and prematurely-terminated proteins (Janssen and Hayes 2012; Gur 2013). Studies in *Caulobacter* in particular have underscored the importance of regulated proteolysis in cell cycle progression and differentiation.

6.1 AAA+ Proteases and Adaptor-Mediated Proteolysis

The ubiquitous AAA+ (ATPases associated with diverse cellular activities) family of proteases mediates the degradation of proteins in a highly controlled manner.

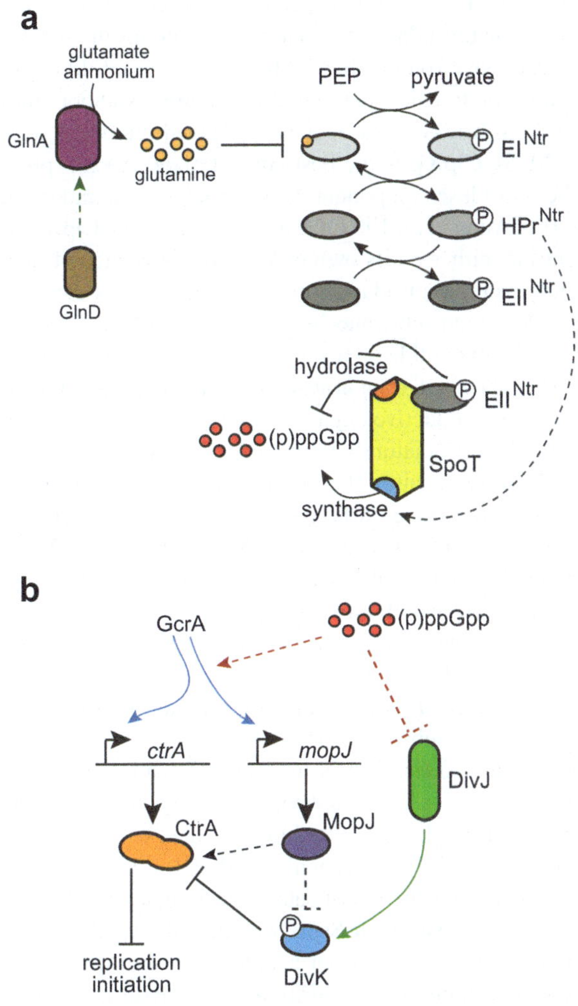

Fig. 3 (**a**) Ammonium limitation stimulates (p)ppGpp synthesis by SpoT via the PTS^{Ntr} system. *Green dashed arrow*, GlnD indirectly stimulates *glnA* transcription and GlnA activity; *black bars*, upstream element inhibits downstream element or process; *black dashed arrow*, HPr^{Ntr} stimulates the (p)ppGpp synthase activity of SpoT by an unknown, indirect mechanism. (**b**) Proposed mechanisms linking (p)ppGpp to CtrA stabilization during ammonium limitation and/or prolonged stationary phase. *Blue arrows*, GcrA-activated transcription; *green arrow*, DivJ phosphorylates DivK; *black bars*, upstream element inhibits downstream element or process; *dashed black bar*, MopJ inhibits DivK~P activity by an unknown mechanism; *dashed black arrow*, MopJ stimulates CtrA activity indirectly via polar localization of DivL; *dashed red arrow*, (p)ppGpp stimulates GcrA-dependent transcription by an unknown mechanism; *dashed red bar*, (p)ppGpp inhibits polar localization of DivJ by an unknown mechanism

AAA+ proteases consist of a barrel-shaped peptidase chamber, the entry of which is gated by an oligomeric unfoldase component that contains the AAA+ domain. The unfoldase uses successive rounds of ATP hydrolysis to unfold the substrate into a linear peptide chain and translocate it through a central axial pore into the peptidase chamber, where the substrate is subsequently cleaved into short peptide fragments (Olivares et al. 2016). ClpXP is the best-characterized AAA+ protease, in which a hexamer of the unfoldase component, ClpX, binds to a tetradecamer of the ClpP peptidase (Fig. 1b) (Wang et al. 1997; Grimaud et al. 1998). ClpX is responsible for substrate recognition, either on its own or with the assistance of dedicated adaptor proteins. The first example of adaptor-mediated proteolysis came from studies of degradation of ssrA-tagged substrates in *E. coli*. The ssrA tag is covalently added to the C-terminus of a polypeptide on a stalled ribosome by transfer-messenger RNA (tmRNA) in a process termed trans-translation, which releases the polypeptide and frees the ribosome for productive engagements (Tu et al. 1995; Keiler et al. 1996). Since the polypeptide is prematurely terminated, the ssrA-tag directs the unfinished product to ClpXP for degradation (Gottesman et al. 1998; Levchenko et al. 2000). Two alanine residues and the α-carboxylate at the C-terminus of the ssrA tag are directly recognized by ClpX (Flynn et al. 2001; Farrell et al. 2007; Martin et al. 2008). Once engaged, ATP hydrolysis induces conformational changes in ClpX that drive substrate translocation and unfolding.

Although ClpXP alone can recognize and degrade ssrA-tagged substrates, degradation is accelerated by the SspB adaptor protein (Gottesman et al. 1998; Levchenko et al. 2000). SspB acts as a tether, binding to both ClpX and the ssrA tag to increase the local substrate concentration (Levchenko et al. 2000; Flynn et al. 2001; Wah et al. 2002). The N-terminus of SspB contains a dimerization domain that binds to the ssrA tag, and this is connected through an unstructured linker to a short C-terminal motif that binds to the N-terminal domain of ClpX (NTD_{ClpX}) (Dougan et al. 2003; Levchenko et al. 2003; Wah et al. 2003). Substrate tethering effectively reduces the K_m of catalysis while negligibly affecting the V_{max} (Levchenko et al. 2000). Although SspB was first shown to facilitate proteolysis of ssrA-tagged substrates, it is also an adaptor for other distinct proteins (Flynn et al. 2004). A simple tethering of substrate to protease is not the only way for an adaptor to stimulate proteolysis. An alternative mechanism is used by the RssB response regulator in *E. coli*, which, when activated by phosphorylation, binds to the stationary phase sigma factor σ^S to induce conformational changes that allow σ^S to be recognized by ClpX (Studemann et al. 2003). In this case, RssB does not contact ClpX directly. Instances of adaptor-mediated proteolysis important for the *Caulobacter* cell cycle and development are explored below.

The *Caulobacter* genome encodes five AAA+ proteases: ClpXP, ClpAP, HslUV, FtsH, and Lon (Nierman et al. 2001). ClpXP, ClpAP, and HslUV are bipartite enzymes, with ATPase and peptidase components encoded by separate genes, while FtsH and Lon contain both domains on a single polypeptide (Lupas et al. 1997). The peptidase ClpP associates with either ClpX or ClpA independently, and these complexes proteolyze distinct substrates (Katayama-Fujimura et al. 1987; Wojtkowiak et al. 1993). So far, ClpXP, ClpAP, and Lon are known to mediate cell cycle-regulated proteolysis in *Caulobacter* (Jenal and Fuchs 1998; Gora et al. 2013; Williams et al. 2014). ClpX, ClpP, and Lon are present at constant levels throughout the cell cycle (Wright et al. 1996; Jenal and Fuchs 1998), but the concentrations of their substrates fluctuate dramatically due to temporal control mechanisms regulating both protein synthesis and degradation (Fig. 2a).

6.2 Cell Cycle Regulation of CtrA Proteolysis

The CtrA response regulator is both dephosphorylated and proteolyzed during swarmer cell differentiation and in the stalked compartment of predivisional cells (Figs. 1a and 2a) to license the initiation of DNA replication by DnaA (Domian et al. 1997; Gorbatyuk and Marczynski 2001; Ryan et al. 2002). CtrA and DnaA bind to overlapping sites in the *Caulobacter* origin of replication (*Cori*), and CtrA has been shown to displace DnaA from its binding sites in vitro (Quon et al. 1998; Taylor et al. 2011). CtrA degradation in these two cell types is accomplished by ClpXP in association with a highly regulated adaptor complex which binds to ClpX (Fig. 2d) (Smith et al. 2014; Joshi et al. 2015). The adaptor complex is composed of three proteins: (1) the single-domain response regulator CpdR (Biondi et al. 2006a; Iniesta et al. 2006), (2) the tethering protein RcdA (McGrath et al. 2006), and (3) the hybrid response regulator-DGC PopA (Duerig et al. 2009). CpdR is phosphorylated on Asp51 by the CckA-ChpT phosphorelay, but only the unphosphorylated form is competent for CtrA degradation (Biondi et al. 2006a; Iniesta et al. 2006). PopA is inactive as a DGC, but binding of cdG to its allosteric "I-site" is required for its function as proteolytic adaptor (Duerig et al. 2009). Efficient, timed degradation of CtrA in vivo absolutely requires the presence and activation of these three proteins, and while none is essential for viability, the omission of any single adaptor component prevents CtrA proteolysis (Iniesta et al. 2006; McGrath et al. 2006; Duerig et al. 2009).

Based on these in vivo findings, it was at first surprising that CtrA could be degraded by ClpXP with ATP in vitro in the absence of any other factors (Chien et al. 2007). The half-life of degradation was estimated at around 5 min, a rate consistent with its clearance during the SW-ST transition (Domian et al. 1997; Chien et al. 2007). In vitro reconstitution experiments revealed that the adaptor proteins work together to stimulate CtrA proteolysis beyond the unassisted rate. Addition of RcdA, PopA, cdG, and unphosphorylated CpdR to reactions with ClpXP, CtrA, and ATP reduces the K_m of proteolysis ten-fold, while having a negligible effect on v_{max} (Smith et al. 2014). Absence of any component of the adaptor complex results in a failure to stimulate CtrA proteolysis beyond the basal rate, indicating a unique and essential role for each component.

Further studies have dissected the roles of individual proteins within the adaptor complex. Using ClpXP substrates that depend on CpdR for degradation, but not on RcdA or PopA, it was shown that CpdR binds directly to NTD_{ClpX} to prime the unfoldase for engagement with substrates (Fig. 2c) (Lau et al. 2015). Interaction with CpdR creates a unique ClpX recruitment interface upon which

CpdR-dependent substrates or additional adaptors bind (Lau et al. 2015). Although CpdR does not interact independently with substrates, it may form part of the primed interface on ClpX where substrates are recognized (Joshi et al. 2015; Lau et al. 2015). This mechanism serves in vivo to limit the degradation of selected proteins to times when CpdR is dephosphorylated.

Priming of ClpX by CpdR is required for the subsequent binding of the adaptor component RcdA (Joshi et al. 2015). All of the *Caulobacter* ClpXP substrates known to require RcdA also need CpdR, but only some substrates additionally require PopA (Joshi et al. 2015). Similar to SspB, RcdA contains an N-terminal dimerization domain and a disordered C-terminal peptide (Taylor et al. 2009). The N-terminal domain of RcdA is thought to bind directly to proteolytic substrates, though this awaits experimental confirmation. Studies in which the C-terminal peptides of SspB and RcdA were exchanged demonstrated that the C-terminus of RcdA interacts with ClpX in a CpdR-dependent manner (Joshi et al. 2015). Both interactions are required for the degradation of RcdA-dependent substrates, suggesting that RcdA can work as a tether, analogous to the function of SspB in the degradation of ssrA-tagged substrates (Fig. 2d).

The final protein in the hierarchical adaptor complex is PopA, which contains two tandem receiver domains followed by a catalytically inactive DGC domain which binds cdG. The N-terminal receiver domain interacts directly with RcdA independent of cdG binding, but only cdG-bound PopA is competent to bind CtrA for delivery to ClpXP (Fig. 2d) (Ozaki et al. 2014; Smith et al. 2014). The $\alpha1$ helix of the CtrA receiver domain contains three amino acids that are critical for the interaction with PopA, but the region of PopA that recognizes CtrA has not been defined (Smith et al. 2014). Because CtrA is both an inhibitor of chromosome replication and an essential transcription factor, its proteolysis must be strictly regulated. The requirement for unphosphorylated CpdR and the cdG-dependency of the PopA-CtrA interaction together ensure that CtrA is only degraded in vivo prior to chromosome replication, when the CckA-ChpT pathway is in phosphatase mode and when cdG levels simultaneously rise (Fig. 2e).

Despite progress in understanding the adaptor complex mechanism, there is a persistent discontinuity between in vitro and in vivo studies of CtrA proteolysis. CtrA can be degraded in vitro by ClpXP and ATP without the addition of any other factors (Chien et al. 2007), and ClpXP is present throughout the *Caulobacter* cell cycle (Jenal and Fuchs 1998), yet pulse-chase assays indicate that CtrA is very stable outside of the short window preceding chromosome replication (Domian et al. 1997). Why does unassisted ClpXP not proteolyze CtrA at other times during the cell cycle? One factor able to protect CtrA is the transcriptional co-regulatory protein SciP. SciP synthesis begins in late PD cells and accumulates to peak levels in SW cells, where it forms a ternary complex with CtrA and DNA at CtrA-binding sequences (Gora et al. 2010; Tan et al. 2010). Most CtrA-activated genes are expressed during the predivisional stage of the cell cycle (Laub et al. 2002), and SciP prevents inappropriate expression of these genes in swarmer cells by blocking the recruitment of RNA polymerase (RNAP) to the ternary complex (Gora et al. 2013). Importantly, SciP increases the affinity of CtrA for its DNA-binding sites, helping to protect CtrA from degradation by unassisted ClpXP in vitro (Gora et al. 2013).

Although SciP and DNA stabilize CtrA in vitro, their contribution to CtrA stability in vivo is unclear. The half-life of CtrA is increased when SciP is overexpressed (Gora et al. 2010, Tan et al. 2010), but the stability of CtrA is unchanged in a Δ*sciP* mutant (Gora et al. 2010). Issues of stoichiometry also limit the number of CtrA molecules that could be protected within CtrA-SciP-DNA complexes. Each *Caulobacter* swarmer cell contains ~9500 molecules of CtrA (Judd et al. 2003), but the chromosome has only ~100 CtrA-dependent promoters (Laub et al. 2002). Even if each promoter bound several CtrA monomers, most CtrA molecules should be excluded from the protective effect of the CtrA-SciP-DNA ternary complex. Finally, because SciP is absent from early predivsional cells (Gora et al. 2010; Tan et al. 2010), a separate protective mechanism for CtrA would be needed at this stage of the cell cycle.

In addition to temporal changes in activity, the CtrA proteolytic complex also dynamically localizes to specific sites within the cell. CtrA, ClpXP, RcdA, CpdR, and PopA are each transiently located at the incipient stalked pole during the swarmer cell development and at the stalked pole in late predivsional cells (Ryan et al. 2002; Iniesta et al. 2006; McGrath et al. 2006; Duerig et al. 2009). The polar organizing factor PopZ, unphosphorylated CpdR, and cdG binding to PopA are all necessary for localization of the protease, adaptor complex, and substrate (Iniesta et al. 2006; Duerig et al. 2009; Holmes et al. 2016). It was originally hypothesized that colocalization of the components, as detected by fluorescence microscopy, was critical for CtrA degradation. However, the in vitro experiments outlined above showed that the adaptor proteins have mechanistic roles beyond substrate localization. Moreover, amino acid substitutions in RcdA were found that prevent its own polar accumulation and that of CtrA, but still support CtrA proteolysis at wild-type rates (Taylor et al. 2009). It may be that proteases and adaptor complexes are located at particular positions chiefly to degrade substrates that are immobilized in large complexes, such as the chemoreceptor array found at the flagellar pole (Alley 2001; Tsai and Alley 2001; Briegel et al. 2008).

6.3 Regulated Proteolysis of Proteins that Modulate CtrA Activity

The PdeA phosphodiesterase contains an N-terminal PAS domain, followed by an inactive DGC domain and a C-terminal EAL domain that hydrolyzes cdG (Christen et al. 2005; Abel et al. 2011; Rood et al. 2012). As described above, PdeA opposes the activity of DgcB in SW cells and prevents a premature increase in cdG levels. Degradation of PdeA during the SW-ST transition triggers the increase in [cdG] that is important both for polar morphogenesis and for the efficient degradation of CtrA.

PdeA is degraded by ClpXP in concert with the adaptor CpdR (Fig. 2c) (Abel et al. 2011). The C-terminal amino acids of PdeA (RG) comprise a weak degradation signal for ClpXP (Rood et al. 2012). Mutation of these residues to DD or addition of a FLAG epitope tag blocks PdeA degradation in vitro and in vivo (Abel et al. 2011, Rood et al. 2012). Underscoring the importance of cdG for swarmer cell development, holdfast production and CtrA proteolysis are delayed in a *pdeA-FLAG* strain, and these effects are exacerbated by simultaneous deletions of either *pleD* or *dgcB* (Abel et al. 2011). Deletion of the PdeA PAS domain blocks degradation. But surprisingly, although the truncated ΔPAS-PdeA variant has PDE activity, it does not produce the same phenotype in vivo as another stable variant with an altered C-terminus, PdeA-DD (Rood et al. 2012). The PAS domain is therefore likely to perform an unknown signaling role in addition to regulating PdeA stability. CpdR reduces the K_M for PdeA degradation three-fold, consistent with a tethering function, but it also increases v_{max} by ~30-fold, suggesting that CpdR can also improve the turnover rate of substrates with intrinsically weak degradation tags (Rood et al. 2012).

As described above, SciP associates with CtrA and DNA in swarmer cells and prevents the inappropriate expression of CtrA-regulated genes that are specifically transcribed in predivisional cells. SciP is itself subject to regulated proteolysis during swarmer cell differentiation, and this process is important for proper cell cycle progression (Gora et al. 2013). Accordingly, expression of a stabilized SciP-M2 variant down-regulates the transcription of CtrA-activated genes and inhibits cell division. SciP is degraded by the Lon protease in vivo and in vitro, and, like CtrA, it is protected from degradation in vitro within the CtrA-SciP-DNA ternary complex (Gora et al. 2013).

6.4 Proteolysis of Substrates that Regulate DNA Replication and Methylation

As in nearly all bacteria, DNA replication in *Caulobacter* is initiated by the highly-conserved AAA+ protein DnaA (Gorbatuyk and Marczynski 2001). When in its active, ATP-bound state, DnaA binds to specific sites within the origin of replication, oligomerizes, and promotes local unwinding of the DNA (Bleichert et al. 2017). These steps permit the subsequent assembly of the replisome and the beginning of new DNA synthesis. A primary mechanism that prevents further, premature replication initiation events is replicatory inactivation of DnaA (RIDA). When bound to the sliding clamp of the replisome on newly synthesized DNA, Hda (homolog of DnaA) contacts DnaA and stimulates its ATP hydrolysis activity to produce the inactive, ADP-bound form of DnaA (Katayama et al. 2017). Unlike *E. coli*, *Caulobacter* replicates its chromosome once and only once per cell division (Marczynski 1999). This periodicity of DNA replication depends upon the essential *Caulobacter* Hda homolog, HdaA, which operates similarly to its *E. coli* counterpart (Collier and Shapiro 2009; Jonas et al. 2011; Fernandez-Fernandez et al. 2013; Wargachuk and Marczynski 2015).

Although HdaA is critical for the timing of replication initiation (Collier and Shapiro 2009), steady-state levels of DnaA also fluctuate somewhat during the *Caulobacter* cell cycle (Fig. 2a) (Collier et al. 2006). DnaA levels increase in swarmer cells and decrease in stalked cells after replication has begun. DnaA is degraded with a half-life of 45–60 min in unstressed conditions (Gorbatuyk and Marczynski 2005; Jonas et al. 2013), and overexpression of *dnaA* causes overinitiation of DNA replication and ultimately death (Jonas et al. 2011). DnaA is chiefly degraded by the Lon protease in vivo (Jonas et al. 2013). In *E. coli*, Lon is known to recognize and degrade unfolded proteins generated by stress or DnaK depletion (Tsilibaris et al. 2006). Consistent with this cellular role, Lon-mediated proteolysis of *Caulobacter* DnaA is stimulated in vitro by the addition of unfolded substrates (Jonas et al. 2013). The allosteric stimulation of Lon activity may only apply to certain substrates, since degradation of SciP is not enhanced by unfolded substrates. The stimulation of DnaA proteolysis by unfolded proteins has physiological consequences in cells exposed to proteotoxic stress. In response to heat shock or loss of the conserved chaperone DnaK, elevated levels of unfolded proteins stimulate Lon to degrade DnaA and consequently inhibit the initiation of chromosome replication (Jonas et al. 2013).

While Lon is the primary protease for DnaA degradation in *Caulobacter* during log phase, ClpAP is required for the complete removal of DnaA observed during stationary phase (Liu et al. 2016). In vitro, ClpAP degrades DnaA, but at a slower rate than Lon. DnaAR357A, a variant that cannot hydrolyze ATP, is degraded much more slowly than wild-type DnaA by Lon, but both substrates are degraded at similar rates by ClpAP (Wargachuk and Marczynski 2015; Liu et al. 2016). These results suggest that ClpAP may contribute significantly to the degradation of active, ATP-bound DnaA when its levels are inappropriately elevated in vivo. Indeed, expression of *dnaAR357A* is much more detrimental to Δ*clpA* cells than to wild-type cells (Liu et al. 2016), indicating that ClpAP-mediated degradation serves as a mechanism to protect *Caulobacter* from an excess of activated DnaA. ClpAP has therefore been designated an auxiliary protease that may help fine-tune the levels of DnaA in the cell or aid in degradation when Lon become saturated.

The Lon protease plays a critical role near the end of the cell cycle, when it is responsible for degrading the essential DNA methyltransferase CcrM (Stephens et al. 1996; Wright et al. 1996). CcrM is restricted to PD cells and catalyzes N^6-methylation of adenine at GAnTC sequences to convert the chromosome from a hemimethylated to a fully methylated state near the end of S-phase (Zweiger et al. 1994; Stephens et al. 1996). Expression of many cell cycle-regulated genes is directly influenced by the methylation state of GAnTC sequences within their promoters (Mohapatra et al. 2014; Haakonsen et al. 2015). Genes nearer the origin of replication become hemimethylated earlier during S-phase and remain so for a longer time than genes nearer the terminus. Thus, genes whose transcription depends upon hemimethylation of promoter sequences are transcribed at different times, depending upon their distance from the origin of replication (Collier et al. 2007). This system of transcriptional regulation does not function correctly if CcrM is present and active throughout S-phase, rapidly methylating all newly synthesized strands of DNA. Thus, constitutive overexpression of *ccrM* or deletion of *lon* nearly eliminates hemimethylated DNA and causes cell morphology defects (Zweiger et al. 1994; Wright et al. 1996). The increase in CcrM abundance at the end of the cell cycle, followed by its rapid degradation, is not believed to be driven by regulated proteolysis. Instead, Lon is thought to degrade CcrM at a constant rate, and a burst of CcrM synthesis late in the cell cycle temporarily overcomes Lon-mediated proteolysis (Zweiger et al. 1994, Wright et al. 1996).

6.5 Degradation of Proteins Involved in Cell Division

The proteolysis of two key regulators of FtsZ dynamics is important for proper assembly of the Z-ring at the midcell prior to division. KidO and GdhZ promote disassembly of the Z-ring in predivisional cells, which facilitates cell division, and in swarmer cells, which prevents premature Z-ring assembly (Radhakrishnan et al. 2010; Beaufay et al. 2015). GdhZ, an NAD-dependent glutamate dehydrogenase that converts glutamate to alpha-ketoglutarate, acts synergistically with KidO to regulate FtsZ (Beaufay et al. 2015). Consistent with these activities, KidO and GdhZ are present in G1-phase swarmer cells and in late predivisional cells but are absent during S-phase (Fig. 2a) (Radhakrishnan et al. 2010, Beaufay et al. 2015). In vivo, ClpXP, CpdR, RcdA, and PopA are each required for the cell cycle-regulated degradation of KidO and GdhZ during swarmer cell differentiation, but it is unknown if these proteins are sufficient for KidO or GdhZ proteolysis in vitro (Radhakrishnan et al. 2010; Beaufay et al. 2015). Constitutive expression of the stabilized variant *kidO-DD* from the *Caulobacter* chromosome inhibits cell division, and overexpression of wild-type *kidO* also disperses FtsZ from Z-rings (Radhakrishnan et al. 2010). Stabilization of GdhZ also results in mild cell elongation, whereas the simultaneous stabilization of GdhZ and the constitutive expression of KidO-DD cause more severe cell filamentation (Beaufay et al. 2015). Collectively, these data indicate that temporally regulated degradation of KidO and GdhZ is important for proper cell division. We discuss below (Sect. 8.2) the possibility that KidO, by binding NADH, could be regulated by changes in the redox status of the *Caulobacter* cytoplasm.

Some components of the divisome itself are proteolyzed during or after septation, including FtsZ, FtsA, and FtsQ (Fig. 2a) (Kelly et al. 1998; Martin et al. 2004). While these proteins are regulated at the level of transcription, constitutive expression throughout the cell cycle still yields oscillations in abundance, consistent with temporally controlled degradation (Kelly et al. 1998; Wang et al. 2001; Martin et al. 2004). FtsZ is degraded by both ClpXP and ClpAP in vivo, and either protease can degrade FtsZ in vitro without any additional proteins (Williams et al. 2014). ClpAP is the primary protease for FtsA degradation in vivo, but the poor solubility of FtsA has precluded a thorough analysis of its degradation in vitro (Williams et al. 2014). Both FtsZ and FtsA are degraded more rapidly in swarmer cells than in stalked cells. Constitutive expression of FtsZ in a Δ*clpA* mutant background results in slightly longer cells with mis-positioned Z-rings, as compared to either mutant individually (Williams et al. 2014), and overexpression of either FtsZ or FtsA results in cell division defects and a decrease in viability (Wang et al. 2001; Martin et al. 2004). However, the physiological consequences of blocking the degradation of divisome components, as opposed to overexpressing them, remain to be characterized. Although unassisted ClpXP or ClpAP can degrade FtsZ in vitro, it is unknown if other factors in vivo govern the cell type-specific degradation of divisome proteins. Further studies are needed to determine if proteolytic clearance of divisome proteins plays a role in cell constriction or in preventing premature divisome assembly in daughter cells.

7 Nutrient Cues Affecting the *Caulobacter* Cell Cycle

The previous sections have described the core network regulating the *Caulobacter* cell cycle and development. These systems have chiefly been studied and described in well-fed, unstressed, exponentially growing cells, but equally important are the ways in which the network is modulated by environmental cues to maximize fitness in changing conditions. In this section we focus on molecular mechanisms that connect nutrient and redox signals to the core network described above.

7.1 (p)ppGpp Is Produced in Response to Starvation in *E. coli*

When *E. coli* cells are starved for various nutrients, they synthesize the "alarmone" signaling molecules guanosine tetraphosphate and guanosine pentaphosphate (here collectively called (p)ppGpp), which trigger the downregulation of macromolecular syntheses (Battesti and Bouveret 2006; Bouveret and Battesti 2011). RelA responds to amino acid starvation by associating with ribosomes and producing (p)ppGpp when an uncharged tRNA molecule enters the A-site of the ribosome (Haseltine and Block 1973). SpoT, which can both synthesize and hydrolyze (p)ppGpp, has been suggested to respond to starvation for several nutrients, including carbon (Xiao et al. 1991), fatty acids (Seyfzadeh et al. 1993), and iron (Vinella et al. 2005). However, only fatty acid starvation has been mechanistically linked to SpoT activation, via the binding of SpoT to holo-acyl carrier protein (Battesti and Bouveret 2006). Increased (p)ppGpp levels directly or indirectly inhibit the synthesis of DNA, RNA, and proteins. For example, by binding to RNAP, (p)ppGpp interferes with the transcription of a subset of RNAs, most importantly rRNAs and tRNAs, which in turn reduces translation (Haugen et al. 2008). (p)ppGpp also inhibits the initiation of chromosome replication in *E. coli* by inhibiting *dnaA* transcription (Chiaramello and Zyskind 1990).

7.2 (p)ppGpp Is a Key Mediator of *Caulobacter* Starvation Responses

It is well established that starvation for carbon or nitrogen blocks the *Caulobacter* cell cycle in the G1 swarmer phase (Gorbatuyk and Marczynski 2005; Lesley and Shapiro 2008; England et al. 2010; Ronneau et al. 2016). Isolated swarmer cells that are released into minimal medium lacking either a carbon or nitrogen source remain motile and do not initiate chromosome replication. In laboratory conditions, synchronized stalked cells that have already initiated chromosome replication are able to complete replication when their nitrogen source is withdrawn (Ronneau et al. 2016). However, cells that complete chromosome replication during carbon or nitrogen starvation may sustain damage that reduces competitive fitness. In addition, it is believed that prolonging the motile phase in response to starvation gives swarmer cells the chance to locate a more favorable environment before committing to a sessile, replicative lifestyle.

At the molecular level, in carbon- or nitrogen-limited swarmer cells, the CtrA protein is stabilized, rather than being degraded during swarmer cell development, and the abundance of DnaA is greatly reduced (Gorbatuyk and Marczynski 2005; Lesley and Shapiro 2008; Boutte et al. 2012). The flagellum is maintained at the pole, and the remodeling of polar signaling proteins is also disrupted. In particular, DivJ does not accumulate at the presumptive stalked pole (Boutte et al. 2012). As in *E. coli*, carbon or nitrogen starvation triggers an increase in intracellular levels of (p)ppGpp (Boutte et al. 2012).

Caulobacter has only one protein capable of synthesizing (p)ppGpp, the bifunctional enzyme SpoT (Nierman et al. 2001). Swarmer cells lacking SpoT do not arrest their cell cycle appropriately when released into medium lacking carbon or nitrogen, suggesting that (p)ppGpp is an important signal of starvation (Lesley and Shapiro 2008; Boutte and Crosson 2011; Boutte et al. 2012). To uncover the cellular effects of (p)ppGpp in the absence of actual starvation, Gonzalez and Collier (2014) expressed a hyperactive (p)ppGpp synthase (RelA′) using an inducible promoter in wild-type *Caulobacter* under nutrient-replete conditions (Gonzalez and Collier 2014). These experiments showed that (p)ppGpp synthesis is sufficient to slow growth, stabilize the CtrA protein, and delay the initiation of chromosome replication and the successive localization of DivJ and release of PleC from the flagellar pole. Given that (p)ppGpp is both necessary and sufficient for these phenotypic effects of starvation, work is ongoing to determine how different types of starvation are sensed by SpoT and how (p)ppGpp modulates cell cycle progression.

7.3 Linkages Between Ammonium Deprivation and (P)ppGpp Synthesis

In contrast to *E. coli*, *Caulobacter* cells do not experience an increase in (p)ppGpp levels during fatty acid starvation (Stott et al. 2015), so *Caulobacter* SpoT is not likely to be regulated in the exact same ways as its homolog in *E. coli*. Recent work, however, has uncovered the mechanism by which *Caulobacter* SpoT senses and responds to ammonium limitation (Fig. 3a) (Ronneau et al. 2016). *Caulobacter* assimilates ammonium exclusively via the glutamine synthetase GlnA, whose transcription and activity are promoted by the general nitrogen sensor GlnD (Reitzer 2003). As expected, the *Caulobacter* Δ*glnD* and Δ*glnA* mutants are auxotrophic for glutamine. Interestingly, however, both mutants grow slowly and accumulate

G1-phase swarmer cells when cultivated in complex PYE medium, which contains a mixture of amino acids (Ely 1991). These growth and cell cycle defects are relieved by adding glutamine to the PYE medium, indicating that low glutamine levels specifically trigger a G1-phase cell cycle delay (Ronneau et al. 2016).

Additional work demonstrated that cellular glutamine levels are communicated to SpoT by a nitrogen-related phosphoenolpyruvate (PEP) phosphotransferase system (PTS^{Ntr}). In canonical PTS systems, which sense and respond to carbon availability, an EI enzyme autophosphorylates using PEP, and the phosphoryl group is transferred sequentially to the HPr and EII proteins, which regulate the uptake of carbohydrates (Deutscher et al. 2014). PTS^{Ntr} systems are comprised of similar components, but respond instead to nitrogen availability. Similar to a PTS^{Ntr} system in *Sinorhizobium meliloti* (Goodwin and Gage 2014), glutamine binding inhibits EI^{Ntr} phosphorylation in *Caulobacter* (Ronneau et al. 2016). Thus, when glutamine is limiting, due to an inability to assimilate ammonium, phosphorylated forms of HPr^{Ntr} and EII^{Ntr} accumulate. Ronneau et al. (2016) found that EII^{Ntr} ~ P binds SpoT directly and inhibits its (p)ppGpp hydrolase activity, while HPr^{Ntr} ~ P indirectly stimulates the (p)ppGpp synthetase activity of SpoT (Fig. 3a). These interactions connect nitrogen limitation with a rise in intracellular (p)ppGpp. Consistent with these findings, deletion of *Caulobacter pstP*, encoding EI^{Ntr}, blocks (p)ppGpp accumulation and G1-phase swarmer cell accumulation during nitrogen starvation (Ronneau et al. 2016).

7.4 Effects of (P)ppGpp in *Caulobacter*

(p)ppGpp appears to affect the function of the CckA-ChpT-CtrA phosphorelay in at least two ways, but no molecular mechanism has yet been described. Mutants lacking either *ptsP* or *spoT* have reduced transcription of *ctrA* and *mopJ*, along with two additional targets of the transcriptional regulator GcrA (Sanselicio and Viollier 2015). GcrA recognizes a specific, methylated DNA motif and binds to the σ^{70} subunit and RNAP core enzyme (Fioravanti et al. 2013; Haakonsen et al. 2015). GcrA is normally essential for *Caulobacter* viability (Holtzendorff et al. 2004), but a Δ*gcrA* mutation can be made in a strain that overproduces (p)ppGpp (Haakonsen et al. 2015). Together, these results suggest that (p)ppGpp participates in the regulation of GcrA-dependent genes, possibly by binding directly to RNAP, or through an indirect mechanism (Fig. 3b). Binding of (p)ppGpp to *Caulobacter* RNAP has not yet been examined, but the amino acid residues that mediate (p)ppGpp binding in *E. coli* are conserved in the respective *Caulobacter* subunits (Ross et al. 2013, 2016).

During abrupt nitrogen or carbon starvation (Gorbatuyk and Marczynski 2005; Lesley and Shapiro 2008) or during glucose exhaustion (Boutte et al. 2012; Leslie et al. 2015), CtrA is maintained at a moderate level, while DnaA levels fall dramatically. The maintenance of CtrA levels during glucose exhaustion requires SpoT (Boutte et al. 2012, Leslie et al. 2015), and (p)ppGpp synthesis by RelA′ in well-fed cells is sufficient to decrease the rate of CtrA proteolysis (Gonzalez and Collier 2014). Together, these results indicate that CtrA is stabilized during nutrient limitation by a process that requires (p)ppGpp. Although it is tempting to speculate that a (p)ppGpp-binding protein directly interferes with CtrA proteolysis, it is perhaps more likely that (p)ppGpp acts upstream in the regulatory network to maintain both CtrA phosphorylation and stability. In support of this idea, glucose exhaustion decreases the fraction of wild-type cells with a polar focus of DivJ, consistent with a delay in swarmer cell differentiation, but glucose exhaustion does not prevent DivJ localization in a Δ*spoT* mutant (Boutte et al. 2012). Further, (p)ppGpp synthesis by RelA′ in well-fed cells delays the localization of DivJ and the delocalization of PleC in developing swarmer cells (Gonzalez and Collier 2014). These results suggest a model in which (p)ppGpp directly or indirectly inhibits the localization of DivJ at the developing flagellar pole, thereby blocking the increases in DivK~P, PleD~P, and cdG levels that would normally occur during differentiation (Fig. 3b). In consequence, CckA remains in kinase mode to activate and stabilize CtrA. If this model is correct, then CckA, CtrA, and CpdR should remain phosphorylated in starved cells, and further studies should reveal a factor involved in or upstream of DivJ localization (Ebersbach et al. 2008; Radhakrishnan et al. 2008; Perez et al. 2017) whose activity is sensitive to (p)ppGpp.

The DNA replication delay observed in starved *Caulobacter* cells could be attributed to CtrA blocking the initiation of chromosome replication, to the rapid clearance of the replication initiation protein DnaA, or to a combination of both. DnaA clearance during nitrogen or carbon starvation was initially attributed to an increase in the rate of DnaA proteolysis (Gorbatuyk and Marczynski 2005; Lesley and Shapiro 2008). However, a subsequent study showed that, while Lon-mediated proteolysis is required to clear DnaA, the rate of DnaA degradation does not increase after exhaustion of the carbon source (Leslie et al. 2015). Instead, the starvation-induced drop in DnaA abundance depends on reduced translation of the *dnaA* message, mediated by an element in its 5′ untranslated region (Leslie et al. 2015). In another discrepancy, one study found SpoT necessary for DnaA clearance during carbon starvation (Lesley and Shapiro 2008), while a different study found that Δ*spoT* cells eliminate DnaA normally upon glucose exhaustion (Leslie et al. 2015). Abrupt withdrawal of glucose from isolated swarmer cells (Lesley and Shapiro 2008) is not identical to the more gradual exhaustion of glucose by an unsynchronized population of cells (Leslie et al. 2015), and it is possible that these subtle differences in experimental design are revealing important nuances in *Caulobacter* responses to nutrient stress.

Yet another similar stress is encountered when a batch culture growing exponentially in rich medium exhausts one or more and enters stationary phase. During the stationary phase transition in *Caulobacter*, CtrA is stabilized, and DnaA is eliminated through a decrease in translation (Leslie et al. 2015). Interestingly, although the Δ*spoT* mutant properly modulates CtrA and DnaA levels during the initial transition into stationary phase (Leslie et al. 2015), SpoT, PtsP, and MopJ are each required to maintain wild-type levels of CtrA after several hours in stationary phase (Fig. 3b) (Sanselicio and Viollier 2015; Sanselicio et al. 2015). *Caulobacter* may therefore use distinct signaling pathways to modulate the cell cycle network in early and prolonged stationary phase. In addition, mutations that cause overproduction of (p)ppGpp increase the transcription of CtrA-dependent genes during prolonged stationary phase (Sanselicio and Viollier 2015). This indicates that (p)ppGpp can upregulate both the abundance and activity of CtrA, in agreement with a model in which it impacts the cell cycle regulatory network at or upstream of CckA.

8 Oscillations in Cellular Redox Status and Their Effects

Bacteria are well known to sense sharp changes in external redox-active compounds and mount appropriate responses to enhance survival (Storz and Spiro 2011). However, more subtle variations in intracellular redox conditions can also function as signals affecting developmental processes such as sporulation and biofilm formation (Sporer et al. 2017). Here we consider the possibility that variations in redox state regulate the *Caulobacter* cell cycle.

Using a derivative of GFP (roGFP2) whose fluorescence is modulated by an intramolecular disulfide bond (Hanson et al. 2004; Bhaskar et al. 2014), Narayanan et al. (2015) were the first to observe cell cycle-dependent variation in the thiol-redox status of the bacterial cytoplasm. *Caulobacter* G1- and early S-phase cells are in a relatively reduced state, followed by a peak of oxidation during S-phase and a slow return to the reduced state in predivisional cells. The causes underlying the redox cycle are presently unknown, but one hypothesis is that increased activity of ribonucleotide reductase during S-phase could temporarily oxidize the cytoplasm, since this enzyme depends on reduced thiol carrier proteins for regeneration (Sporer et al. 2017). To date, two systems have emerged as possible links between the cytoplasmic redox status and cell cycle progression.

8.1 Redox-Sensitive NstA Regulates Topoisomerase IV

Topoisomerase IV (topo IV) is an essential enzyme responsible for decatenating linked circular chromosomes just prior to cell division (Espeli et al. 2003). The small protein NstA binds to topo IV and inhibits its decatenating activity, and constitutive expression of NstA blocks *Caulobacter* chromosome segregation and cell division (Narayanan et al. 2015). The active form of NstA, a disulfide-linked dimer, is present during S-phase, when the cytoplasm is relatively oxidized, but is absent during G1-phase, when the cytoplasm is relatively reduced. Redox-sensitive disulfide bond formation may thereby limit the inhibitory activity of NstA to S-phase, which would release topo IV from inhibition when it is needed late in the cell cycle (Badrinarayanan et al. 2015). It remains to be demonstrated, however, that the inhibitory NstA dimer is absent from late predivisional cells (Narayanan et al. 2015). The deletion of *nstA* causes no reported phenotypic consequences in *Caulobacter* (Narayanan et al. 2015), indicating either that a redundant mechanism inhibits topo IV activity in S-phase cells or that temporary inhibition of topo IV is not critical for cell cycle progression in laboratory conditions.

8.2 NADH-Producing and NADH-Binding Proteins Regulate Z-Ring Assembly

The divisome is a potential target of the cytoplasmic redox state, because two modulators of the Z-ring, KidO and GdhZ, both interact with NAD(H) (Radhakrishnan et al. 2010; Beaufay et al. 2015). KidO is similar to NAD(P)H-dependent oxidoreductases and binds NAD(H), but it lacks critical catalytic residues (Radhakrishnan et al. 2010). When bound to NADH, KidO inhibits Z-ring formation by preventing FtsZ filament bundling (Beaufay et al. 2015). GdhZ is an NAD-dependent glutamate dehydrogenase that oxidizes glutamate, yielding α-ketoglutarate and NADH (Beaufay et al. 2015). This reaction is required for the catabolism of specific amino acids, such as glutamate and glutamine, but not for the catabolism of sugars such as glucose and xylose (Minambres et al. 2000). GdhZ stimulates the GTPase activity of FtsZ, which inhibits filament polymerization (Beaufay et al. 2015). GdhZ may also indirectly inhibit Z-ring formation by colocalizing with KidO on Z-rings and providing the NADH cofactor for KidO activity (Radhakrishnan et al. 2010; Beaufay et al. 2015).

Because it is specifically the NADH-bound form of KidO that inhibits Z-rings, and because the cytoplasm is relatively reduced in swarmer and predivisional cells, an attractive hypothesis is that the ratio of NADH to NAD^+ in the cytoplasm controls KidO activity. However, if cytoplasmic redox status is involved, it is not the only regulator of KidO. As described above (Sect. 6.5), KidO and GdhZ are both regulated by proteolysis, such that their levels are high in swarmer and late predivisional cells but low during S-phase (Fig. 2a). Expression of a stabilized variant of KidO during S-phase results in FtsZ mislocalization and cell elongation (Radhakrishnan et al. 2010), indicating that the relatively oxidized state of the cytoplasm during S-phase is not sufficient to inhibit KidO activity and preserve Z-rings.

During growth on complex PYE medium, where GdhZ is necessary for the catabolism of amino acids (Ely 1991), KidO and GdhZ interact with each other and with the Z-ring (Radhakrishnan et al. 2010; Beaufay et al. 2015). In contrast, GdhZ is delocalized from the Z-ring when glucose is added to the medium, suggesting that it doesn't regulate FtsZ in this growth condition (Beaufay et al. 2015). Mutants lacking either KidO or GdhZ have irregular cell sizes and altered Z-ring dynamics when grown in PYE medium, but addition of glucose to the medium suppresses these defects in the *ΔgdhZ* mutant (Radhakrishnan et al. 2010; Beaufay et al. 2015). Taken together, these results suggest that during growth on glucose, either GdhZ's direct inhibition of the Z-ring is unnecessary, or GdhZ is not an important source of the NADH cofactor for KidO, or both. Mutants in which the catalytic activity of GdhZ (producing NADH) is separated from its ability to stimulate the GTPase activity of FtsZ would help to distinguish between these models. Furthermore, the redox status of the *Caulobacter* cytoplasm may be different during growth on amino acids, which is unknown, than on glucose (Narayanan et al. 2015). If the cytoplasmic pool of NAD(H) is more reduced during growth on glucose than on amino acids, then KidO may be able to obtain NADH and regulate Z-ring assembly without the need for GdhZ. It is also possible, however, that an unknown division regulator cooperates with KidO during growth on glucose.

9 Outlook

Although the *Caulobacter* cell cycle signaling machinery has been intensively studied, fundamental questions continue to arise. Several non-canonical interactions between two-component proteins have been uncovered in the *Caulobacter* cell cycle network, and we expect that future work will continue to describe the molecular details of signal transduction in these novel systems. *Caulobacter* was one of the first bacteria in which signaling proteins, proteases, and proteolytic adaptors were found to be dynamically, subcellularly localized. We expect that future studies in *Caulobacter* will continue to tackle the challenging problems of observing and explaining at a molecular level cases where a protein performs distinct activities in different subcellular locations. Studies of cdG-dependent processes in *Caulobacter* have revealed new classes of proteins that bind cdG, and additional effector proteins are likely to be discovered during efforts to link cdG to stalk and pilus biosynthesis. Studies in other systems suggest that protein–protein interactions between a DGC or PDE enzyme and a cdG-dependent effector can generate spatially regulated signals, where the cdG produced (or degraded) by an enzyme only affects one or a small number of cellular processes (Lindenberg et al. 2013; Dahlstrom et al. 2016). Such regulatory mechanisms may underlie the specific effects of particular DGC enzymes on cell motility and attachment (Abel et al. 2013) or the flagellar-pole specific protection of CckA from cdG (Lori et al. 2015). With respect to proteolysis, we expect that additional recognition mechanisms and adaptor proteins will be discovered that target substrates to proteases other than ClpXP. It will also be important to reconcile in vitro and in vivo approaches to determine how substrates that are degraded in an unregulated manner in vitro are proteolyzed under rigorous cell cycle control in vivo. Finally, although the unstressed, exponential-phase cell cycle has been the main focus of the field in the past, there is now keen interest in understanding exactly how (p)ppGpp modulates the cell cycle network during nutrient stress and stationary phase.

Acknowledgements The authors wish to thank Charlie Huang, Sneha Jani, and Lisa Bowers for helpful discussions and comments on the manuscript.

References

Abel S, Chien P, Wassmann P et al (2011) Regulatory cohesion of cell cycle and cell differentiation through interlinked phosphorylation and second messenger networks. Mol Cell 43:550–560. https://doi.org/10.1016/j.molcel.2011.07.018

Abel S, Bucher T, Nicollier M et al (2013) Bi-modal distribution of the second messenger c-di-GMP controls cell fate and asymmetry during the *Caulobacter* cell cycle. PLoS Genet 9:e1003744. https://doi.org/10.1371/journal.pgen.1003744

Aldridge P, Jenal U (1999) Cell cycle-dependent degradation of a flagellar motor component requires a novel-type response regulator. Mol Microbiol 32:379–391

Alley MR (2001) The highly conserved domain of the *Caulobacter* McpA chemoreceptor is required for its polar localization. Mol Microbiol 40:1335–1343

Angelastro PS, Sliusarenko O, Jacobs-Wagner C (2010) Polar localization of the CckA histidine kinase and cell cycle periodicity of the essential master regulator CtrA in *Caulobacter*. J Bacteriol 192:539–552. https://doi.org/10.1128/JB.00985-09

Badrinarayanan A, Le TBK, Laub MT (2015) Bacterial chromosome organization and segregation. Annu Rev Cell Dev Biol 31:171–199. https://doi.org/10.1146/annurev-cellbio-100814-125211

Barchinger SE, Ades SE (2013) Regulated proteolysis: control of the *Escherichia coli* σ(E)-dependent cell envelope stress response. Subcell Biochem 66:129–160. https://doi.org/10.1007/978-94-007-5940-4_6

Battesti A, Bouveret E (2006) Acyl carrier protein/SpoT interaction, the switch linking SpoT-dependent stress response to fatty acid metabolism. Mol Microbiol 62:1048–1063. https://doi.org/10.1111/j.1365-2958.2006.05442.x

Beaufay F, Coppine J, Mayard A et al (2015) A NAD-dependent glutamate dehydrogenase coordinates metabolism with cell division in *Caulobacter crescentus*. EMBO J 34:1786–1800. https://doi.org/10.15252/embj.201490730

Bhaskar A, Chawla M, Mehta M et al (2014) Reengineering redox sensitive GFP to measure mycothiol redox potential of *mycobacterium tuberculosis* during infection. PLoS Pathog 10:e1003902. https://doi.org/10.1371/journal.ppat.1003902

Biondi EG, Reisinger SJ, Skerker JM et al (2006a) Regulation of the bacterial cell cycle by an integrated genetic circuit. Nature 444:899–904. https://doi.org/10.1038/nature05321

Biondi EG, Skerker JM, Arif M et al (2006b) A phosphorelay system controls stalk biogenesis during cell cycle progression in *Caulobacter crescentus*. Mol Microbiol 59:386–401. https://doi.org/10.1111/j.1365-2958.2005.04970.x

Bleichert S, Botchan MR, Berger JM (2017) Mechanisms for initiating cellular DNA replication. Science 355:eaah6317. https://doi.org/10.1126/science.aah6317

Bodenmiller D, Toh E, Brun YV (2004) Development of surface adhesion in *Caulobacter crescentus*. J Bacteriol 186:1438–1447

Bourret RB (2010) Receiver domain structure and function in response regulator proteins. Curr Opin Microbiol 13:142–149. https://doi.org/10.1016/j.mib.2010.01.015

Boutte CC, Crosson S (2011) The complex logic of stringent response regulation in *Caulobacter crescentus*: starvation signalling in an oligotrophic environment. Mol Microbiol 80:695–714. https://doi.org/10.1111/j.1365-2958.2011.07602.x

Boutte CC, Henry JT, Crosson S (2012) ppGpp and polyphosphate modulate cell cycle progression in *Caulobacter crescentus*. J Bacteriol 194:28–35. https://doi.org/10.1128/JB.05932-11

Bouveret E, Battesti A (2011) The stringent response. In: Storz G, Hengge R (eds) Bacterial stress responses. ASM Press, Washington, DC, pp 231–250

Briegel A, Ding HJ, Li Z et al (2008) Location and architecture of the *Caulobacter crescentus* chemoreceptor array. Mol Microbiol 69:30–41. https://doi.org/10.1111/j.1365-2958.2008.06219.x

Chen YE, Tsokos CG, Biondi EG et al (2009) Dynamics of two phosphorelays controlling cell cycle progression in *Caulobacter crescentus*. J Bacteriol 191:7417–7429. https://doi.org/10.1128/JB.00992-09

Chen YE, Tropini C, Jonas K et al (2011) Spatial gradient of protein phosphorylation underlies replicative asymmetry in a bacterium. Proc Natl Acad Sci U S A 108:1052–1057. https://doi.org/10.1073/pnas.1015397108

Chiaramello AE, Zyskind JW (1990) Coupling of DNA replication to growth rate in *Escherichia coli*: a possible role for guanosine tetraphosphate. J Bacteriol 172:2013–2019

Chien P, Perchuk BS, Laub MT et al (2007) Direct and adaptor-mediated substrate recognition by an essential AAA+ protease. Proc Natl Acad Sci U S A 104:6590–6595. https://doi.org/10.1073/pnas.0701776104

Childers WS, Xu Q, Mann TH et al (2014) Cell fate regulation governed by a repurposed bacterial histidine kinase. PLoS Biol 12:e1001979. https://doi.org/10.1371/journal.pbio.1001979

Chou SH, Galperin MY (2016) Diversity of cyclic di-GMP-binding proteins and mechanisms. J Bacteriol 198:32–46. https://doi.org/10.1128/JB.00333-15

Christen M, Christen B, Folcher M et al (2005) Identification and characterization of a cyclic di-GMP-specific phosphodiesterase and its allosteric control by GTP. J Biol Chem 280:30829–30837. https://doi.org/10.1074/jbc.M504429200

Christen M, Christen B, Allan MG et al (2007) DgrA is a member of a new family of cyclic diguanosine monophosphate receptors and controls flagellar motor function in *Caulobacter crescentus*. Proc Natl Acad Sci U S A 104:4112–4117. https://doi.org/10.1073/pnas.0607738104

Christen M, Kulasekara HD, Christen B et al (2010) Asymmetrical distribution of the second messenger c-di-GMP upon bacterial cell division. Science 328:1295–1297. https://doi.org/10.1126/science.118865

Collier J, Shapiro L (2009) Feedback control of DnaA-mediated replication initiation by replisome-associated HdaA protein in *Caulobacter*. J Bacteriol 191:5706–5716. https://doi.org/10.1128/JB.00525-09

Collier J, Murray SR, Shapiro L (2006) DnaA couples DNA replication and the expression of two cell cycle master regulators. EMBO J 25:346–356. https://doi.org/10.1038/sj.emboj.7600927

Collier J, McAdams HH, Shapiro L (2007) A DNA methylation ratchet governs progression through a bacterial cell cycle. Proc Natl Acad Sci U S A 104:17111–17116. https://doi.org/10.1073/pnas.0708112104

Dahlstrom KM, Giglio KM, Sondermann H et al (2016) The inhibitory site of a diguanylate cyclase is a necessary element for interaction and signaling with an effector protein. J Bacteriol 198:1595–1603. https://doi.org/10.1128/JB.00090-16

Davis NJ, Cohen Y, Sanselicio S et al (2013) De- and repolarization mechanism of flagellar morphogenesis during a bacterial cell cycle. Genes Dev 27:2049–2062. https://doi.org/10.1101/gad.222679.113

Deutscher J, Ake FM, Derkaoui M et al (2014) The bacterial phosphoeholpyruvate:carbohydrate phosphotransferase system: regulation by protein phosphorylation and phosphorylation-dependent protein-protein interactions. Microbiol Mol Biol Rev 78:231–256. https://doi.org/10.1128/MMBR.00001-14

Domian IJ, Quon KC, Shapiro L (1997) Cell type-specific phosphorylation and proteolysis of a transcriptional regulator controls the G1-to-S transition in a bacterial cell cycle. Cell 90:415–424

Dougan DA, Weber-Ban E, Bukau B (2003) Targeted delivery of an ssrA-tagged substrate by the adaptor protein SspB to its cognate AAA+ protein ClpX. Mol Cell 12:373–380

Dubey BN, Lori C, Ozaki S et al (2016) Cyclic di-GMP mediates a histidine kinase/phosphatase switch by noncovalent domain cross-linking. Sci Adv 2:e1600823. https://doi.org/10.1126/sciadv.1600823

Duerig A, Abel S, Folcher M et al (2009) Second messenger-mediated spatiotemporal control of protein degradation regulates bacterial cell cycle progression. Genes Dev 23:93–104. https://doi.org/10.1101/gad.502409

Ebersbach G, Briegel A, Jensen GJ et al (2008) A self-associating protein critical for chromosome attachment, division, and polar organization in *Caulobacter*. Cell 134:956–968. https://doi.org/10.1016/j.cell.2008.07.016

Ely B (1991) Genetics of *Caulobacter crescentus*. Methods Enzymol 204:372–384

England JC, Perchuk BS, Laub MT et al (2010) Global regulation of gene expression and cell differentiation in *Caulobacter crescentus* in response to nutrient availability. J Bacteriol 192:819–833

Espeli O, Levine C, Hassing H et al (2003) Temporal regulation of topoisomerase IV activity in *E. coli*. Mol Cell 11:189–201

Farrell CM, Baker TA, Sauer RT (2007) Altered specificity of a AAA+ protease. Mol Cell 25:161–166. https://doi.org/10.1016/j.molcel.2006.11.018

Fernandez-Fernandez C, Grosse K, Sourjik V et al (2013) The β-sliding clamp directs the localization of HdaA to the replisome in *Caulobacter crescentus*. Microbiology 159:2237–2248. https://doi.org/10.1099/mic.0.068577-0

Fioravanti A, Fumeaux C, Mohaptra SS et al (2013) DNA binding of the cell cycle transcriptional regulator GcrA depends on N6-adenosine methylation in *Caulobacter crescentus* and other *Alphaproteobacteria*. PLoS Genet 9:e1003541. https://doi.org/10.1371/journal.pgen.1003541

Flynn JM, Levchenko I, Seidel M et al (2001) Overlapping recognition determinants within the ssrA degradation tag allow modulation of proteolysis. Proc Natl Acad Sci U S A 98:10584–10589. https://doi.org/10.1073/pnas.191375298

Flynn JM, Levchenko I, Sauer RT et al (2004) Modulating substrate choice: the SspB adaptor delivers a regulator of the extracytoplasmic-stress response to the AAA+ protease ClpXP for degradation. Genes Dev 18:2292–2301. https://doi.org/10.1101/gad.1240104

Gao R, Stock AM (2010) Molecular strategies for phosphorylation-mediated regulation of response regulator activity. Curr Opin Microbiol 13:160–167. https://doi.org/10.1016/j.mib.2009.12.009

Gonzalez D, Collier J (2014) Effects of (p)ppGpp on the progression of the cell cycle of *Caulobacter crescentus*. J Bacteriol 196:2514–2525. https://doi.org/10.1128/JB.01575-14

Goodwin RA, Gage DJ (2014) Biochemical characterization of a nitrogen-type phosphotransferase system reveals that enzyme EI(Ntr) integrates carbon and nitrogen signaling in *Sinorhizobium meliloti*. J Bacteriol 196:1901–1907. https://doi.org/10.1128/JB.01489-14

Gora KG, Tsokos CG, Chen YE et al (2010) A cell-type-specific protein-protein interaction modulates transcriptional activity of a master regulator in *Caulobacter crescentus*. Mol Cell 39:45–467. https://doi.org/10.1016/j.molcel.2010.06.024

Gora KG, Cantin A, Wohlever M et al (2013) Regulated proteolysis of a transcription factor complex is critical to cell cycle progression in *Caulobacter crescentus*. Mol Microbiol 87:1277–1289. https://doi.org/10.1111/mmi.12166

Gorbatuyk B, Marczynski GT (2001) Physiological consequences of blocked *Caulobacter crescentus dnaA* expression, an essential DNA replication gene. Mol Microbiol 40:485–497

Gorbatuyk B, Marczynski GT (2005) Regulated degradation of chromosome replication proteins DnaA and CtrA in *Caulobacter crescentus*. Mol Microbiol 55:1233–1245

Gottesman S, Roche E, Zhou Y et al (1998) The ClpXP and ClpAP proteases degrade proteins with carboxy-terminal peptide tails added by the SsrA-tagging system. Genes Dev 12:1338–1347

Grimaud R, Kessel M, Beuron F et al (1998) Enzymatic and structural similarities between the *Escherichia coli* ATP-dependent proteases, ClpXP and ClpAP. J Biol Chem 273:12476–12481

Grunenfelder B, Tawfilis S, Gehrig S et al (2004) Identification of the protease and the turnover signal responsible for cell cycle-dependent degradation of the *Caulobacter* FliF motor protein. J Bacteriol 186:4960–4971. https://doi.org/10.1128/JB.186.15.4960-4971.2004

Gur E (2013) The Lon AAA+ protease. Subcell Biochem 66:35–51. https://doi.org/10.1007/978-94-007-5940-4_2

Haakonsen DL, Yuan AH, Laub MT (2015) The bacterial cell cycle regulator GcrA is a σ70 cofactor that drives gene expression from a subset of methylated promoters. Genes Dev 29:2272–2286. https://doi.org/10.1101/gad.270660.115

Hanson GT, Aggeler R, Oglesbee D et al (2004) Investigating mitochondrial redox potential with redox-sensitive green fluorescent protein indicators. J Biol Chem 279:13044–13053. https://doi.org/10.1074/jbc.M312846200

Haseltine WA, Block R (1973) Synthesis of guanosine tetra- and pentaphosphate requires the presence of a codon-specific uncharged transfer ribonucleic acid in the acceptor site of ribosomes. Proc Natl Acad Sci U S A 70:1564–1568

Haugen SP, Ross W, Gourse RL (2008) Advances in bacterial promoter recognition and its control by factors that do not bind DNA. Nat Rev Microbiol 6:507–519. https://doi.org/10.1038/nrmicro1912

Hecht GB, Lane T, Ohta N et al (1995) An essential single domain response regulator required for normal cell division and differentiation in *Caulobacter crescentus*. EMBO J 14:3915–3924

Holmes JA, Follett SE, Wang H et al (2016) *Caulobacter* PopZ forms an intrinsically disordered hub in organizig bacterial cell poles. Proc Natl Acad Sci U S A 113:12490–12495. doi:https://doi.org/10.1073/pnas.1602380113

Holtzendorff J, Hung D, Brende P et al (2004) Oscillating global regulators control the genetic circuit driving a bacterial cell cycle. Science 304:983–987. https://doi.org/10.1126/science.1095191

Hug I, Deshpande S, Sprecher KS et al (2017) Second messenger-mediated tactile response by a bacterial rotary motor. Science 358:531–534. https://doi.org/10.1126/science.aan5353

Hung D, Shapiro L (2002) A signal transduction protein cues proteolytic events critical to *Caulobacter* cell cycle progression. Proc Natl Acad Sci U S A 99:13160–13165. https://doi.org/10.1073/pnas.202495099

Huynh TN, Stewart V (2011) Negative control in two-component signal transduction by transmitter phosphatase activity. Mol Microbiol 82:275–286. https://doi.org/10.1111/j.1365-2958.2011.07829.x

Iniesta AA, McGrath PT, Reisenauer A et al (2006) A phospho-signaling pathway controls the localization and activity of a protease complex critical for bacterial cell cycle progression. Proc Natl Acad Sci U S A 103:10935–10940. https://doi.org/10.1073/pnas.0604554103

Iniesta AA, Hillson NJ, Shapiro L (2010) Cell pole-specific activation of a critical bacterial cell cycle kinase. Proc Natl Acad Sci U S A 107:7012–7017. https://doi.org/10.1073/pnas.1001767107

Jacobs C, Domian IJ, Maddock JR et al (1999) Cell cycle-dependent polar localization of an essential bacterial histidine kinase that controls DNA replication and cell division. Cell 97:111–120

Jacobs C, Hung D, Shapiro L (2001) Dynamic localization of a cytoplasmic signal transduction response regulator controls morphogenesis during the *Caulobacter* cell cycle. Proc Natl Acad Sci U S A 98:4095–4100. https://doi.org/10.1073/pnas.051609998

Jacobs C, Ausmees N, Cordwell SJ et al (2003) Functions of the CckA histidine kinase in *Caulobacter* cell cycle control. Mol Microbiol 47:1279–1290

Janssen BD, Hayes CS (2012) The tmRNA ribosome-rescue system. Adv Protein Chem Struct Biol 86:151–191. https://doi.org/10.1016/B978-0-12-386497-0.00005-0

Jenal U, Fuchs T (1998) An essential protease involved in bacterial cell-cycle control. EMBO J 17:5658–5669. https://doi.org/10.1093/emboj/17.19.5658

Jenal U, Reinders A, Lori C (2017) Cyclic di-GMP: second messenger extraordinaire. Nat Rev Microbiol 15:271–284. https://doi.org/10.1038/nrmicro.2016.190

Jonas K, Chen YE, Laub MT (2011) Modularity of the bacterial cell cycle enables independent spatial and temporal control of DNA replication. Curr Biol 21:1092–1101. https://doi.org/10.1016/j.cub.2011.05.040

Jonas K, Liu J, Chien P et al (2013) Proteotoxic stress induces a cell-cycle arrest by stimulating Lon to degrade the replication initiator DnaA. Cell 154:623–636. https://doi.org/10.1016/j.cell.2013.06.034

Joshi KK, Berge M, Radhakrishnan SK et al (2015) An adaptor hierarchy regulates proteolysis during a bacterial cell cycle. Cell 163:419–431. https://doi.org/10.1016/j.cell.2015.09.030

Judd EM, Ryan KR, Moerner WE et al (2003) Fluorescence bleaching reveals asymmetric compartment formation prior to cell division in *Caulobacter*. Proc Natl Acad Sci U S A 100:8235–8240. https://doi.org/10.1073/pnas.1433105100

Katayama T, Kasho K, Kawakami H (2017) The DnaA cycle in *Escherichia coli*: activation, function, and inactivation of the initiator protein. Front Microbiol 8:2496. https://doi.org/10.3389/fmicb.2017.02496

Katayama-Fujimura Y, Gottesman S, Maurizi MR (1987) A multiple-component, ATP-dependent protease from *Escherichia coli*. J Biol Chem 262:4477–4485

Keiler KC, Waller PR, Sauer RT (1996) Role of a peptide tagging system in degradation of proteins synthesized from damaged messenger RNA. Science 271:990–993

Kelly AJ, Sackett MJ, Din N et al (1998) Cell cycle-dependent transcriptional and proteolytic regulation of FtsZ in *Caulobacter*. Genes Dev 12:880–893

Kirkpatrick CL, Viollier PH (2012) Decoding *Caulobacter* development. FEMS Microbiol Rev 36:193–205. https://doi.org/10.1111/j.1574-6976.2011.00309.x

Lau J, Hernandez-Alicea L, Vass RH et al (2015) A phosphosignaling adaptor primes the AAA+ protease ClpXP to drive cell cycle-regulated proteolysis. Mol Cell 59:104–116. https://doi.org/10.1016/j.molcel.2015.05.014

Laub MT, McAdams HH, Feldblyum T et al (2000) Global analysis of the genetic network controlling a bacterial cell cycle. Science 290:2144–2148

Laub MT, Chen SL, Shapiro L et al (2002) Genes directly controlled by CtrA, a master regulator of the *Caulobacter* cell cycle. Proc Natl Acad Sci U S A 99:4632–4637. https://doi.org/10.1073/pnas.062065699

Lesley JA, Shapiro L (2008) SpoT regulates DnaA stability and initiation of DNA replication in carbon-starved *Caulobacter crescentus*. J Bacteriol 190:6867–6880. https://doi.org/10.1128/JB.00700-08

Leslie DJ, Heinen C, Shcramm FD et al (2015) Nutritional control of DNA replication initiation through the proteolysis and regulated translation of DnaA. PLoS Genet 11:e1005342. https://doi.org/10.1371/journal.pgen.1005342

Levchenko I, Seidel M, Sauer RT et al (2000) A specificity-enhancing factor for the ClpXP degradation machine. Science 289:2354–2356

Levchenko I, Grant RA, Wah DA et al (2003) Structure of a delivery protein for an AAA+ protease in complex with a peptide degradation tag. Mol Cell 12:365–372

Levi A, Jenal U (2006) Holdfast formation in motile swarmer cells optimizes surface attachment during *Caulobacter crescentus* development. J Bacteriol 188:5315–5318. https://doi.org/10.1128/JB.01725-05

Li G, Brown PJ, Tang JX et al (2012) Surface contact stimulates the just-in-time deployment of bacterial adhesins. Mol Microbiol 83:41–51. https://doi.org/10.1111/j.1365-2958.2011.07909.x

Lindenberg S, Klauck G, Pesavento C et al (2013) The EAL domain protein YciR acts as a trigger enzyme in a c-di-GMP signalling cascade in *E. coli* biofilm control. EMBO J 32:2001–2014. https://doi.org/10.1038/emboj.2013.120

Liu J, Francis LI, Jonas K et al (2016) ClpAP is an auxiliary protease for DnaA degradation in *Caulobacter crescentus*. Mol Microbiol 102:1075–1085. https://doi.org/10.1111/mmi.13537

Lori C, Ozaki S, Steiner S et al (2015) Cyclic di-GMP acts as a cell cycle oscillator to drive chromosome replication. Nature 523:236–239. https://doi.org/10.1038/nature14473

Lupas A, Flanagan JM, Tamura T et al (1997) Self-compartmentalizing proteases. Trends Biochem Sci 22:399–404

Maddock JR, Shapiro L (1993) Polar localization of the chemoreceptor complex in the *Escherichia coli* cell. Science 259:1717–1723

Mann TH, Childers WS, Blair JA et al (2016) A cell cycle kinase with tandem sensory PAS domains integrates cell fate cues. Nat Commun 7:11454. https://doi.org/10.1038/ncomms11454

Marczynski GT (1999) Chromosome methylation and measurement of faithful, once and only once per cell cycle chromosome replication in Caulobacter crescentus. J Bacteriol 181:1984–1993

Martin ME, Trimble MJ, Brun YV (2004) Cell cycle-dependent abundance, stability and localization of FtsA and FtsQ in *Caulobacter crescentus*. Mol Microbiol 54:60–74. https://doi.org/10.1111/j.1365-2958.2004.04251.x

Martin A, Baker TA, Sauer RT (2008) Diverse pore loops of the AAA+ ClpX machine mediate unassisted and adaptor-dependent recognition of ssrA-tagged substrates. Mol Cell 29:441–450. https://doi.org/10.1016/j.molcel.2008.02.002

Matroule J-Y, Lam H, Burnette DT et al (2004) Cytokinesis monitoring during development: rapid pole-to-pole shuttling of a signaling protein by localized kinase and phosphatase in *Caulobacter*. Cell 118:579–590. https://doi.org/10.1016/j.cell.2004.08.019

McGrath PT, Iniesta AA, Ryan KR et al (2006) A dynamically localized protease complex and a polar specificity factor control a cell cycle master regulator. Cell 124:535–547. https://doi.org/10.1016/j.cell.2005.12.033

Minambres B, Olivera ER, Jensen RA et al (2000) A new class of glutamate dehydrogenases (GDH). Biochemical and genetic characterization of the first member, the AMP-requiring NAD-specific GDH of *Streptomyces clavuligerus*. J Biol Chem 275:39529–39542. https://doi.org/10.1074/jbc.M005136200

Mohapatra SS, Fioravanti A, Biondi EG (2014) DNA methylation in *Caulobacter* and other Alphaproteobacteria during cell cycle progession. Trends Microbiol 22:528–535. https://doi.org/10.1016/j.tim.2014.05.003

Moliere N, Turgay K (2013) General and regulatory proteolysis in *Bacillus subtilis*. Subcell Biochem 66:73–103. https://doi.org/10.1007/978-94-007-5940-4_4

Narayanan S, Janakiraman B, Kumar L et al (2015) A cell cycle-controlled redox switch regulates the topoisomerase IV activity. Genes Dev 29:1175–1187. https://doi.org/10.1101/gad.257030.114

Nesper J, Reinders A, Glatter T et al (2012) A novel capture compound for the identification and analysis of cyclic di-GMP binding proteins. J Proteome 75:4874–4878. https://doi.org/10.1016/j.jprot.2012.05.033

Nesper J, Hug I, Kato S et al (2017) Cyclic di-GMP differentially tunes a bacterial flagellar motor through a novel class of CheY-like regulators. eLife 6:e28842. https://doi.org/10.7554/eLife.28842

Nierman WC, Feldblyum TV, Laub MT et al (2001) Complete genome sequence of *Caulobacter crescentus*. Proc Natl Acad Sci U S A 98:4136–4141. https://doi.org/10.1073/pnas.061029298

Olivares AO, Baker TA, Sauer RT (2016) Mechanistic insights into bacterial AAA+ proteases and protein-remodelling machines. Nat Rev Microbiol 14:33–44. https://doi.org/10.1038/nrmicro.2015.4

Ozaki S, Schlach-Moser A, Zumthor L et al (2014) Activation and polar sequestration of PopA, a c-di-GMP effector protein involved in *Caulobacter crescentus* cell cycle control. Mol Microbiol 94:580–594. https://doi.org/10.1111/mmi.12777

Paul R, Weiser S, Amiot NC et al (2004) Cell cycle-dependent dynamic localization of a bacterial response regulator with a novel di-guanylate cyclase output domain. Genes Dev 18:715–727. https://doi.org/10.1101/gad.289504

Paul R, Abel S, Wassmann P et al (2007) Activation of the diguanylate cyclase PleD by phosphorylation-mediated dimerization. J Biol Chem 282:29170–29177. https://doi.org/10.1074/jbc.M704702200

Paul R, Jaeger T, Abel S et al (2008) Allosteric regulation of histidine kinases by their cognate response regulator determines cell fate. Cell 133:452–461. https://doi.org/10.1016/j.cell.2008.02.045

Perez AM, Mann TH, Lasker K et al (2017) A localized complex of two protein oligomers controls the orientation of cell polarity. mBio 8:e02238-16. https://doi.org/10.1128/mBio.02238-16

Petters T, Zhang X, Nesper J et al (2012) The orphan histidine protein kinase SgmT is a c-di-GMP receptor and regulates composition of the extracellular matrix together with the orphan DNA binding response regulator DigR in *Myxococcus xanthus*. Mol Microbiol 84:147–165. https://doi.org/10.1111/j.1365-2958.2012.08015.x

Pierce DL, O'Donnol D, Allen RC et al (2006) Mutations in DivL and CckA rescue a *divJ* null mutant of *Caulobacter crescentus* by reducing the activity of CtrA. J Bacteriol 188:2473–2482. https://doi.org/10.1128/JB.188.7.2473-2482.2006

Poindexter JS (1964) Biological properties and classification of the *Caulobacter* group. Bacteriol Rev 28:231–295

Quon KC, Marczynski GT, Shapiro L (1996) Cell cycle control by an essential bacterial two-component signal transduction protein. Cell 84:83–93

Quon KC, Yang B, Domian IJ et al (1998) Negative control of bacterial DNA replication by a cell cycle regulatory protein that binds at the chromosome origin. Proc Natl Acad Sci U S A 95:120–125

Radhakrishnan SK, Thanbichler M, Viollier PH (2008) The dynamic interplay between a cell fate determinant and a lysozyme homolog drives the asymmetric division cycle of *Caulobacter crescentus*. Genes Dev 22:212–225. https://doi.org/10.1101/gad.1601808

Radhakrishnan SK, Pritchard S, Viollier PH (2010) Coupling prokaryotic cell fate and division control with a bifunctional and oscillating oxidoreductase homolog. Dev Cell 18:90–101. https://doi.org/10.1016/j.devcel.2009.10.024

Reisinger SJ, Huntwork S, Viollier PH et al (2007) DivL performs critical cell cycle functions in *Caulobacter crescentus* independent of kinase activity. J Bacteriol 189:8308–8320. https://doi.org/10.1128/JB.00868-07

Reitzer L (2003) Nitrogen assimilation and global regulation in *Escherichia coli*. Annu Rev Microbiol 57:155–176. https://doi.org/10.1146/annurev.micro.57.030502.090820

Ronneau S, Petit K, De Bolle X et al (2016) Phosphotransferase-dependent accumulation of (p)ppGpp in response to glutamine deprivation in *Caulobacter crescentus*. Nat Commun 7:11423. https://doi.org/10.1038/ncomms11423

Rood KL, Clark NE, Stoddard PR et al (2012) Adaptor-dependent degradation of a cell-cycle regulator uses a unique substrate architecture. Structure 20:1223–1232. https://doi.org/10.1016/j.str.2012.04.019

Ross W, Vrentas CE, Sanchez-Vazquez P et al (2013) The magic spot: a ppGpp binding site on *E. coli* RNA polymerase responsible for regulation of transcription initiation. Mol Cell 50:420–429. https://doi.org/10.1016/j.molcel.2013.03.021

Ross W, Sanchez-Vazquez P, Chen AY et al (2016) ppGpp binding to a site at the RNAP-DskA interface accounts for its dramatic effects on transcription initiation during the stringent response. Mol Cell 62:811–823. https://doi.org/10.1016/j.molcel.2016.04.029

Ryan KR, Judd EM, Shapiro L (2002) The CtrA response regulator essential for *Caulobacter crescentus* cell-cycle progression requires a bipartite degradation signal for temporally controlled proteolysis. J Mol Biol 324:443–455

Sanselicio S, Viollier PH (2015) Convergence of alarmone and cell cycle signaling from trans-encoded sensory domains. mBio 6:e01415-01415. https://doi.org/10.1128/mBio.01415-15

Sanselicio S, Berge M, Theraulaz L et al (2015) Topological control of the *Caulobacter* cell cycle circuitry by a polarized single-domain PAS protein. Nat Commun 6:7005. https://doi.org/10.1038/ncomms8005

Sciochetti SA, Ohta N, Newton A (2005) The role of polar localization in the function of an essential *Caulobacter crescentus* tyrosine kinase. Mol Microbiol 56:1467–1480. https://doi.org/10.1111/j.1365-2958.2005.04652.x

Seyfzadeh M, Keener J, Nomura M (1993) *spoT*-dependent accumulation of guanosine tetraphosphate in response to fatty acid starvation in *Escherichia coli*. Proc Natl Acad Sci U S A 90:11004–11008

Shapiro L, Maizel JV Jr (1973) Synthesis and structure of *Caulobacter crescentus* flagella. J Bacteriol 113:478–485

Silversmith RE (2010) Auxiliary phosphatases in two-component signal transduction. Curr Opin Microbiol 13:177–183. https://doi.org/10.1016/j.mib.2010.01.004

Smit J, Agabian N (1982) *Caulobacter crescentus* pili: analysis of production during development. Dev Biol 89:237–247

Smith SC, Joshi KK, Zik JJ et al (2014) Cell cycle-dependent adaptor complex for ClpXP-mediated proteolysis directly integrates phosphorylation and second messenger signals. Proc Natl Acad Sci U S A 111:14299–14234. https://doi.org/10.1073/pnas.1407862111

Sommer JM, Newton A (1989) Turning off flagellum rotation requires the pleiotropic gene *pleD*: *pleA*, *pleC*, and *pleD* define two morphogenic pathways in *Caulobacter crescentus*. J Bacteriol 171:392–401

Sporer AJ, Kahl LJ, Price-Whelan A et al (2017) Redox-based regulation of bacterial development and behavior. Annu Rev Biochem 86:777–797. https://doi.org/10.1146/annurev-biochem-061516-044453

Sprecher KS, Hug I, Nesper J et al (2017) Cohesive properties of the *Caulobacter crescentus* holdfast adhesin are regulated by a novel c-di-GMP effector protein. mBio 8:e00294–e00217. https://doi.org/10.1128/mBio.00294-17

Stephens C, Reisenauer A, Wright R et al (1996) A cell cycle-regulated bacterial DNA methyltransferase is essential for viability. Proc Natl Acad Sci U S A 93:1210–1214

Storz G, Spiro S (2011) Sensing and responding to reactive oxygen and nitrogen species. In: Storz G, Hengge R (eds) Bacterial stress responses. ASM Press, Washington, DC, pp 157–173
Stott KV, Wood SM, Blair JA et al (2015) (p)ppGpp modulates cell size and the initiation of DNA replication in *Caulobacter crescentus* in response to a block in lipid biosynthesis. Microbiology 161:553–564. https://doi.org/10.1099/mic.0.000032
Studemann A, Noirclerc-Savoye M, Klauck E et al (2003) Sequential recognition of two distinct sites in sigma(S) by the proteolytic targeting factor RssB and ClpX. EMBO J 22:4111–4120. https://doi.org/10.1093/emboj/cdg411
Subramanian K, Paul MR, Tyson JJ (2015) Dynamical localization of DivL, and PleC in the asymmetric division cycle of *Caulobacter crescentus*: a theoretical investigation of alternative models. PLoS Comp Biol 11:e1004348. https://doi.org/10.1371/journal.pcbi.1004348
Tan MH, Kozdon JB, Shen X et al (2010) An essential transcription factor, SciP, enhances robustness of *Caulobacter* cell cycle regulation. Proc Natl Acad Sci U S A 107:18985–18990. https://doi.org/10.1073/pnas.1014395107
Taylor JA, Wilbur JD, Smith SC et al (2009) Mutations that alter RcdA surface residues decouple protein localization and CtrA proteolysis in *Caulobacter crescentus*. J Mol Biol 394:46–60. https://doi.org/10.1016/j.jmb.2009.08.076
Taylor JA, Ouimet M-C, Wargachuk R et al (2011) The *Caulobacter crescentus* chromosome replication origin evolved two classes of weak DnaA binding sites. Mol Microbiol 82:312–326. https://doi.org/10.1111/j.1365-2958.2011.07785.x
Tropini C, Huang KC (2012) Interplay between the localization and kinetics of phosphorylation in flagellar pole development of the bacterium *Caulobacter crescentus*. PLoS Comp Biol 8:e1002602. https://doi.org/10.1371/journal.pcbi.1002602
Tsai JW, Alley MR (2001) Proteolysis of the *Caulobacter* McpA chemoreceptor is cell cycle regulated by a ClpX-dependent pathway. J Bacteriol 183:5001–5007
Tsilibaris V, Maenhaut-Michel G, Van Melderen L (2006) Biological roles of the Lon ATP-dependent protease. Res Microbiol 157:701–713. https://doi.org/10.1016/j.resmic.2006.05.004
Tsokos CG, Perchuk BS, Laub MT (2011) A dynamic complex of signaling proteins uses polar localization to regulate cell-fate asymmetry in *Caulobacter crescentus*. Dev Cell 20:329–341. https://doi.org/10.1016/j.devcel.2011.01.007
Tu GF, Reid GE, Zhang JG et al (1995) C-terminal extension of truncated recombinant proteins in *Escherichia coli* with a 10Sa RNA decapeptide. J Biol Chem 270:9322–9326
Vinella D, Albrecht C, Cashel M et al (2005) Iron limitation induces SpoT-dependent accumulation of ppGpp in *Escherichia coli*. Mol Microbiol 56:958–970. https://doi.org/10.1111/j.1365-2958.2005.04601.x
Wah DA, Levchenko I, Baker TA et al (2002) Characterization of a specificity factor for an AAA+ ATPase: assembly of SspB dimers with ssrA-tagged proteins and the ClpX hexamer. Chem Biol 9:1237–1245
Wah DA, Levchenko I, Rieckhof GE et al (2003) Flexible linkers leash the substrate binding domain of SspB to a peptide module that stabilizes delivery complexes with the AAA+ ClpXP protease. Mol Cell 12:355–363
Wang J, Hartling JA, Flanagan JM (1997) The structure of ClpP at 2.3 Å resolution suggests a model for ATP-dependent proteolysis. Cell 91:447–456
Wang Y, Jones BD, Brun YV (2001) A set of *ftsZ* mutants blocked at different stages of cell division in *Caulobacter*. Mol Microbiol 40:347–360
Wargachuk R, Marczynski GT (2015) The *Caulobacter crescentus* homolog of DnaA (HdaA) also regulates the proteolysis of the replication initiator protein DnaA. J Bacteriol 197:3521–3532. https://doi.org/10.1128/JB.00460-15
Wheeler RT, Shapiro L (1999) Differential localization of two histidine kinases controlling bacterial cell differentiation. Mol Cell 4:683–694
Williams B, Bhat N, Chien P et al (2014) ClpXP and ClpAP proteolytic activity on divisome substrates is differentially regulated following the *Caulobacter* asymmetric cell division. Mol Microbiol 93:853–866. https://doi.org/10.1111/mmi.12698

Wojtkowiak D, Georgopoulos C, Zylicz M (1993) Isolation and characterization of ClpX, a new ATP-dependent specificity component of the Clp protease of *Escherichia coli*. J Biol Chem 268:22609–22617

Wright R, Stephens C, Zweiger G et al (1996) *Caulobacter* Lon protease has a critical role in cell-cycle control of DNA methylation. Genes Dev 10:1532–1542

Wu J, Ohta N, Zhao J-L et al (1999) A novel bacterial tyrosine kinase essential for cell division and differentiation. Proc Natl Acad Sci U S A 96:13068–13073

Xiao H, Kalman M, Ikehara K et al (1991) Residual guanosine 3′,5′-bispyrophosphate synthetic activity of *relA* null mutants can be eliminated by *spoT* null mutations. J Biol Chem 266:5980–5990

Zschiedrich C, Keidel V, Szurmant H (2016) Molecular mechanisms of two-component signal transduction. J Mol Biol 428:3752–3775. https://doi.org/10.1016/j.jmb.2016.08.003

Zweiger G, Marczynski GT, Shapiro L (1994) A *Caulobacter* DNA methyltransferase that functions only in the predivisional cell. J Mol Biol 235:472–485. https://doi.org/10.1006/jmbi.1994.1007

Cell Division in *Caulobacter crescentus*: A Molecular-Scale Model

Patrick J. Lariviere and Erin D. Goley

Abstract

During bacterial division, a single mother cell undergoes a stereotyped shape change to split into two daughter cells. The division process is particularly well characterized in the model organism *Caulobacter crescentus*, which has numerous mechanisms in place to spatially and temporally control division. The shape changes that underpin division are physically mediated by remodeling of peptidoglycan, the building block of the cell wall, which ultimately leads to constriction of the entire bacterial envelope. The cytokinetic Z-ring, a dynamic ring-like structure formed by the concentration of FtsZ protofilaments at midcell, is the central hub for the peptidoglycan remodeling enzymes and the rest of the division machinery, collectively referred to as the divisome. Divisome proteins have diverse functions within the division process, but they all contribute to at least one of the stages of division. Following Z-ring formation at the division site, cellular elongation at midcell begins. Subsequently, we propose that FtsZ acts as a "dynamic activator" of constriction, whereby it signals upstream of the constriction machinery to initiate constriction. Throughout the division process, numerous factors ensure that envelope integrity is maintained to ensure that the cellular barrier to the environment remains intact as envelope shape and composition are altered. At the end of division, the daughter cells finally

P. J. Lariviere
Department of Integrative Biology, University of Texas at Austin, Austin, TX, USA

E. D. Goley (✉)
Department of Biological Chemistry, Johns Hopkins University School of Medicine, Baltimore, MD, USA
e-mail: egoley1@jhmi.edu

E. Biondi (ed.), *Cell Cycle Regulation and Development in Alphaproteobacteria*,
https://doi.org/10.1007/978-3-030-90621-4_4

separate and the cell poles undergo morphological organization. In this chapter, we present a comprehensive molecular-scale model for division in *Caulobacter*, with a focus on spatiotemporal regulation of division factors, and additionally highlight key aspects of division that remain unresolved.

1 An Overview of Division

For a replicating bacterium, the objective of cell division is relatively simple: split into two viable cells that are themselves capable of growing and dividing. Despite the simplicity of this goal, the mechanism for achieving successful division in *Caulobacter crescentus* (hereafter, *Caulobacter*) is in fact exceedingly sophisticated. The first layer of complexity involves proper timing of division initiation, such that the start of division is coordinated with the other events of the cell cycle. *Caulobacter* cells begin their life cycle as motile, flagellated swarmer cells, which grow, but do not themselves divide (Terrana and Newton 1975; Aaron et al. 2007). In their natural habitat, swarmer cells are believed to swim around in search of a nutrient-rich zone. Upon finding such an environment, swarmer cells undergo a cell-type transition in which they lose their flagella and grow a long, thin extension of the cell body called the stalk. At this time, stalked cells begin to replicate their DNA, grow, and initiate the pre-constriction phase of the cell division program. At some point after DNA replication has begun, cells begin to constrict, eventually dividing into two daughter cells.

Division in *Caulobacter* requires that cells undergo a complex series of synchronized shape changes in the cell envelope. In Gram-negative cells including *Caulobacter*, the envelope is the multilayered casing around the cell made up of an inner membrane (IM), an outer membrane (OM), and a cell wall in between the two, occupying the space known as the periplasm. The cell wall in *Caulobacter* is a semirigid structure made of the macromolecule peptidoglycan (PG), which helps resist turgor pressure and maintain cell shape (Cabeen and Jacobs-Wagner 2005; Woldemeskel and Goley 2017). PG itself is made of glycan strands of repeating disaccharides that run along the circumference of the cell and are attached to each other by pentapeptide crosslinks (Hayhurst et al. 2008; Gan et al. 2008; Huang et al. 2008). Incorporation of new PG material into the cell wall (also referred to as PG or cell wall remodeling) drives both constriction, the inward invagination of the envelope, and elongation, the longitudinal extension of the envelope that occurs before and during constriction in *Caulobacter*.

The next layer of the complexity of division lies within the sheer number of factors involved in facilitating cell wall remodeling and other events during the division process. The components of the cell division machinery are collectively known as the divisome, with each protein playing one or more specialized role(s) in division. The structural core of the divisome, FtsZ, forms a cytokinetic ring referred to as the Z-ring. In *Caulobacter*, the Z-ring marks the site of division, which is located approximately at midcell. There, the Z-ring acts as a scaffold for the rest of the division machinery, including the enzymes that insert new PG into the cell wall. Such a scaffold nucleates an intricate network of interactions among the various divisome proteins that is crucial for allowing division to occur.

The last level of complexity in *Caulobacter* division lies in the requirement for tight coordination of the activities of the numerous members of the divisome in time and space, such that the envelope can undergo the correct shape changes at the right time. Here we present a description of division in *Caulobacter*. We have divided division into five concrete steps, which we will explain in brief here and in greater detail in the following sections. Note that we introduce these events each in turn for simplicity, but they are not strictly separate. Instead, they are tightly coordinated and overlapping in time and space. These steps are as follows: (1) Formation of the Z-ring, (2) Cell elongation at midcell, (3) Envelope maintenance, (4) Activation of the constriction machinery and constriction, and (5) Cell separation and polar organization (Fig. 1). Z-ring formation marks the beginning of the cell division program in *Caulobacter* and consists of localizing the Z-ring to midcell at the correct point in time in the cell cycle with the help of other divisome proteins and additional factors. Following assembly of the Z-ring, proteins involved in cell elongation and envelope maintenance localize to the ring to initiate growth at midcell, and preserve integrity of the cell wall. While cells elongate, additional division proteins gradually localize to midcell in a stepwise fashion. We will refer to these later arriving division proteins as the constriction machinery, as they are more directly responsible for either activating or effecting constriction. Once all of the constriction machinery has assembled, constriction initiates, whereby the entire envelope starts to invaginate. When cells are close to finishing division, that is, they are characterized by deep constrictions, polarity determining factors arrive at midcell and factors responsible for separating the two future daughter cells help execute the last stages of the division process. Finally, the cell splits into two, thus completing division. Having introduced the major steps of division, we will go into each in more detail in the rest of this chapter. First, however, we will introduce the Z-ring and describe what is known about its structure and dynamics.

2 Z-Ring Composition, Structure, and Dynamics

The Z-ring is an essential apparatus required for division in most bacteria, including *Caulobacter*, though the full range of its activities are still being investigated. A guiding principle in this endeavor has been that the structure of the Z-ring dictates

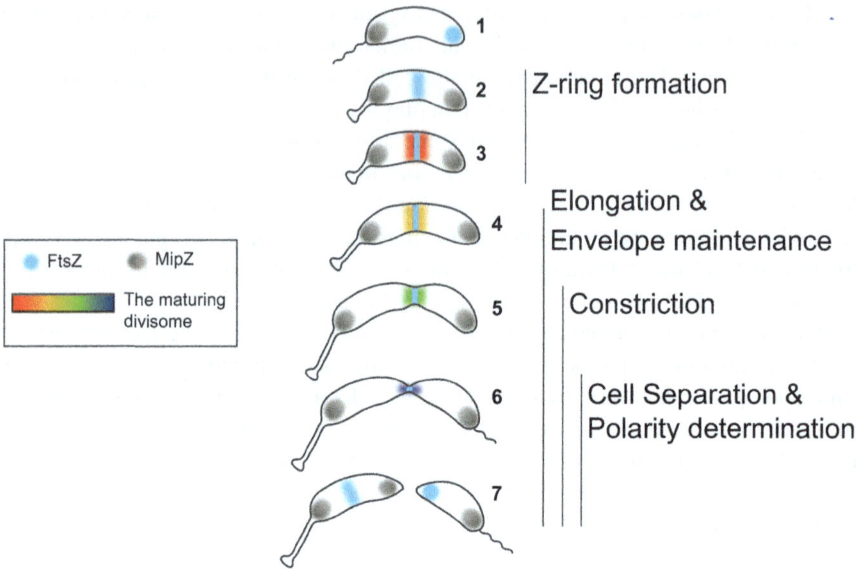

Fig. 1 Overview of division in *Caulobacter. Caulobacter* cells undergo a series of events during division, facilitated by maturation of the divisome. (1) Cells begin in the swarmer phase, with unipolar MipZ and FtsZ at opposite poles. (2) During the swarmer to stalked cell transition, MipZ localizes to both poles to initiate Z-ring formation, forcing FtsZ to localize in an unfocused band at midcell. (3) Z-ring formation factors arrive (red) to help focus the Z-ring and tether FtsZ to the membrane. (4) The elongation machinery (yellow) arrives and cells begin to elongate by PG insertion at midcell; at the same time, the envelope maintenance machinery (also yellow) arrives to ensure envelope integrity is preserved throughout the division process. (5) The constriction machinery (green) arrives and constriction begins. (6) As constriction progresses, factors involved in cell separation and polarity determination (dark blue) arrive at midcell to help terminate division and mark the future site of the new pole. (7) Division terminates when the cell finally splits apart, forming two new daughter cells

its functions within the cell, and so a detailed understanding of its architecture has been a longstanding goal. The fundamental building block of the Z-ring is the GTPase FtsZ, a tubulin-like protein that polymerizes into head-to-tail assemblies of monomers to form protofilaments. In vivo, these FtsZ protofilaments assemble into a dynamic, annular superstructure—the Z-ring—along the circumference of the cell's short axis at midcell. Formation of this structure is crucial for FtsZ to serve as a scaffold for the rest of the divisome, a concept that we will explore in a later section. In order to lay the groundwork for understanding the functions of the Z-ring during division, this section will focus first on the composition of the Z-ring, then on its structure and dynamics.

2.1 FtsZ: The Building Block of the Z-Ring

FtsZ is a widely conserved protein (Vaughan et al. 2004) that is essential for division in *Caulobacter* and most other bacterial species. First characterized in *E. coli*, *ftsZ* was identified by complementation of a mutation that conferred cell filamentation at restrictive temperature, indicating that it is required for division. Subsequent studies established FtsZ as a crucial positive regulator of division by demonstrating its overexpression rescued various division mutant phenotypes (Lutkenhaus et al. 1986; Bi and Lutkenhaus 1990). FtsZ was then found to localize to midcell at the site of division, first in *E. coli*, then in *Bacillus subtilis*, forming a ring-like structure now known as the Z-ring (Bi and Lutkenhaus 1991; Wang and Lutkenhaus 1993; Ma et al. 1996).

FtsZ is considered to be a tubulin homolog through amino acid sequence and structural comparison (Mukherjee et al. 1993; Löwe and Amos 1998; Löwe 1998; Vaughan et al. 2004). Like tubulin, FtsZ is a GTPase, capable of binding and hydrolyzing GTP in its GTPase domain (de Boer et al. 1992; RayChaudhuri and Park 1992; Mukherjee et al. 1993). However, unlike tubulin, *Caulobacter* FtsZ contains two additional motifs: a C-terminal conserved (CTC) peptide and a disordered C-terminal linker (CTL), which connects the GTPase domain to the CTC (Sundararajan et al. 2015).

Like tubulin, FtsZ monomers form protofilaments in a nucleotide-dependent manner in vitro (Mukherjee and Lutkenhaus 1994), with GTP binding and hydrolysis influencing polymer assembly and disassembly dynamics (Mukherjee and Lutkenhaus 1998, 1999). FtsZ binds GTP to initiate polymerization, with FtsZ-GTP subunits assembling in an end-to-end fashion (Mukherjee and Lutkenhaus 1994; Löwe and Amos 1999). GTP is quickly hydrolyzed into GDP, which subsequently destabilizes the filaments (Scheffers and Driessen 2002; Chen and Erickson 2005). Finally, subunits fall off the protofilament and exchange GDP for GTP to begin the cycle anew (Scheffers and Driessen 2002; Chen and Erickson 2005). By electron microscopy (EM), FtsZ protofilaments form straight or gently curved polymers in the presence of GTP, but they can also form highly curved mini-ring-shaped assemblies under strongly stabilizing conditions (i.e., on DEAE-dextran) with GDP present (Mukherjee and Lutkenhaus 1994, 1998, 1999; Erickson et al. 1996; Goley et al. 2010b). Additional structural and in vitro data indicate that individual FtsZ protofilaments can associate with one another further to form higher-order structures, depending on in vitro conditions. Specifically, FtsZ protofilaments can interact laterally to form small filament bundles (made up of as few as two filaments), larger multifilament bundles, or structured sheets (Erickson et al. 1996; Yu and Margolin 1997; Mukherjee and Lutkenhaus 1999; Löwe and Amos 1999; Oliva et al. 2004; Milam and Erickson 2013).

2.2 Z-Ring Structure and Dynamics

In vivo, by conventional fluorescence microscopy (FM), FtsZ localizes to midcell to form the band-like Z-ring (Ma et al. 1996). Super-resolution microscopy has provided additional insight into the architecture of the Z-ring. Viewing cells from a top-down perspective, photoactivated localization microscopy (PALM) shows that FtsZ is tightly focused at midcell, with a mean longitudinal width of 71–108 nm in *Caulobacter* (Biteen et al. 2012; Holden et al. 2014; Woldemeskel et al. 2017). Imaging *Caulobacter* cells in cross-section by PALM confirms the suspected ring-like structure, with FtsZ most concentrated near the membrane (Holden et al. 2014), similar to the ring-like organization observed in other organisms (Fu et al. 2010; Strauss et al. 2012). *Caulobacter* Z-rings were found to have a diameter roughly corresponding with cell diameter [150 nm (near the z resolution limit) to 650 nm, depending on what stage of division the cell is in] and a mean radial thickness of around 65 nm (Biteen et al. 2012; Holden et al. 2014). *Caulobacter* Z-rings, similar to other organisms, have a clustered distribution of FtsZ suggesting the *Caulobacter* Z-ring is patchy and discontinuous (Strauss et al. 2012; Holden et al. 2014; Coltharp et al. 2016; Lyu et al. 2016).

Whole-cell cryo-electron tomography (cryo-ET) has provided a higher resolution view of the *Caulobacter* Z-ring. In one study, curved filamentous FtsZ structures have been observed at midcell roughly 16 nm from the membrane (Li et al. 2007). Overexpression of FtsZ or expression of a less dynamic form of FtsZ increased the number of these structures, suggesting they are in fact made of FtsZ (Li et al. 2007). These structures are similar in width to single FtsZ protofilaments (5 nm wide by cryo-ET vs. 4 nm wide FtsZ structure by crystallography), with neighboring structures being either approximately parallel or slightly overlapping (Li et al. 2007). Moreover, these structures are relatively short at 80–160 nm in length and are heterogeneously distributed at midcell, providing additional evidence for a discontinuous Z-ring (Li et al. 2007). A follow-up study showed that in the initial stages of constriction, FtsZ filaments were often found localized to one side of the cell at a site of invagination, suggesting the heterogeneous distribution of the Z-ring begins early in the division process (Yao et al. 2017). However, in similar experiments from another group, the Z-ring instead appears as a single connected, continuous filament (Szwedziak et al. 2014). Since cryo-ET does not use labeling, not all of the FtsZ in a cell may be visualized, which may account for discrepancies between the two sets of experiments. Questions therefore still remain about the continuous nature of the *Caulobacter* Z-ring, though evidence presented in subsequent sections, as well as the super-resolution light microscopy data discussed above, support the discontinuous ring model.

Although the Z-ring displays a relatively ordered architecture, it is highly dynamic at multiple timescales. On a longer timescale, the structure of the Z-ring is dynamic through the cell cycle: FtsZ moves from the pole to midcell during the swarmer to stalked cell transition, assembles into a focused ring, then decreases in diameter throughout the constriction process (Goley et al. 2011; Holden et al. 2014).

On a much shorter timescale, fluorescence recovery after photobleaching (FRAP) experiments from *E. coli* suggest that the Z-ring exchanges with cytoplasmic FtsZ on the order of seconds, with the speed of this process depending on FtsZ's GTPase rate (Stricker et al. 2002; Anderson et al. 2004). In addition to subunit exchange, seminal work in other species suggests the Z-ring exhibits another form of dynamics: treadmilling. In both *E. coli* and *B. subtilis*, by three-dimensional structured-illumination microscopy (3D-SIM) or total internal reflection fluorescence microscopy (TIRFM), FtsZ polymers/clusters appear to move around the circumference of the cell (Bisson-Filho et al. 2017; Yang et al. 2017). However, FtsZ molecules are actually motionless, and the polymers themselves likely do not "move" either (Bisson-Filho et al. 2017; Yang et al. 2017). Yang et al. and Bisson-Filo et al. instead propose that FtsZ clusters treadmill, potentially achieving dynamics through addition of FtsZ subunits (or oligomers or polymers) at the leading edge of the cluster and loss of FtsZ from the lagging edge (Bisson-Filho et al. 2017; Yang et al. 2017), allowing FtsZ clusters to have the appearance of movement. The speed of this movement is dependent on the GTPase rate of FtsZ, indicating that polymerization and depolymerization dictate FtsZ dynamics within the cell (Bisson-Filho et al. 2017; Yang et al. 2017). Experiments to assess FtsZ treadmilling in *Caulobacter* have not yet been reported, likely because its smaller diameter complicates resolution of the Z-ring. However, since FtsZ treadmilling is conserved in both Gram-positive and Gram-negative bacteria, we can infer that it occurs in *Caulobacter* as well.

2.3 Open Questions

- Is the Z-ring discontinuous? Are there filaments within the Z-ring that have not been resolved by cryo-ET?
- Does FtsZ treadmill in *Caulobacter*? If so, what is the function of treadmilling?
- How is treadmilling regulated in *Caulobacter*?

3 Building the Z-Ring: Spatiotemporal Regulation of Z-Ring Formation

Having detailed the architecture of the Z-ring, we will now consider *how* the Z-ring forms. The timing of Z-ring formation must be tightly controlled since precise timing of the *Caulobacter* cell cycle is necessary to ensure that cells balance replication speed with fitness of both mother and daughter cells. Z-ring formation is prevented in swarmer cells both by transcriptional and posttranslational limitation of FtsZ levels, and by inhibition of FtsZ polymerization. Removal of these blocks occurs only when swarmer cells are ready to transition into the division-competent stalked cells, thus allowing the Z-ring to assemble. Spatial regulation of Z-ring formation is also crucial for the division process, ensuring that division yields two properly sized daughters. As cells enter the stalked phase, the site of Z-ring formation is set at approximately midcell in a process that is tightly coupled to the cell cycle. Shortly after the midcell division plane is established, FtsZ rapidly assembles into a ring-like structure. Additional factors are required to "focus" the ring to a defined width. Concurrently, at least one other factor is thought to attach FtsZ polymers to the membrane in order to bring the Z-ring proximal to its site of action. The result is a highly ordered super-structure that is now ready to serve as the backbone for the rest of the divisome for the remainder of the division process.

3.1 Cell Cycle Control of FtsZ Protein Levels

The initiation of division is regulated, at least in part, through temporal control of division protein concentrations. As such, FtsZ protein levels in *Caulobacter* fluctuate over the course of the cell cycle (Quardokus and Brun 2002); FtsZ is absent or in low abundance in swarmer cells, with its expression increasing around the time of the swarmer to stalked cell transition (Quardokus and Brun 2002). FtsZ levels peak in pre-divisional cells, before finally dropping significantly after division is complete (Quardokus and Brun 2002). At least two mechanisms are responsible for the control of FtsZ protein levels: transcriptional regulation and proteolysis. *ftsZ* mRNA levels were found to fluctuate over the course of the cell cycle in a manner dependent on the cell cycle regulator CtrA (Kelly et al. 1998; Laub et al. 2000). Depletion of CtrA leads to a corresponding increase in *ftsZ* expression, whereas overexpression of CtrA reduces *ftsZ* expression (Kelly et al. 1998). CtrA, therefore, represses *ftsZ* expression during the swarmer phase by directly binding to its promoter (Laub et al. 2002), and as its own levels decrease upon transition to the stalked cell phase, *ftsZ* levels go up. Other cell cycle-associated transcriptional regulators—DnaA, GcrA, and CcrM—have been implicated in directly activating *ftsZ* expression, allowing for its transcription throughout the majority of the division process (Hottes et al. 2005; McAdams and Shapiro 2009; Gonzalez and Collier 2013; Haakonsen et al. 2015). Additionally, transcripts of other division genes (*ftsA*, *ftsQ*, *ftsW*, *ftsI*, *ftsK*, *fzlA*, *ftsB*, *kidO*, and *murG*) have also been found to fluctuate over the course of the cell cycle (Laub et al. 2000; Goley et al. 2011). Multiple transcriptional regulators probably control these transcript levels, as *kidO*, *fzlA*, and *ftsK* have CtrA binding motifs and *ftsA*, *ftsQ*, and *ftsB* have both CtrA and CcrM motifs.

The levels of FtsZ and at least a few other divisome proteins are also regulated posttranslationally. When either *ftsZ*, *ftsQ*, or *ftsA* is constitutively transcribed, levels of the corresponding protein still change over the cell cycle similar to WT cells (Kelly et al. 1998; Martin et al. 2004), suggesting cell cycle-linked protein degradation by proteolysis. The AAA+ protease ClpXP, which was first found to degrade FtsZ in *E. coli* (Camberg et al. 2009), was determined to be partially responsible for the cell cycle-dependent proteolysis of FtsZ in *Caulobacter* (Bhat et al. 2013; Williams et al. 2014). Another protease, ClpAP, also degrades both *Caulobacter* FtsZ and FtsA in a cell cycle-dependent manner (Williams et al. 2014). Deletion of ClpA or inactivation of ClpX leads to accumulation of FtsZ in swarmer cells (Williams et al. 2014), indicating that these proteases are critical for preventing premature FtsZ accumulation and Z-ring formation. Initiation of the cell division program is therefore inhibited during G1 phase and only begins once FtsZ (and FtsA) transcript and protein levels increase sufficiently to support Z-ring formation upon transition to S phase.

3.2 Z-Ring Site Selection

As cells enter S phase, and FtsZ levels increase, the Z-ring is now ready to assemble. Formation of the Z-ring is not, however, merely dependent on the presence of FtsZ. In fact, despite multiple forms of negative regulation of FtsZ accumulation during G1 phase, a small amount of FtsZ is actually present in cells localized to the pole opposite the flagellum, indicating a mechanism is in place to regulate FtsZ localization. It is only upon transition to S phase, concurrent with the increase in FtsZ protein levels, that FtsZ relocates from the pole to midcell (Figs. 1 and 3). How, then, does FtsZ exhibit dynamic localization and how is localization eventually limited to midcell? Additionally, how is FtsZ localization regulated temporally? A negative regulator of FtsZ polymerization, the dimeric ATPase MipZ, is primarily responsible for regulating localization of the Z-ring in time and space (Thanbichler and Shapiro 2006). MipZ is technically nonessential, though its deletion (or depletion) causes severe perturbations in division including cell filamentation and loss of nearly all constriction sites (Thanbichler and Shapiro 2006; Radhakrishnan et al. 2010). In swarmer cells, MipZ localizes to the flagellated pole (opposite from the FtsZ-localized pole). Coincident with FtsZ's relocation to midcell, MipZ becomes bipolar in cells that have recently undergone the swarmer to stalked cell transition (Figs. 1 and 3), forming a gradient with maxima at the poles and a minimum at midcell (Thanbichler and Shapiro 2006; Goley et al. 2011). Overexpressing *mipZ*, which causes it to become diffuse throughout the cytoplasm, makes FtsZ localize to the poles instead of midcell. MipZ, therefore, dictates FtsZ localization, with FtsZ present at the minimum of the MipZ gradient. What is the mechanism of this regulation? In vitro, MipZ has been shown to destabilize FtsZ filaments and promote polymer turnover. MipZ is thus proposed to negatively regulate FtsZ polymerization in vivo through its formation of a gradient, spatially limiting where FtsZ can localize and polymerize. This regulation is necessary for allowing FtsZ polymers to condense into a focused Z-ring, since depletion of MipZ leads to the formation of FtsZ puncta instead of rings (Thanbichler and Shapiro 2006).

If MipZ is able to dictate FtsZ's localization, how then is its own localization determined? The DNA partitioning protein ParB, which binds to the centromeric DNA locus *parS* to help segregate the chromosome in a ParA-dependent manner, also directly interacts with and colocalizes with MipZ (Thanbichler and Shapiro 2006). MipZ localization is dependent on ParB, becoming diffuse upon ParB depletion. MipZ's localization is therefore determined by the localization of the ParB-*parS* complex. During G1 phase, ParB-*parS* localizes to the flagellated pole;

upon DNA replication in S phase, one copy of ParB-*parS* is segregated to the opposite cell pole. Localization of ParB-*parS* to the poles, therefore, underlies MipZ gradient formation (Thanbichler and Shapiro 2006). In light of these localization dependencies, it is therefore unsurprising that proper FtsZ localization to midcell requires initiation of DNA replication (Quardokus and Brun 2002). When the DNA replication initiator DnaA is depleted, FtsZ forms constrictions at subpolar regions of the cell (Quardokus and Brun 2002). Interestingly, FtsZ also appears to become punctate in some cells blocked for DNA replication (Quardokus and Brun 2002), reminiscent of MipZ depletion. Further study is required to flesh out the links between DNA replication and the localization patterns of MipZ and FtsZ.

MipZ's dimerization state also influences both its own and FtsZ's localization within the cell, as indicated by a study involving MipZ mutants with aberrant ATP binding, hydrolysis, and/or dimerization (Kiekebusch et al. 2012). Polar ParB recruits ADP-bound MipZ monomers to the poles. MipZ then exchanges ADP for ATP, causing MipZ to dimerize and release ParB. Dimeric MipZ begins to diffuse from the pole, but its diffusion is limited by sequence non-specific interactions with the chromosome. This dimeric form of MipZ is also able to inhibit FtsZ polymerization. MipZ eventually hydrolyzes ATP and becomes monomeric again, releasing from the chromosome and going back to the poles due to its high affinity for ParB. These dynamics allow for formation of a MipZ gradient in the cell with the poles serving as the points of highest concentration. Since a large concentration of dimeric MipZ is still close to the poles, FtsZ is forced to polymerize near midcell where the concentration of MipZ is the lowest (Kiekebusch et al. 2012). More accurately, however, the Z-ring does not actually localize to the exact midpoint between the poles—rather, it tends to assemble slightly closer to the non-stalked pole. Shtylla proposes that this is due to the presence of more free ParB at the stalked pole, allowing slightly more MipZ to localize to this pole, pushing the MipZ minimum closer to the other pole (Shtylla 2017).

In addition to chromosome segregation, cell polarity also helps MipZ regulate Z-ring placement spatially and temporally. Disrupting proper cell polarity through deletion of *tipN*, a polarity determinant, has multiple effects on MipZ activity, which in turn adversely affects Z-ring localization (Schofield et al. 2010). TipN directly influences ParA localization, both spatially and temporally, impacting the timing and location of ParB, and finally MipZ localization (Schofield et al. 2010). Accordingly, in *tipN* deleted cells, MipZ establishes its bipolar gradient both less robustly, with weaker MipZ maxima at each pole, and also later in the cell cycle, which causes the Z-ring to form later (Schofield et al. 2010). Additionally, the MipZ gradient was flipped in TipN deleted cells, with a higher concentration of MipZ found at the non-stalked pole instead of the stalked pole, causing reversed Z-ring asymmetry (Schofield et al. 2010). Since proper MipZ localization depends on the timing of ParAB-mediated DNA segregation, itself a tightly timed process, MipZ-mediated Z-ring formation is intimately linked to the cell cycle both temporally and spatially.

3.3 Cell Cycle Control of FtsZ Self-Interaction: A Putative Model

Another mode of Z-ring assembly inhibition, which is still poorly understood, entails temporal regulation of FtsZ self-interaction. Two nonessential FtsZ-binding proteins, GdhZ, an NAD+-dependent glutamate hydrolase, and KidO, an NADH binding protein, are proposed to negatively regulate FtsZ self-association in a cell cycle-dependent manner (Radhakrishnan et al. 2010; Beaufay et al. 2015). In vivo, KidO and GdhZ are present in high concentrations during G1 phase and are diffusely localized, but their protein levels decrease during S phase in a CtrA-dependent manner. Intriguingly, qualitative observations suggest that deletion of either protein may cause the Z-ring to assemble earlier than normal (Radhakrishnan et al. 2010; Beaufay et al. 2015). However, a more quantitative analysis would be useful in confirming this hypothesis. In vitro, GdhZ (which can also convert NAD+ into NADH) inhibits FtsZ polymerization in an NAD+- or a glutamate-dependent manner, while KidO inhibits FtsZ polymer bundling in an NADH-dependent manner (Radhakrishnan et al. 2010; Beaufay et al. 2015). Beaufay et al. have proposed a model by which GdhZ and KidO regulate Z-ring formation (Beaufay et al. 2015): GdhZ and KidO inhibit self-interaction of what little FtsZ is present in G1 phase, preventing Z-ring formation (Beaufay et al. 2015). Upon a downshift in *gdhZ* and *kidO* transcript levels during S phase, FtsZ self-interaction is no longer inhibited, and it is now able to form the Z-ring (Beaufay et al. 2015). These negative regulators are proposed to act in synergy, with GdhZ supplying the NADH required for KidO's activity (Beaufay et al. 2015). Finally, GdhZ's glutamate hydrolysis activity may link FtsZ polymerization state to metabolism (Beaufay et al. 2015), though the implications of this are still unclear. Though Beaufay et al. present a tantalizing model to explain Z-ring assembly regulation, more data are required to fully validate it.

3.4 Focusing the Z-Ring

Although MipZ is necessary for localizing FtsZ to the future division site, it is not sufficient for forming a coherent Z-ring. Additional factors are indeed required to help focus the Z-ring into a narrow band that can efficiently promote constriction. Key players in this process in *Caulobacter* are the nonessential, cytoplasmic, coiled-coil proteins ZapA and ZauP (Fig. 2), which are recruited to midcell by FtsZ shortly after FtsZ itself arrives (Fig. 3) (Goley et al. 2011; Woldemeskel et al. 2017). ZapA directly interacts with both ZauP and FtsZ, and mediates recruitment of ZauP to midcell. Similar to deletion of *zapA* in *E. coli*, deletion of *Caulobacter zapA*, *zauP*, or both causes the Z-ring to become more dispersed and less tightly focused along the longitudinal axis. The mechanism through which this occurs may be distinct from *E. coli*, however, based on a few divergent observations. The most recent cellular scale model in *E. coli* suggests that ZapA's ability to focus the Z-ring depends on connections to the chromosome mediated by two proteins, MatP and

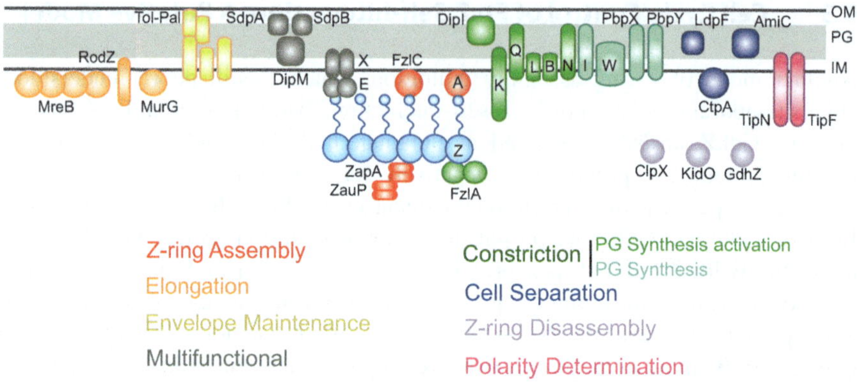

Fig. 2 Architecture of the *Caulobacter* divisome. Depiction of all of the known members of the *Caulobacter* divisome at the envelope at midcell (*IM* inner membrane, *PG* peptidoglycan, *OM* outer membrane). Members of the divisome have been placed into functional groups: Z-ring assembly factors (red); elongation machinery (orange); envelope maintenance machinery (yellow); multifunctional factors (gray); constriction machinery, subdivided into PG synthesis activation factors (green) and PG synthesis enzymes (turquoise); cell separation factors (dark purple); Z-ring disassembly factors (lavender); polarity determination factors (magenta). Note that the divisome is a dynamic structure and not all of its components are necessarily localized to midcell at the same time, as shown here. See Fig. 3 for divisome localization timing. Adapted from Goley et al. (2011)

ZapB, which form a ring concentric to ZapA (Buss et al. 2015, 2017). However, *Caulobacter* lacks homologs to these factors, and ZauP, which forms a ring with the same diameter as ZapA, does not bind DNA. Like ZapB in *E. coli*, ZauP midcell foci persist upon FtsZ depletion, suggesting that ZauP may form a stable scaffold, albeit independent of the chromosome. The observation that *zapB* and *zauP* homologs are mutually exclusive further indicates that there is a separate mode of Z-ring focusing. The details of this proposed ZauP-dependent process are still unknown, however. On the molecular scale, *E. coli* ZapA's ability to bundle FtsZ was initially thought to be important for its activity in vivo, though this has been called into question (Buss et al. 2013). Therefore, the finding that *Caulobacter* ZapA does not affect FtsZ polymerization or bundling suggests an alternate mechanism may indeed be more plausible. *E. coli* ZapA has also been shown to crosslink FtsZ, which may instead contribute to its activity in vivo (Dajkovic et al. 2010). Further testing will be required to determine if *Caulobacter* ZapA also crosslinks FtsZ.

Additional divisome proteins also help to focus the Z-ring. Genetic perturbation of the membrane anchor FzlC or the putative membrane anchor FtsEX, a complex of the proteins FtsE (a cytoplasmic ATPase) and FtsX (a transmembrane protein), has effects on Z-ring focusing (Meier et al. 2016, 2017). FzlC and FtsE each interact directly with FtsZ, as demonstrated in *Caulobacter* and *E. coli*, respectively (Corbin et al. 2007; Meier et al. 2016). Overexpression of either *ftsEX* or *fzlC* causes Z-rings to become broader, and overexpression of *ftsE* alone results in punctate FtsZ. Deletion of *fzlC* alone does not impact FtsZ width, but deletion of *ftsE* alone or *fzlC*

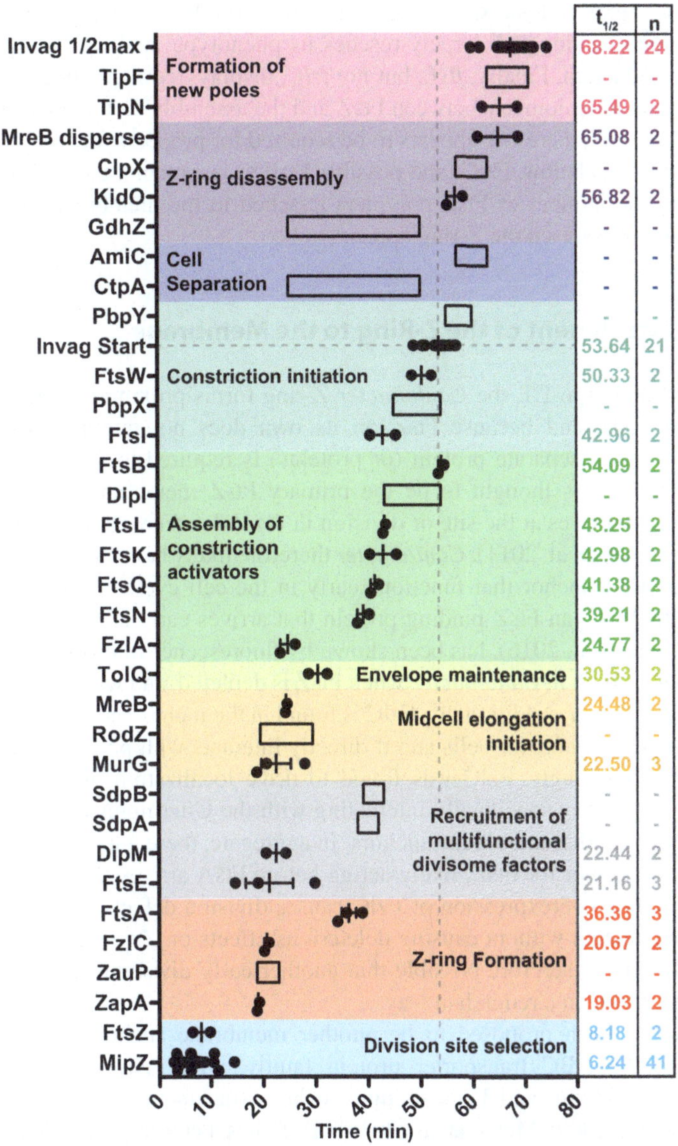

Fig. 3 Timing of divisome localization. Midcell localization timing and order of the members of the divisome. The half-maximal midcell localization time of each protein (the time at which half of all of a protein has localized to midcell; $t_{1/2}$) was either calculated from measured data (black circles, reported in Goley et al. 2011) or predicted (empty boxes). Measured data were taken from Goley et al. (2011) in which timing of localization of 19 different divisome proteins were assessed in parallel. For proteins where $t_{1/2}$ was predicted, localization timing was extracted from the literature. When possible, the timing of protein localization was compared to the timing of localization of another, already characterized protein. Absent such information, a best guess was made (i.e., a protein was predicted to arrive around the same time as interaction partners or before/after a major event, such as constriction initiation). Adapted from Goley et al. (2011)

and *ftsE* together yields dispersed Z-rings. Interestingly, overexpression of *fzlC* in a background lacking *ftsE* largely rescues its phenotype, making Z-rings become more focused again. Finally, *ftsE*, but not *fzlC*, displays synthetic interactions with *zapA*. Proper stoichiometry between FtsZ and the assembly factors FtsE, FzlC, and potentially ZapA, therefore appears to be required for properly focused Z-rings. As will be discussed below, FzlC (and possibly FtsEX) is a membrane anchor for FtsZ, so varying the number of FtsZ polymers attached to the membrane likely has an impact on how focused the Z-ring appears to be.

3.5 Attachment of the Z-Ring to the Membrane

By PALM and cryo-ET, the *Caulobacter* Z-ring forms proximal to the membrane in stalked cells, and because FtsZ on its own does not interact strongly with the membrane, a separate protein (or proteins) is required to anchor FtsZ to the membrane. FtsA is thought to be the primary FtsZ membrane anchor in many species, but it arrives at the site of division in *Caulobacter* well after initial Z-ring assembly (Goley et al. 2011). *Caulobacter* therefore likely has at least one additional FtsZ membrane anchor that functions early in the cell cycle, with one candidate being FzlC. FzlC, an FtsZ binding protein that arrives early at the Z-ring (Figs. 2 and 3) (Meier et al. 2016), has been shown by fluorescence microscopy to localize to membranes both in *Caulobacter* when FtsZ is depleted, and when heterologously expressed in *E. coli*. Additionally, FzlC is found in the membrane fraction of whole-cell lysate of *Caulobacter* cells and it directly interacts with phospholipid vesicles in vitro. Compellingly, FzlC was found to drive localization of FtsZ to vesicle membranes in vitro, specifically interacting with the C-terminus of FtsZ, a known binding site for other membrane anchors. In aggregate, these data implicate FzlC as a membrane anchor for FtsZ, likely acting before FtsA arrives (Meier et al. 2016). However, while overexpression of FzlC causes division defects, it is not essential and can be deleted without causing deleterious effects on Z-ring assembly (Meier et al. 2016). It is therefore possible that another early divisome protein is able to anchor the Z-ring in a redundant way.

FtsEX has been proposed to be another membrane anchor for FtsZ. FtsEX belongs to the ABC transporter protein family, with FtsE predicted to be a cytoplasmic ATPase, and FtsX thought to be a transmembrane protein (Fig. 2) (Arends et al. 2009; Meier et al. 2016). FtsE has been shown to interact with FtsZ in *E. coli* by co-immunoprecipitation, suggesting FtsEX could link FtsZ to the membrane (Corbin et al. 2007). In *Caulobacter*, FtsE arrives at the division plane early (Fig. 3), around the same time as FzlC (Goley et al. 2011). Overexpression or deletion of FtsE causes the Z-ring to become punctate, consistent with a direct interaction between FtsE and FtsZ (Meier et al. 2017). In addition, FtsE genetically interacts with FzlC. Deletion of both gives a synthetic sick phenotype, whereas overexpression of *fzlC* reduces the severity of an *ftsE* deletion phenotype (Meier et al. 2016). Taken together, these data present a case that FtsEX may serve as a membrane anchor, though biochemical evidence of this activity will be required to advance this hypothesis.

The last known FtsZ membrane anchor to arrive at the divisome is FtsA (Figs. 2 and 3), an essential division protein in *Caulobacter* that is widely conserved across bacteria (Sackett et al. 1998; Rothfield et al. 1999; Pichoff and Lutkenhaus 2005; Goley et al. 2011). FtsA has been shown to anchor FtsZ to the membrane in a number of in vitro and in vivo systems (Pichoff and Lutkenhaus 2005; Szwedziak et al. 2012, 2014), an activity facilitated by its FtsZ-interacting domain (Pichoff and Lutkenhaus 2007) and a C-terminal amphipathic helix capable of binding to the membrane (Pichoff and Lutkenhaus 2005; Szwedziak et al. 2012). An actin homolog, FtsA is an ATPase capable of both forming protofilaments and binding to membranes in vitro, with polymerization being required for membrane attachment (Szwedziak et al. 2012; Krupka et al. 2014). Cryo-ET of liposomes containing *Thermotoga maritima* FtsZ and FtsA demonstrate that, at least in vitro, FtsA protofilaments can form a ring-like structure sandwiched between the membrane and FtsZ protofilaments (Szwedziak et al. 2014). *E. coli* FtsA has been shown to stimulate FtsZ dynamics at the membrane in vitro (Loose and Mitchison 2014), indicating FtsA is not simply a passive membrane anchor. Interestingly, genetic data suggests that FtsA functions in a distinct manner compared to one of the other *Caulobacter* FtsZ membrane anchors, FzlC (Meier et al. 2016). Overexpression of *fzlC* in *Caulobacter* cells causes Z-rings to become wider and cells to become slightly longer, whereas overexpression of *ftsA* induces broad Z-rings and patchy FtsZ localization, as well as severe cell filamentation (Meier et al. 2016). The ability of FtsA to regulate FtsZ dynamics could account for these phenotypic differences, although it has not been tested if FzlC also possesses this activity. Alternatively, or in addition, FtsA is believed to be involved in signaling to the PG synthetic machinery in the divisome, whereas FzlC has genetic interactions with factors involved in PG hydrolysis, further highlighting functional differences between the two.

3.6 Open Questions

- Do Z-rings form earlier in *gdhZ*/*kidO* deletion strains? Quantitative temporal analysis of Z-ring localization will be required to address this question.
- How is metabolism coordinated with Z-ring formation? Do glutamate levels play a role, as suggested by GdhZ's activity?
- Why are there multiple membrane anchors? (How) do they interact with each other? Do they have distinct functions?
- What do membrane anchor dynamics look like in vivo? Do they treadmill with FtsZ?
- How does FtsEX affect Z-ring localization? Is FtsEX a bona fide membrane anchor? If so, to which region of FtsZ does FtsEX bind?
- What are the mechanisms by which ZapA and ZauP focus the Z-ring? Is there a conserved mechanism of focusing via ZapA between *E. coli* and *Caulobacter*? Does the presence of different FtsZ site selection systems (MipZ vs. Min/SulA) mean that *Caulobacter* does not require linking to the nucleoid via a MatP-like system?

4 The Z-Ring: A Scaffold for the Rest of the Divisome

Having explored how the Z-ring is assembled, we are now ready to consider the function of the Z-ring in the context of division. Itself a well-defined and tightly focused structure, the Z-ring helps to ensure that the division site is also well-defined by spatially constraining the site of envelope remodeling. The Z-ring achieves this feat by acting as a scaffold for the divisome, precisely positioning the division machinery near midcell in the longitudinal axis.

What comprises the division machinery? The divisome is made up of over 30 proteins that are involved in remodeling the envelope during division, which we have categorized into functional groups: Z-ring formation factors (ZapA, ZauP, FzlC, FtsA), discussed in the previous section; elongation machinery (MreB, MurG, RodZ); constriction machinery (FzlA, FtsN, FtsQ, FtsI, FtsK, FtsL, FtsW, FtsB, PbpX, PbpY, DipI); envelope maintenance machinery (the Tol-Pal complex); cell separation factors (AmiC, LdpF, CtpA); Z-ring disassembly factors (KidO, GdhZ, ClpXP); polarity determining factors (TipN, TipF); and multifunctional factors (FtsEX, DipM, SdpA, SdpB) (Fig. 2). Note that we have categorized a handful of factors as multifunctional since they contribute to multiple processes, which will be discussed throughout the chapter.

The order and timing of midcell localization for most of these proteins have been determined, with timing of functional group localization corresponding well to the activity associated with that group (Goley et al. 2011). In general, proteins within each functional group localize to the Z-ring in close temporal proximity to one another (Fig. 3). The Z-ring formation factors are the first to arrive to midcell, which ensure proper focusing and membrane attachment. Shortly after this, the elongation and envelope maintenance machineries both begin to localize to the future site of division. The constriction machinery assembles to the Z-ring in a more drawn out fashion, with the first factor appearing early, and the subsequent proteins arriving up until constriction starts. The factors involved in cell separation, Z-ring disassembly, and polarity determination localize to midcell last, only after constriction has initiated. All members of the divisome require FtsZ for localization and there is a nonlinear hierarchy of assembly (Goley et al. 2011), with numerous protein–protein and protein–envelope interactions likely important for divisome assembly.

Having introduced the concept that the Z-ring serves as a scaffold for the divisome, we are now ready to describe the next steps in division and the role that each divisome protein plays in more detail. First, we will discuss divisome-associated elongation, followed by constriction. We will then examine the importance of envelope maintenance in the division process. Finally, we will consider cell separation and the completion of division.

4.1 Open Questions

- What additional factors, if any, are recruited to the Z-ring?
- How is the timing of divisome assembly established?
- How is each divisome component recruited to the division site?

5 Divisome-Associated Elongation

Caulobacter cells begin to elongate shortly after assembling the Z-ring, and both before and during constriction, suggesting a potential link between the division machinery and elongation. In a landmark study, Aaron et al. showed that *Caulobacter* elongation occurs primarily via insertion of PG at the midcell, colocalized with the Z-ring (Aaron et al. 2007). Insertion of PG was found to start in relatively young stalked cells, before the initiation of constriction. Additionally, this process was found to be FtsZ-dependent since loss of FtsZ leads to PG incorporation throughout the entire length of the cell (Aaron et al. 2007). Intriguingly, links have been found between FtsZ and the elongation machinery, also known as the elongasome. Soon after Z-ring assembly, at least two members of the elongasome arrive at midcell (Goley et al. 2011). MreB, an actin homolog located in the cytoplasm that scaffolds the elongasome and MurG, a PG precursor synthesizing enzyme also found in the cytoplasm, localize to the future site of division in an FtsZ-dependent manner (Fig. 3) (Aaron et al. 2007; Goley et al. 2011). FtsZ is therefore thought to at least passively regulate elongation (Aaron et al. 2007), though the details of this mechanism are still hazy.

Two additional elongasome proteins, MraY (a PG precursor synthesis enzyme) and RodZ (a transmembrane MreB binding protein, required for its circumferential motion in *E. coli*), have also been shown to localize to midcell either concurrent with constriction in *E. coli* (MraY) or prior to constriction in *Cauobacter* (RodZ) (Figs. 2 and 3) (Alyahya et al. 2009; Morgenstein et al. 2015; Liu et al. 2017). However, it is unclear if the rest of the elongasome localizes to midcell. On the one hand, patchy localization has been reported for numerous other elongasome proteins, including PG modifying enzymes, with localization to midcell only occurring after exposure to cell shape stressors such as osmolarity upshift or MreB inhibition (Divakaruni et al. 2007; White et al. 2010; Hocking et al. 2012). On the other hand, MreB localization itself is either patchy or primarily midcell localized, depending on the stage of the cell cycle (Figge et al. 2004; Gitai et al. 2004; Aaron et al. 2007; Goley et al. 2011). A comprehensive study assessing elongasome localization throughout the cell cycle will be required to address these concerns.

Inversely, some proteins associated with the elongasome also play a role in division. MurG and, probably MraY, contribute to cell wall remodeling during constriction, since division also requires the synthesis of PG precursors. It has also been proposed in *E. coli* that MreB somehow "transfers" PG synthesis enzymes to FtsZ prior to constriction (Fenton and Gerdes 2013), though this hypothesis relies heavily on two-hybrid data and has not been validated in *Caulobacter*. Additional work will be required to clarify the relationship between the *Caulobacter* elongasome and divisome.

5.1 Open Questions

- Where are PG precursors synthesized?
- Do all the members of the elongasome localize to midcell or only some? Is their midcell localization important for constriction?
- What directs elongation at midcell before and then during constriction? MreB? FtsZ? A combination? Is elongation before constriction similar to or different from the type of elongation that occurs during constriction?
- How is the balance between elongation and constriction during division maintained? Directional insertion of PG? Types of reactions catalyzed, such as glycosylation vs. transpeptidation?

6 Constriction: A PG Synthesis-Driven Process

Constriction in *Caulobacter* is marked by the simultaneous increase in length and decrease in width at a particular spot. Like elongation, constriction is thought to be facilitated by the insertion of new PG into the cell wall at midcell. Vital to this process is the constriction machinery, made up of the constriction-specific PG synthesis enzymes and their regulators, and of course, the Z-ring, which plays a crucial role in recruiting, scaffolding, and potentially activating the constriction machinery at midcell. The members of the constriction machinery localize to the Z-ring in an ordered fashion, with the first protein, FzlA, arriving around the same time that elongation starts (Goley et al. 2011). Subsequently, the rest of the constriction machinery factors (FtsA, FtsN, FtsQ, FtsI, FtsK, FtsL, FtsW, FtsB) gradually assemble at the Z-ring, with the arrival of FtsB coinciding with the initiation of constriction (Goley et al. 2011). In our model for constriction initiation, we propose that FtsZ activates the constriction-specific PG synthases through its dynamics, physical signals, and/or interaction partners (Fig. 4). A number of downstream constriction activators likely help transduce the activation signal from FtsZ to the PG synthases, potentially occupying multiple pathways.

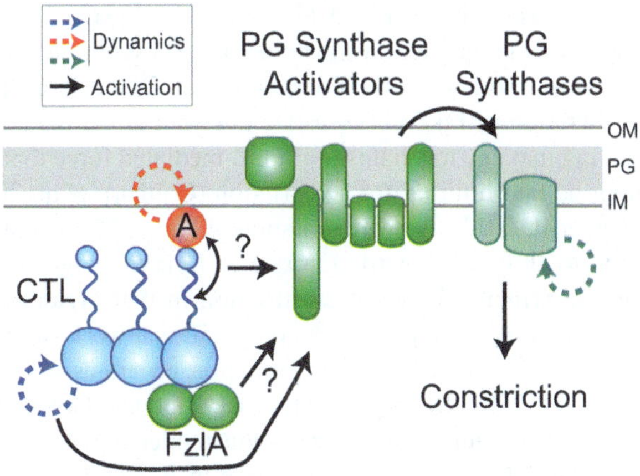

Fig. 4 A dynamic activator model for FtsZ-mediated constriction initiation. Model for FtsZ-directed activation of constriction in *Caulobacter*. FtsZ is proposed to signal through the PG synthase activators (e.g., FtsQLB, FtsN, and FtsK) to PG synthases (e.g., FtsWI), initiating PG remodeling and potentially impacting PG synthase dynamics (turquoise dashed line), to start constriction. There are a number of putative mechanisms by which FtsZ may signal to the PG synthase activators (black arrows, as shown) or potentially directly to the PG synthases (not shown): Through FtsZ's CTL, which may influence protein–protein interactions directly or indirectly, possibly through mechanical signaling and/or through an interaction with FtsA (double-headed black arrow); through a FzlA-mediated protein–protein interaction; and/or through FtsZ dynamics (blue dashed line), which may influence protein–protein interactions indirectly. FtsA dynamics (red dashed line) may also play a role in constriction activation, though precisely how is still unclear

6.1 Force Generation Through PG Synthesis

One of the major questions concerning not only *Caulobacter* division, but bacterial division in general, is what *drives* constriction? That is, what generates the force needed to push or pull the cell envelope inward at the site of division? Early models implicated FtsZ as the primary driver of constriction, since FtsZ alone was able to facilitate liposome constriction in vitro (Osawa et al. 2008; Osawa and Erickson 2013). FtsZ-mediated constriction was proposed to work either through maximization of lateral bonds through Z-ring condensation or through FtsZ curvature-mediated bending of the IM (Lan et al. 2009; Erickson et al. 2010; Xiao and Goley 2016; Erickson and Osawa 2017; Coltharp and Xiao 2017). The second model is supported by evidence that FtsZ protofilaments can deform membranes with directional specificity (Osawa et al. 2009), with FtsZ filament curvature likely driving membrane bending in a GTP-hydrolysis independent manner (Erickson and Osawa 2017; Osawa and Erickson 2018). However, these FtsZ-centric models have been disfavored in the past few years for a few major reasons. The force required to counter turgor pressure and constrict the IM at the division site is predicted to be 400 pN, whereas the Z-ring is calculated to only produce between 8 and 100 pN of force (Lan et al. 2007, 2009; Allard and Cytrynbaum 2009; Paez et al. 2009; Xiao and

Goley 2016; Coltharp and Xiao 2017). Additionally, FtsZ leaves the divisome in *E. coli* before completion of constriction and before the PG synthases leave, indicating FtsZ is not required for this later step (Söderström et al. 2014, 2016; Xiao and Goley 2016). Finally, alteration of FtsZ GTPase rate does not affect the constriction rate in *E. coli*, indicating that any mechanism of FtsZ-mediated force that requires GTP hydrolysis cannot be rate-limiting for constriction (Coltharp et al. 2016; Xiao and Goley 2016; Yang et al. 2017). Instead, mutation of FtsI, a PG synthase highlighted below, was found to slow the constriction rate in *E. coli*, suggesting that midcell PG synthesis is the primary driver of constriction in that organism (Coltharp et al. 2016). This hypothesis is supported by computational work suggesting that the chemical energy released upon PG bond formation generates sufficient force for constriction (Lan et al. 2007; Banerjee et al. 2016; Xiao and Goley 2016; Coltharp and Xiao 2017). A PG-centric force generation model may hold not across all species, however. Alteration of FtsZ GTPase rate in *B. subtilis does* influence the constriction rate and inhibition of FtsZ in *S. aureus* blocks constriction initiation (but not its completion), indicating that GTP hydrolysis can be rate-limiting during constriction in other organisms (Bisson-Filho et al. 2017; Monteiro et al. 2018). Regardless of these differences, it is clear that both the PG synthesis machinery and the Z-ring play central roles in constriction in *Caulobacter*. The following sections will therefore explore the contributions of *Caulobacter*'s PG synthases and their regulators, including FtsZ, to constriction.

6.2 Constriction-Specific PG Synthesis Machinery

During constriction, the cell wall is remodeled at the division plane through the addition of new PG (Aaron et al. 2007; Divakaruni et al. 2007). As described above, PG consists of glycan strands attached to each other by peptide crosslinks. Addition to this network requires that these glycan strands are extended or polymerized de novo through glycosyltransferase (GTase) enzymatic activity and that new peptide crosslinks are made via transpeptidase (TPase) activity. The enzymes responsible for this activity, the constriction-specific GTases and TPases, together will be referred to as the PG synthetic machinery in this text from this point on. This machinery is thought to be crucial for facilitating efficient constriction, though the activities of its individual components and their collective mechanism of action are still poorly defined in *Caulobacter* and other organisms.

Perhaps the best understood constriction-specific PG synthetic enzyme in *Caulobacter* is the TPase FtsI, also known as PBP3. FtsI plays an essential role in constriction, with loss of functional FtsI in *Caulobacter* causing cell filamentation (Costa et al. 2008). FtsI is similarly essential for constriction in *E. coli* (Goehring and Beckwith 2005; Typas et al. 2012), with partial inhibition of function slowing the rate of constriction (Coltharp et al. 2016) and full knockdown completely blocking division (Spratt 1977; Pogliano et al. 1997). A single-pass transmembrane protein, FtsI contains a small cytoplasmic motif and periplasmic transpeptidase domain capable of catalyzing the PG transpeptidation reaction (Wissel and Weiss

2004; Sauvage et al. 2014). FtsI localizes to the IM and, along with the rest of the constriction machinery, is recruited to midcell in an FtsZ-dependent manner before the initiation of constriction (Figs. 2 and 3) (Costa et al. 2008; Goley et al. 2011). FtsI localization at the *Caulobacter* midcell was found to be highly dynamic by FRAP (Costa et al. 2008). Higher-resolution imaging in *E. coli* and *B. subtilis* showed that their division-specific TPases (FtsI in *E. coli*; Pbp2B in *B. subtilis*) move around the circumference of the division site in a manner dependent on FtsZ treadmilling speed (Bisson-Filho et al. 2017; Yang et al. 2017), insinuating that *Caulobacter* FtsI may demonstrate similar dynamics. Further work on *Caulobacter* FtsI will be required to test this hypothesis.

Another essential member of the constriction machinery in *Caulobacter* is FtsW, which has been shown to bind to FtsI in *E. coli* (Karimova et al. 2005; Fraipont et al. 2011). FtsW, whose depletion leads to filamentation and loss of constriction initiation (Goley et al. 2011), is a multipass transmembrane protein that is part of the SEDS (shape, elongation, division, and sporulation) family (Meeske et al. 2016). Located at the IM (Fig. 2) (Fraipont et al. 2011), FtsW arrives at midcell just before the start of constriction (Fig. 3) (Goley et al. 2011). FtsW's precise role in division, however, has been the subject of much debate. One hypothesis supported by in vitro work contends that *E. coli* FtsW is a flippase for the PG precursor lipid II, transferring lipid II from cytoplasmic face of the inner membrane to the periplasmic side (Mohammadi et al. 2011, 2014). However, in vivo data has failed to demonstrate FtsW's flippase activity in *E. coli* (Sham et al. 2014). An alternative hypothesis suggests that FtsW is instead a GTase (Cho et al. 2016). Not only are the SEDS family proteins topologically analogous to GTases, but the *E. coli* SEDS protein RodA was directly shown to have GTase activity in vitro (Meeske et al. 2016), suggesting FtsW may have GTase activity as well. It is, therefore, worth examining FtsW's GTase activity in *Caulobacter* in order to gain a better understanding of its true function. As an aside, it will also be important to identify the lipid II flippase in *Caulobacter*, as this is likely a key player in all processes requiring PG synthesis, including division. The leading candidate is MurJ, which was shown to be required for flipping lipid II in *E. coli* (Sham et al. 2014) and has a homolog in *Caulobacter*.

Caulobacter has additional putative PG synthases that may contribute to constriction. Pbp1A, PbpC, PbpX, PbpY, and PbpZ are bifunctional penicillin-binding proteins (PBPs) located at the IM, each containing a transmembrane domain, a TPase domain, and a GTase domain (Yakhnina and Gitai 2013; Strobel et al. 2014). Each of these PBPs is nonessential and, with the exception of PbpZ, each individually is sufficient for supporting growth and/or division (Yakhnina and Gitai 2013; Strobel et al. 2014). However, deletion of *pbpX* leads to a mild elongation phenotype and deletion of all five together is lethal (Yakhnina and Gitai 2013; Strobel et al. 2014). PbpX and PbpY localize along the length of the cell at the IM prior to constriction, with PbpX relocating to midcell at or just prior to constriction initiation and PbpY arriving shortly after (Figs. 2 and 3) (Yakhnina and Gitai 2013; Strobel et al. 2014). Additionally, both PbpX and PbpY have been found to interact with the divisome proteins FtsN, FtsL, and DipM by bacterial two-hybrid analysis, strengthening the claim that these PBPs are involved in division (Strobel et al. 2014). Intriguingly, PbpC and Pbp1A do not localize to midcell (PbpZ localization could not be determined) (Yakhnina and Gitai 2013), so enrichment of a bifunctional PBP at midcell is not absolutely required for division. Altogether, the sequence homology, localization, and interaction data strongly suggest that PbpX and PbpY play an auxiliary role in division, possibly as a secondary function to general PG maintenance. The nature of the roles that PbpZ, PbpC, and Pbp1A play in division is still unknown.

6.3 FtsZ-Mediated PG Synthase Localization and Activation

So what role does FtsZ play in constriction, then? FtsZ of course is necessary for localizing the constriction machinery to the site of division, and it now seems that treadmilling is important for directing movement of PG synthases at the division site. As described in previous sections, FtsZ in *E. coli* and *B. subtilis* has been shown to treadmill around the circumference of the cell, with its GTPase rate dictating treadmilling speed (Bisson-Filho et al. 2017; Yang et al. 2017). FtsI in *E. coli* (or Pbp2B in *B. subtilis*) moves around the cell at a speed that is dependent on FtsZ treadmilling (Bisson-Filho et al. 2017; Yang et al. 2017). In *B. subtilis*, but not *E. coli*, treadmilling speed also correlates with the rate of PG synthesis (Bisson-Filho et al. 2017; Yang et al. 2017). Interestingly, in *S. aureus*, it was shown that FtsZ treadmilling is required for an early phase of constriction, but is dispensable later on (Monteiro et al. 2018). *S. aureus* FtsZ may therefore activate PG synthases early in division, but is not required for their continued regulation, highlighting mechanistic similarities to both *E. coli* and *B. subtilis* FtsZ (Monteiro et al. 2018). Altogether, these data suggest *Caulobacter* FtsZ probably treadmills too and is therefore important for spatial regulation of PG insertion. However, they do not provide any coherent insight as to whether FtsZ regulates the rate of PG synthesis, which may well vary from species to species.

Based on the FtsZ treadmilling studies, to what extent FtsZ regulates PG remodeling is still in question, with evidence from *E. coli* pointing toward a passive role for FtsZ and evidence from *B. subtilis* indicating a more active role for FtsZ. Studies from our lab have helped to shed light on this point, suggesting that FtsZ may more actively regulate PG synthase activity in *Caulobacter*. FtsZ's CTL, the flexible linker between the GTPase domain and C-terminal peptide, has been shown to be essential for proper PG synthesis in *Caulobacter* (Sundararajan et al. 2015). Cells producing an FtsZ variant lacking its CTL (FtsZ-ΔCTL) form bulges instead of constrictions at midcell, similar to the cell shape changes induced by β-lactam treatment (Sundararajan et al. 2015). Muropeptide analysis revealed that PG composition in these cells is altered, indicating a defect in the activity of cell wall remodeling enzyme(s) downstream of FtsZ (Sundararajan et al. 2015). Because the rest of the divisome appears to still form properly in the presence of FtsZ-ΔCTL, it was proposed that FtsZ regulates PG synthesis in a CTL-dependent manner, possibly by facilitating mechanical signaling to the PG synthesis machinery (Sundararajan et al. 2015). Another pathway by which *Caulobacter* FtsZ influences PG synthase rate is through the FtsZ-bending protein, FzlA. An early localizing FtsZ binding protein (Fig. 3), FzlA is cytoplasmic (Fig. 2) and has been shown to curve FtsZ protofilaments in vitro (Goley et al. 2010b). Mutations in FzlA that resulted in weakened interaction with FtsZ also caused a decrease in constriction rate relative to elongation rate, indicating FzlA, through its interaction with FtsZ, influences PG insertion rate at the site of division (Lariviere et al. 2018). FzlA may therefore act as an intermediary through which FtsZ regulates PG synthase activity. Additional work here entails characterizing the upstream and downstream factors of this pathway, as well as probing the effect of FzlA mutation on FtsZ dynamics and PG synthase activity in vivo (Lariviere et al. 2018).

Based on the above data, we propose two models by which FtsZ could regulate PG remodeling during division, whereby the Z-ring acts as either a "dynamic scaffold" or a "dynamic activator." In the dynamic scaffold model, which is in line with FtsZ function in *E. coli*, FtsZ treadmilling is required only for distributing the PG synthases at the site of division, not for affecting their activity. According to the dynamic activator model, which more accurately describes FtsZ function in *B. subtilis*, FtsZ treadmilling is needed to both distribute and activate the PG synthases at midcell. We, therefore, propose that *Caulobacter* FtsZ is also a dynamic activator (Fig. 4). We can envision several possible mechanisms through which *Caulobacter* FtsZ might activate PG synthesis according to a dynamic activator model, which need not be mutually exclusive: mechanical signaling to PG synthases requiring force transduction through the CTL; FzlA-dependent signaling to PG synthases; and treadmilling-dependent distribution and/or activation of PG synthases (Fig. 4). FtsZ does not directly bind to FtsW or FtsI, therefore each of these proposed mechanisms likely requires that signal transduction from FtsZ to the PG synthases goes through one or more intermediary factors (Fig. 4). Substantial future work will be required to determine the precise mechanisms and proteins through which FtsZ activates PG synthesis in *Caulobacter*.

6.4 Stress Response-Mediated Constriction Regulation

Downstream of FtsZ, additional factors are likely required to regulate the PG synthetic machinery activity, ensuring that constriction occurs at the correct time. However, before discussing this regulation under normal conditions, we will consider how the PG synthases can be deactivated (leading to a constriction block) under stress, as this will provide us with valuable insight into the activation process. The best-studied stress that leads to a late-stage constriction block in *Caulobacter* is DNA damage. Induction of DNA damage results in the upregulation of numerous genes involved in DNA repair, as well as two genes responsible for inhibiting division in *Caulobacter, sidA* and *didA*. *sidA* is a member of the canonical DNA damage response SOS regulon, with its expression driven by cleavage of the SOS repressor LexA (Modell et al. 2011). *didA*, on the other hand, is not part of the SOS regulon, with its expression driven instead by a poorly understood transcriptional activator, DriD (Modell et al. 2014). Normally, when cells sense DNA damage, they halt division to allow time for DNA repair. However, when *sidA* or *didA* are deleted, division occurs more quickly and leads to more growth defects in the presence of DNA damage (Modell et al. 2011, 2014). Deletion of both genes together greatly enhances this effect, indicating SidA and DidA are both needed for proper division inhibition in response to DNA damage (Modell et al. 2011, 2014). Overproduction of either protein alone in the absence of DNA damage results in division inhibition, indicating each protein is sufficient to block division (Modell et al. 2011, 2014). SidA and DidA each localize to midcell, with SidA binding FtsW and both binding FtsN (Modell et al. 2011, 2014). FtsN is an essential transmembrane divisome protein (Figs. 2 and 3), which binds to PG via a periplasmic SPOR domain (Möll and Thanbichler 2009). FtsN likely binds to FtsW and FtsI, putatively forming an FtsWIN complex (Karimova et al. 2005; Alexeeva et al. 2010; Fraipont et al. 2011; Modell et al. 2011, 2014). Because neither SidA nor DidA recruitment to midcell results in delocalization of any of the divisome proteins tested, including FtsW, FtsI, or FtsN, Modell et al. proposed that they inactivate the FtsWIN complex (Modell et al. 2014). In support of this hypothesis, mutations in FtsW and FtsI which have been shown to increase constriction rate under WT conditions (Lambert et al. 2018), as well as a mutation in FtsN proposed to hyperactivate constriction under WT conditions, suppress the ability of SidA or DidA to block division (Modell et al. 2014). These data suggest FtsWIN may occupy either an inactive state or an active state (Modell et al. 2014), though it is still unclear precisely how SidA and DidA alter the transition from one to the other.

6.5 Downstream PG Synthase Activation Pathways

Having discussed regulation of the PG synthesis machinery in response to stress, we now have a better understanding of how PG synthase regulation downstream to FtsZ may occur under normal conditions. FtsN is thought to activate the PG synthases FtsW and FtsI, since mutation of FtsN can suppress a constriction block caused by SidA and DidA overproduction, thus allowing FtsW and FtsI to function (Modell et al. 2014). Because FtsN forms a complex with FtsW and FtsI, this activation is likely direct (Modell et al. 2014). However, data from *E. coli* suggests FtsN may also activate FtsW and FtsI via an indirect route as well. FtsN is proposed signal to the PG synthases through the FtsQLB complex, comprising the IM transmembrane proteins FtsQ, FtsL, and FtsB (Figs. 2 and 3) (Goley et al. 2011; Tsang and Bernhardt 2015; Liu et al. 2015; Glas et al. 2015). This may occur through direct interactions with FtsQLB in the periplasm and/or through FtsA in the cytoplasm. This latter assertion not only suggests that FtsA is more than a passive membrane anchor for FtsZ, but also implicates a parallel, overlapping pathway in constriction initiation (Tsang and Bernhardt 2015; Liu et al. 2015). Thus, by analogy to *E. coli*, there are potentially a number of ways in which FtsN activates FtsWI in *Caulobacter*, all of which require validation before we can propose a fully fleshed-out model.

In addition to the above factors, several other proteins have been implicated in constriction activation in *Caulobacter.* FtsK is a DNA translocase, with transmembrane and cytoplasmic portions (Fig. 2), that localizes to midcell just prior to constriction (Fig. 3) and is essential for proper chromosome segregation (Wang et al. 2006; Grainge 2010; Goley et al. 2011; Wolfe et al. 2014). In *E. coli*, FtsK binds members of the FtsWI and FtsQLB complexes, and a temperature sensitive mutant allows for partial suppression of deletion of FtsQ, FtsB, FtsN, and FtsA (Geissler and Margolin 2005; Grainge 2010). FtsK has therefore been proposed to play a regulatory function in the divisome, possibly licensing constriction initiation after DNA segregation has occurred, though no mechanism has been described (Grainge 2010). FzlA is another essential activator of constriction in *Caulobacter* as described earlier. Yet another essential protein required for division in *Caulobacter*, DipI, was recently discovered. Believed to reside in the periplasm (Fig. 2), DipI localizes to midcell just prior to constriction initiation (Fig. 3), and is thought to bind FtsQLB (Osorio et al. 2017). Osorio et al., therefore, propose that DipI helps FtsQLB to activate the PG synthesis machinery, though the details of this mechanism are unresolved (Osorio et al. 2017).

As a note, we have characterized the above pathways as downstream of FtsZ in our dynamic activator model, in which they relay activating signals from FtsZ to PG synthases (Fig. 4). However, at this time it is unclear how biophysical signals from FtsZ are relayed to FtsQLB, FtsN, FtsK, and/or DipI, or if any of these pathways are parallel to FtsZ activation signals instead of downstream. Future work will focus on characterizing how these constriction activators interact genetically with one another and how they transduce an activation signal from FtsZ to PG synthases.

6.6 Open Questions

- Does PG synthesis generate force? This still has not been directly demonstrated in vitro. What are the relative contributions to the constrictive force from FtsZ and PG?
- Does *Caulobacter* FtsZ's GTPase rate influence the rate of PG synthesis and constriction?
- How is the rate of PG synthesis determined? Treadmilling speed? Intrinsic activity of PG synthases? Amount of activating signal?
- Does the speed of PG synthase movement correspond with catalysis rate?
- Does FtsZ activate the PG synthetic machinery and if so, how? Do FtsZ dynamics, force transduction, and/or downstream protein interactions contribute to this activation? What is the most upstream constriction initiation signal?
- How do the downstream activators promote constriction? What is the nature of the interactions between these factors and what do the pathways that they comprise look like?

- How are the relative amounts of TPase vs. GTase activity maintained? Is this ratio important for division?
- What is the function of the bifunctional PBPs? Do they play a proofreading role?
- Does FtsW have transglycosylase activity? Does MurJ have flippase activity?

7 Envelope Maintenance

Addition of new PG to the cell wall is not sufficient for efficient division to occur in *Caulobacter*. The integrity of the envelope needs to be maintained during division in order to ensure that the entire *Caulobacter* envelope invaginates in a concerted manner and that cells do not lyse in the process of division. The IM and OM, separated by a distance of 31.5 nm in *Caulobacter* (Goley et al. 2010a), must be physically linked in order to maintain envelop integrity and to ensure equal rates of constriction of both membranes. The Tol-Pal complex is a multi-protein system that spans the entire length of the envelope and serves precisely such a function, with TolA, TolQ, and TolR located at the inner membrane, TolB at the periplasm, and Pal at the outer membrane (Fig. 2) (Gerding et al. 2007; Yeh et al. 2010). Given the location of each protein within the envelope, TolA is thought to bind Pal in the periplasm, physically tethering the IM to the OM. Accordingly, by cryo-EM, the OM to PG distance increases in cells depleted of Pal and the PG to IM distance increases in cells depleted of TolA (Yeh et al. 2010), indicating loss of a functional Tol-Pal system weakens the IM to OM connection. Depletion of TolA or Pal leads to membrane blebbing throughout the cell, including at the division plane, with loss of TolB causing blebbing at the division plane as well (Yeh et al. 2010). Because blebbing results from loss of envelope integrity and impaired IM-PG-OM attachment, its occurrence at the site of division prevents efficient and concerted constriction of the envelope. However, while this envelope maintenance function is important for constriction, it is unlikely to be specific to it. TolQ, followed by TolA, localizes to midcell well before constriction is initiated, around the time that the elongation machinery assembles (Fig. 3) (Yeh et al. 2010; Goley et al. 2011); the rest of the Tol-Pal machinery is presumed to follow suit. This early assembly, in conjunction with the presence of cell-wide shape defects in Tol-Pal depletion strains, suggests that the Tol-Pal system is likely active during elongation as well.

Another envelope maintenance protein well characterized in *Caulobacter* is the LytM and LysM domain-containing protein, DipM. Located in the periplasm (Fig. 2), DipM is proposed to be involved in PG hydrolysis, causing cell lysis when it is overexpressed (Goley et al. 2010a; Möll et al. 2010; Poggio et al. 2010). Like the Tol-Pal proteins, deletion of *dipM* causes membrane blebbing at the site of division and throughout the rest of the cell (Goley et al. 2010a; Möll et al. 2010). And similar to TolQ, DipM localizes to midcell early, around the time that the elongation machinery arrives (Fig. 3) (Goley et al. 2010a, 2011; Möll et al. 2010; Poggio et al. 2010). Intriguingly, cells depleted of DipM were found to have thicker PG throughout the entire length of the cell, indicating that DipM is important for regulating PG width (Goley et al. 2010a). Further, it has been hypothesized that proper PG width is required to maintain contact between TolA and Pal in the periplasm, with thicker PG pushing TolA and Pal out of reach of each other (Goley et al. 2010a). This results in loss of IM-OM contact and would explain the OM blebbing present in Δ*dipM* cells.

DipM is likely not a PG hydrolase itself, however. While zymogram analysis shows that DipM can bind PG, this assay cannot differentiate PG binding from hydrolysis (Uehara et al. 2010; Möll et al. 2010). By RBB-labeled sacculus dye release assay, a more reliable method for measuring PG hydrolysis, DipM does not show hydrolytic activity on its own (Meier et al. 2017). Instead, DipM's crucial function is likely to recruit and/or activate other PG hydrolases in order to drive PG breakdown (Meier et al. 2017; Zielińska et al. 2017). SdpA and SdpB are two such putative lytic transglycoslyases, which reside in the periplasm and localize to midcell early in the division process in a FtsN- and DipM-dependent manner (Figs. 2 and 3) (Zielińska et al. 2017). Deletion of *sdpA* and *sdpB* causes membrane blebbing, highlighting their importance in envelope maintenance (Zielińska et al. 2017). Further deletion of *dipM* in a background lacking *sdpA* and *sdpB* results in a synthetic sick/lethal phenotype, indicating DipM may have other downstream targets (Zielińska et al. 2017). What lies upstream of DipM? Interestingly, FtsN has been shown to interact with DipM by bacterial two-hybrid analysis, and is also required for recruitment of DipM, SpdA, and SdpB to midcell (Möll et al. 2010; Zielińska et al. 2017). Although it is unclear precisely how the late arriving FtsN recruits the early arriving DipM, the localization dependencies and interaction data indicate a potential link between the constriction and envelope maintenance machineries, suggesting that FtsN may help coordinate their activities (Möll et al. 2010). Finally, while DipM, SdpA, and SdpB are all involved in envelope maintenance during constriction, like Tol-Pal, they are likely also important during elongation due to their early midcell localization and cell-wide envelope defects.

7.1 Open Questions

- Do SdpA and SdpB have lytic transglycosylase activity?
- Does DipM have other downstream targets involved in envelope maintenance?
- How is envelope maintenance coupled to elongation and constriction? Does FtsN serve as a hub for coordinating envelope maintenance during constriction?
- How is the PG hydrolase activity from envelope maintenance balanced with the PG synthase activity from elongation and constriction?
- How does DipM interact genetically with Tol-Pal? Is limited PG thickness in fact required to maintain Tol-Pal contact, and does this influence OM-IM linkage?

8 Cell Separation, Polar Organization, and the Final Events of Division

Following the initiation and continuation of constriction, the final stage of division is marked by a few major events. During cell separation, the IMs fuse, compartmentalizing the cytoplasm into two separate daughters; PG hydrolases cleave PG connecting the daughter cells in a process we will refer to as terminal PG hydrolysis; and finally, the OMs fuse. Prior to or concurrent with these events, polarity factors are recruited to the constriction site to mark the creation of a new pole. Toward the end of division, MreB disperses and the Z-ring begins to be dismantled. Finally, the completion of division results in two new daughter cells, with two new poles, shaped by the constriction process, in place of the division plane.

8.1 Cell Separation

A number of factors have been implicated in cell separation that function in at least three pathways: 1. The FtsEX-LdpF pathway, 2. The AmiC pathway, and 3. The DipM-SdpAB pathway (Meier et al. 2017) (Fig. 5a). FtsEX, which we introduced as a putative FtsZ membrane anchor, has also been implicated in terminal PG hydrolysis (Meier et al. 2017). LdpF is a LytM-domain containing protein that is predicted to be periplasmic, and like DipM, LdpF is thought to lack the endopeptidase activity typically associated with LytM proteins due to the absence of key active site residues (Meier et al. 2017; Zielińska et al. 2017). (Note that while LdpF is thought to be active at midcell, it localizes diffusely throughout the cell, so it is not included in Fig. 2 (Meier et al. 2017).) AmiC is an N-acetylmuramoyl-L-alanine amidase, responsible for cleaving the bond between MurNac and the pentapeptide stem in PG (Meier et al. 2017; Zielińska et al. 2017; Dubey and Priyadarshini 2018). AmiC is predicted to reside in the periplasm and localizes to midcell shortly after constriction starts (Figs. 2 and 3) (Meier et al. 2017; Zielińska et al. 2017). Finally, DipM, as previously mentioned, is another early recruit involved in envelope maintenance, that is a periplasmic PG binding protein (Goley et al. 2010a; Möll et al. 2010; Poggio et al. 2010). The soluble lytic transglycosylases (SLTs) SdpA and SdpB, which have putative lytic transglycosylase activity, are dependent on DipM for localization for midcell (Zielińska et al. 2017).

Genetic perturbation of the cell separation factors has informed functional grouping according to phenotype (Fig. 5b). Loss of *ftsE*, *amiC*, or *ldpF* alone or together results in a similar chaining phenotype with varying severity (Meier et al. 2017; Zielińska et al. 2017). Specifically, cells lacking one or more of these factors constrict to a very late point, then extend thin connections between daughter cells that remain unresolved. In an *ftsE amiC* double mutant, unfused inner membrane connections were observed in these thin cell–cell connections, implicating these genes in inner membrane fusion (Fig. 5b). Deletion of *dipM*, *sdpA*, or *sdpB*, on the other hand, yields short, blunt, septum-like connections, in which the OM has not

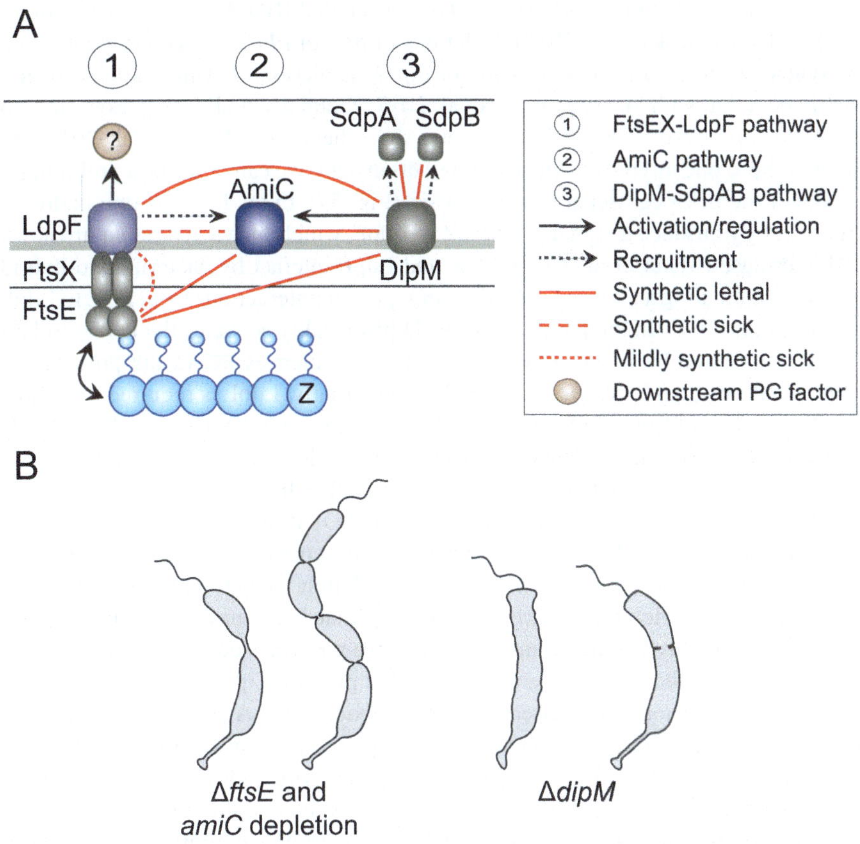

Fig. 5 Cell separation pathways. (**a**) Model for cell separation pathways in *Caulobacter*. (1) In the FtsEX-LdpF pathway, FtsEX and LdpF are proposed to signal to an unidentified PG hydrolase to initiate cell separation. FtsZ, which interacts with FtsEX, may serve as an upstream activator for this pathway. (2) In the AmiC pathway, LdpF recruits AmiC to midcell, and DipM subsequently activates AmiC's PG hydrolase activity. (3) In the DipM-SdpAB pathway, DipM recruits the putative PG hydrolases SdpA and SdpB to midcell, which then begin PG hydrolysis. Synthetic interactions between genes (red solid, dashed, and dotted lines) indicate varying degrees of highlight the complexity of the interactions among and between pathways. Adapted from Meier et al. (2017). (**b**) Deletion or depletion phenotypes of *ftsE*, *amiC*, and *dipM* in *Caulobacter*. Deletion of *ftsE* and depletion of AmiC yields a heterogeneous population of chained cells, with some connected by a long, skinny connection and others connected by a much shorter connection. Deletion of *dipM* leads to filamentation, envelope defects, blunter poles, and septum-like constriction sites. The gene deletions correspond with the cell separation pathways depicted in panel A and highlight the heterogeneity of roles that each protein/pathway plays in cell separation

constricted fully, but the IMs have fused (Goley et al. 2010a; Möll et al. 2010; Meier et al. 2017; Zielińska et al. 2017). Deletions of *ftsE* or *ldpF* are each synthetic sick or synthetic lethal with loss of *amiC* or *dipM*; depletion of AmiC is additionally synthetic lethal with *ΔdipM* (Meier et al. 2017; Zielińska et al. 2017). Loss of both *ftsE* and *ldpF* results in a mild synthetic sick phenotype (Meier et al. 2017). In light of these phenotypic and genetic interaction data, we can describe a preliminary model of the three cell separation pathways (Fig. 5a). In the FtsEX-LdpF pathway, FtsEX may transduce a signal from FtsZ to LdpF, which then activates an unknown PG hydrolase (Meier et al. 2017). FtsX and LdpF interact by bacterial two-hybrid analysis, indicating a series of direct protein–protein interactions from FtsZ to LdpF (Meier et al. 2017). In the AmiC pathway, DipM weakly activates the PG hydrolase AmiC, supported by in vitro evidence that AmiC cleaves PG in the presence of DipM (but not LdpF) (Meier et al. 2017). Interestingly, LdpF is required for AmiC localization to midcell (Zielińska et al. 2017), suggesting the activity of the AmiC pathway may be partially dependent on the FtsEX-LdpF pathway. As for the DipM-SdpAB pathway, DipM likely activates the putative hydrolases SdpA and SdpB (Zielińska et al. 2017). We would like to note here that the membrane anchor FzlC also interacts genetically with a number of cell separation factors (Meier et al. 2016). Loss of *dipM*, *ftsE*, or *amiC* in a *fzlC* deletion background yields synthetic sick phenotypes (Meier et al. 2016), indicating that FzlC may be involved in cell separation, possibly helping to integrate the three pathways described above.

In addition to the above factors, an additional protein, CtpA, is implicated in cell separation in *Caulobacter*. Essential in *Caulobacter*, CtpA is a putative tyrosine phosphatase that resides on or near the IM (Fig. 2) and localizes to midcell before constriction starts (Fig. 3) (Shapland et al. 2011). Although CtpA does not have detectable activity in vitro, it has homology to other tyrosine phosphatases, and mutation of its predicted catalytic site leads to growth inhibition (Shapland et al. 2011). Depletion of CtpA not only causes cell chaining and inefficient cell separation, but also results in OM blebbing throughout the cell body (Shapland et al. 2011). CtpA is therefore proposed to regulate a factor/factors involved in OM maintenance, cell separation, or both (Shapland et al. 2011). Further work is required to identify CtpA's specific targets.

8.2 Recruitment of Polarity Determining Factors

The last proteins recruited to the divisome are the polarity determining factors, TipN and TipF (Huitema et al. 2006; Lam et al. 2006; Goley et al. 2011). Predicted to be transmembrane proteins at the IM (Fig. 2), TipN and TipF are responsible for marking the new cell pole that will be formed after division, a function necessary for properly localized flagellar development, chromosome segregation, and daughter cell length determination (Huitema et al. 2006; Lam et al. 2006; Ptacin et al. 2010). TipN arrives at midcell in an FtsZ-dependent fashion after constriction is well underway, and is itself necessary for recruitment of TipF to midcell (Fig. 3) (Huitema et al. 2006; Goley et al. 2011). TipN recruitment to midcell also depends on the Tol-Pal system, with which it was shown to interact by co-immunoprecipitation (Yeh et al. 2010).

8.3 Z-Ring Disassembly and MreB Dispersal

Upon completion of division, the former constriction site becomes the new poles, FtsZ levels decrease in both daughter cells (Williams et al. 2014), and the Z-ring is disassembled. In the new swarmer cell, a small amount of FtsZ remains at the new pole, whereas in the new stalked cell, FtsZ immediately leaves the new pole for the new midcell (Goley et al. 2011; Williams et al. 2014). Several mechanisms are proposed to facilitate these changes. The cytoplasmic proteases ClpXP and ClpAP, which were previously shown to limit FtsZ levels in G1 phase through degradation, are hypothesized to also be involved in Z-ring disassembly. Although ClpA remains diffuse throughout the entire *Caulobacter* cell cycle, ClpX localizes to midcell just before completion of division (Figs. 2 and 3) (Williams et al. 2014). ClpXP may therefore assist in Z-ring disassembly by reducing FtsZ concentration at the old constriction site/new poles to a point below its critical concentration and thereby favoring depolymerization. GdhZ and KidO, cytoplasmic inhibitors of FtsZ self-interaction in G1 phase, may also contribute to Z-ring disassembly. Expression of GdhZ and KidO is downregulated during S phase, but increases again starting in pre-divisional cells (Radhakrishnan et al. 2010; Beaufay et al. 2015). KidO subsequently localizes to midcell shortly after the initiation of constriction, whereas GdhZ appears to arrive sometime before constriction starts (Figs. 2 and 3) (Radhakrishnan et al. 2010; Goley et al. 2011; Beaufay et al. 2015). Around the time that constriction finishes, KidO and GdhZ are thought to help disassemble the Z-ring by blocking FtsZ self-interaction (Radhakrishnan et al. 2010), though it is unclear how they delay their activity until completion of division. Finally, the other negative regulator of Z-ring assembly, MipZ, likely plays a role in disassembly of FtsZ foci at newborn cell poles. MipZ localizes to the new pole in stalked cells as soon as chromosome segregation has occurred (Thanbichler and Shapiro 2006; Goley et al. 2011), suggesting it helps to quickly displace FtsZ from the last division site. Altogether, Z-ring disassembly from the old division site appears to be due to a combination of decreased FtsZ levels and localized inhibition of self-interaction.

It is also worth mentioning that MreB disperses from the division plane toward the end of division, after constriction has initiated and before the cell has completed separation (Fig. 3) (Goley et al. 2011). Precisely why or how MreB leaves midcell at this time is not currently known, though we can speculate that if MreB contributes to elongation earlier during division, its departure may allow for an increased focus of PG synthesis on constriction.

8.4 Cell Pole Shape Determination

Following cell separation, a single *Caulobacter* cell splits into two, and the constriction site becomes two new poles. Because the shape of the poles can be influenced by numerous factors during constriction, analysis of its shape is sometimes useful for assessing the efficiency of different processes during division. Cell pole shape can be characterized according to two metrics: "pointiness," a measure of how tapered or blunt the pole looks, and envelope topology, which indicates if the OM, PG, and IM are still properly attached. As pole pointiness is dictated by the events of division, it is logically determined by the ratio of the elongation rate to the constriction rate, such that cells elongating faster relative to constriction will have pointier poles. This has been shown to be the case in *E. coli*, where a mutation in FtsI causes cells to start to develop pointy poles during division (Taschner et al. 1988; Costa et al. 2008). In *Caulobacter*, however, it has previously been shown that poles become pointier well after division has finished (Aaron et al. 2007), so it is less clear to what extent division also plays a role in pole shape determination in this species. Consideration of division protein mutants that alter cell pole pointiness has helped to shed some light on this matter. Mutation of FzlA has been shown to yield pointy poles, likely because it slows the rate of constriction relative to elongation (Lariviere et al. 2018). Overexpression of FzlC also leads to pointy poles, though the reason for this is still unclear (Meier et al. 2016). Finally, deletion of DipM leads to blunter poles, though again it is not clear if DipM deletion impacts constriction and/or elongation rate (Möll et al. 2010). A properly functioning divisome is therefore required for WT pole pointiness, though studies with finer temporal resolution will be required to determine if DipM and FzlC influence pole pointiness during or following division.

The other metric for pole shape, envelope topology, is determined by members of the envelope maintenance machinery described in previous sections, including DipM and the Tol-Pal system. Loss of envelope integrity due to DipM or Tol-Pal deletion/depletion leads to OM blebbing during constriction, which is often still present at the new poles following division, either in the form of a small bulge or a larger bleb (Goley et al. 2010a; Möll et al. 2010; Yeh et al. 2010). All in all, the determination of cell pole shape is highly sensitive to the interplay amongst multiple division processes.

8.5 Open Questions

- What are the relative contributions of the three characterized cell separation pathways? Why are there multiple pathways and why is there some crosstalk? Does each pathway differentially contribute to IM fusion, terminal PG hydrolysis, and OM fusion?
- Does PG metabolism contribute to inner membrane fusion? If so, how? If not, why do *ftsE-amiC* mutants fail to fuse their inner membranes?

- What are the PG hydrolases downstream of LdpF and DipM?
- Is DipM's contribution to cell separation distinct from its role in envelope maintenance?
- What role does FzlC play in cell separation? Does it help to integrate multiple cell separation pathways?
- Does CtpA have phosphatase activity? What are its downstream effectors? Is it involved in any of the cell separation pathways that we have described?
- To what extent do KidO and GdhZ help disassemble the Z-ring? (Why) is Z-ring disassembly necessary?
- Why and how does MreB disassemble from midcell toward the end of division?
- How do DipM and FzlC determine cell pole pointiness? Are their effects on pole shape immediate (during division) or delayed (after division)?

9 Conclusions and Outlook

In this chapter, we have outlined what is currently known about division in *Caulobacter*. The Z-ring is a central player in division, helping to coordinate temporally overlapping processes. Multiple assembly factors regulate Z-ring formation in both time and space, defining the division plane at midcell after chromosome replication has begun. The Z-ring subsequently recruits a host of divisome proteins, the first of which contribute to cellular elongation from the midcell. Upon arrival of the constriction machinery, FtsZ is hypothesized to activate PG synthesis through one or multiple putative mechanisms, including dynamics-, mechanical-, or protein interaction-based signal transduction, involving up to a dozen potential additional regulators. Since both elongation and constriction involve cell wall remodeling, maintenance of envelope integrity is crucial for effective division. As a cell nears the end of division, PG hydrolysis drives cell separation, the Z-ring begins to disassemble, and the cell finally splits into two daughter cells.

Though numerous studies in recent years have greatly informed our understanding of division in *Caulobacter*, we would like to highlight some of the remaining fundamental unanswered questions. To what extent do FtsZ and PG remodeling contribute to the forces required for constriction to occur? Direct evidence that PG remodeling generates a constrictive force is still lacking. What is the nature of FtsZ's role in activating PG synthesis in *Caulobacter*? How is this activation mechanistically facilitated and to what degree, if at all, do dynamics, force transduction, and downstream protein signaling each play a role in PG synthase regulation? How do the constriction machinery activators (FtsN, FtsQ, FtsL, etc.) interact with one another and affect constriction? Is the PG remodeling that facilitates elongation during division distinct from or related to the PG remodeling that drives constriction? How are the activities of constriction-specific PG synthases and PG hydrolases regulated with respect to one another? How are the activities of the factors involved in cell separation coordinated to drive OM fusion, terminal PG hydrolysis, and IM fusion? Finally, how is division influenced by metabolic processes?

In order to address these questions and gain a better understanding of division in *Caulobacter*, the use of emerging tools will be invaluable. Advanced imaging techniques such as PALM and cryo-ET have proved useful for gaining a higher resolution view of microscopic processes. Computational analyses of single-cell experiments have allowed for the rapid and easy generation of high-quality datasets. Additionally, the declining cost of whole genome sequencing and the proliferation of other genomics techniques, such as transposon-sequencing and CRISPR-based editing, should make genetic analysis faster and more accessible. The use of cutting-edge technologies to complement tried-and-true approaches like genetics and in vitro biochemistry will drive future discoveries in the field. Our hope is that their implementation in the study of *Caulobacter* will provide insight into the fundamental nature of division.

References

Aaron M, Charbon G, Lam H et al (2007) The tubulin homologue FtsZ contributes to cell elongation by guiding cell wall precursor synthesis in Caulobacter crescentus. Mol Microbiol 64:938–952. https://doi.org/10.1111/j.1365-2958.2007.05720.x

Alexeeva S, Gadella TWJ, Verheul J et al (2010) Direct interactions of early and late assembling division proteins in Escherichia coli cells resolved by FRET. Mol Microbiol 77:384–398. https://doi.org/10.1111/j.1365-2958.2010.07211.x

Allard JF, Cytrynbaum EN (2009) Force generation by a dynamic Z-ring in Escherichia coli cell division. Proc Natl Acad Sci U S A 106:145–150. https://doi.org/10.1073/pnas.0808657106

Alyahya SA, Alexander R, Costa T et al (2009) RodZ, a component of the bacterial core morphogenic apparatus. Proc Natl Acad Sci U S A 106:1239–1244. https://doi.org/10.1073/pnas.0810794106

Anderson DE, Gueiros-Filho FJ, Erickson HP (2004) Assembly dynamics of FtsZ rings in Bacillus subtilis and Escherichia coli and effects of FtsZ-regulating proteins. J Bacteriol 186:5775–5781. https://doi.org/10.1128/JB.186.17.5775-5781.2004

Arends SJR, Kustusch RJ, Weiss DS (2009) ATP-binding site lesions in FtsE impair cell division. J Bacteriol 191:3772–3784. https://doi.org/10.1128/JB.00179-09

Banerjee S, Scherer NF, Dinner AR (2016) Shape dynamics of growing cell walls. Soft Matter 12:3442–3450. https://doi.org/10.1039/C5SM02991K

Beaufay F, Coppine J, Mayard A et al (2015) A NAD-dependent glutamate dehydrogenase coordinates metabolism with cell division in Caulobacter crescentus. EMBO J 34:1786–1800. https://doi.org/10.15252/embj.201490730

Bhat NH, Vass RH, Stoddard PR et al (2013) Identification of ClpP substrates in Caulobacter crescentus reveals a role for regulated proteolysis in bacterial development. Mol Microbiol 88:1083–1092. https://doi.org/10.1111/mmi.12241

Bi E, Lutkenhaus J (1990) FtsZ regulates frequency of cell division in Escherichia coli. J Bacteriol 172:2765–2768

Bi EF, Lutkenhaus J (1991) FtsZ ring structure associated with division in Escherichia coli. Nature 354:161–164. https://doi.org/10.1038/354161a0

Bisson-Filho AW, Hsu Y-P, Squyres GR et al (2017) Treadmilling by FtsZ filaments drives peptidoglycan synthesis and bacterial cell division. Science 355:739–743. https://doi.org/10.1126/science.aak9973

Biteen JS, Goley ED, Shapiro L, Moerner WE (2012) Three-dimensional super-resolution imaging of the midplane protein FtsZ in live Caulobacter crescentus cells using astigmatism. ChemPhysChem 13:1007–1012. https://doi.org/10.1002/cphc.201100686

Buss J, Coltharp C, Huang T et al (2013) In vivo organization of the FtsZ-ring by ZapA and ZapB revealed by quantitative super-resolution microscopy. Mol Microbiol 89:1099–1120. https://doi.org/10.1111/mmi.12331

Buss J, Coltharp C, Shtengel G et al (2015) A multi-layered protein network stabilizes the Escherichia coli FtsZ-ring and modulates constriction dynamics. PLoS Genet 11:e1005128. https://doi.org/10.1371/journal.pgen.1005128

Buss JA, Peters NT, Xiao J, Bernhardt TG (2017) ZapA and ZapB form an FtsZ-independent structure at midcell. Mol Microbiol 104:652–663. https://doi.org/10.1111/mmi.13655

Cabeen MT, Jacobs-Wagner C (2005) Bacterial cell shape. Nat Rev Microbiol 3:601–610. https://doi.org/10.1038/nrmicro1205

Camberg JL, Hoskins JR, Wickner S (2009) ClpXP protease degrades the cytoskeletal protein, FtsZ, and modulates FtsZ polymer dynamics. Proc Natl Acad Sci U S A 106:10614–10619. https://doi.org/10.1073/pnas.0904886106

Chen Y, Erickson HP (2005) Rapid in vitro assembly dynamics and subunit turnover of FtsZ demonstrated by fluorescence resonance energy transfer. J Biol Chem 280:22549–22554. https://doi.org/10.1074/jbc.M500895200

Cho H, Wivagg CN, Kapoor M et al (2016) Bacterial cell wall biogenesis is mediated by SEDS and PBP polymerase families functioning semi-autonomously. Nat Microbiol 1:16172. https://doi.org/10.1038/nmicrobiol.2016.172

Coltharp C, Xiao J (2017) Beyond force generation: why is a dynamic ring of FtsZ polymers essential for bacterial cytokinesis? BioEssays 39. https://doi.org/10.1002/bies.201600179

Coltharp C, Buss J, Plumer TM, Xiao J (2016) Defining the rate-limiting processes of bacterial cytokinesis. Proc Natl Acad Sci U S A 113:E1044–E1053. https://doi.org/10.1073/pnas.1514296113

Corbin BD, Wang Y, Beuria TK, Margolin W (2007) Interaction between cell division proteins FtsE and FtsZ. J Bacteriol 189:3026–3035. https://doi.org/10.1128/JB.01581-06

Costa T, Priyadarshini R, Jacobs-Wagner C (2008) Localization of PBP3 in Caulobacter crescentus is highly dynamic and largely relies on its functional transpeptidase domain. Mol Microbiol 70:634–651. https://doi.org/10.1111/j.1365-2958.2008.06432.x

Dajkovic A, Pichoff S, Lutkenhaus J, Wirtz D (2010) Cross-linking FtsZ polymers into coherent Z rings. Mol Microbiol 78:651–668. https://doi.org/10.1111/j.1365-2958.2010.07352.x

de Boer P, Crossley R, Rothfield L (1992) The essential bacterial cell-division protein FtsZ is a GTPase. Nature 359:254–256. https://doi.org/10.1038/359254a0

Divakaruni AV, Baida C, White CL, Gober JW (2007) The cell shape proteins MreB and MreC control cell morphogenesis by positioning cell wall synthetic complexes. Mol Microbiol 66:174–188. https://doi.org/10.1111/j.1365-2958.2007.05910.x

Dubey A, Priyadarshini R (2018) Amidase activity is essential for medial localization of AmiC in *Caulobacter crescentus*. Curr Genet 64(3):661–675. https://doi.org/10.1007/s00294-017-0781-9

Erickson HP, Osawa M (2017) FtsZ constriction force – curved Protofilaments bending membranes. Subcell Biochem 84:139–160. https://doi.org/10.1007/978-3-319-53047-5_5

Erickson HP, Taylor DW, Taylor KA, Bramhill D (1996) Bacterial cell division protein FtsZ assembles into protofilament sheets and minirings, structural homologs of tubulin polymers. Proc Natl Acad Sci U S A 93:519–523

Erickson HP, Anderson DE, Osawa M (2010) FtsZ in bacterial cytokinesis: cytoskeleton and force generator all in one. Microbiol Mol Biol Rev 74:504–528. https://doi.org/10.1128/MMBR.00021-10

Fenton AK, Gerdes K (2013) Direct interaction of FtsZ and MreB is required for septum synthesis and cell division in Escherichia coli. EMBO J 32:1953–1965. https://doi.org/10.1038/emboj.2013.129

Figge RM, Divakaruni AV, Gober JW (2004) MreB, the cell shape-determining bacterial actin homologue, co-ordinates cell wall morphogenesis in Caulobacter crescentus. Mol Microbiol 51:1321–1332. https://doi.org/10.1111/j.1365-2958.2003.03936.x

Fraipont C, Alexeeva S, Wolf B et al (2011) The integral membrane FtsW protein and peptidoglycan synthase PBP3 form a subcomplex in Escherichia coli. Microbiol Read Engl 157:251–259. https://doi.org/10.1099/mic.0.040071-0

Fu G, Huang T, Buss J et al (2010) In vivo structure of the E. coli FtsZ-ring revealed by photoactivated localization microscopy (PALM). PLoS One 5:e12682. https://doi.org/10.1371/journal.pone.0012680

Gan L, Chen S, Jensen GJ (2008) Molecular organization of gram-negative peptidoglycan. Proc Natl Acad Sci U S A 105:18953–18957. https://doi.org/10.1073/pnas.0808035105

Geissler B, Margolin W (2005) Evidence for functional overlap among multiple bacterial cell division proteins: compensating for the loss of FtsK. Mol Microbiol 58:596–612. https://doi.org/10.1111/j.1365-2958.2005.04858.x

Gerding MA, Ogata Y, Pecora ND et al (2007) The trans-envelope Tol-pal complex is part of the cell division machinery and required for proper outer-membrane invagination during cell constriction in E. coli. Mol Microbiol 63:1008–1025. https://doi.org/10.1111/j.1365-2958.2006.05571.x

Gitai Z, Dye N, Shapiro L (2004) An actin-like gene can determine cell polarity in bacteria. Proc Natl Acad Sci U S A 101:8643–8648. https://doi.org/10.1073/pnas.0402638101

Glas M, van den Berg van Saparoea HB, McLaughlin SH et al (2015) The soluble periplasmic domains of Escherichia coli cell division proteins FtsQ/FtsB/FtsL form a trimeric complex with Submicromolar affinity. J Biol Chem 290:21498–21509. https://doi.org/10.1074/jbc.M115.654756

Goehring NW, Beckwith J (2005) Diverse paths to midcell: assembly of the bacterial cell division machinery. Curr Biol 15:R514–R526. https://doi.org/10.1016/j.cub.2005.06.038

Goley ED, Comolli LR, Fero KE et al (2010a) DipM links peptidoglycan remodelling to outer membrane organization in Caulobacter. Mol Microbiol 77:56–73. https://doi.org/10.1111/j.1365-2958.2010.07222.x

Goley ED, Dye NA, Werner JN et al (2010b) Imaging-based identification of a critical regulator of FtsZ protofilament curvature in Caulobacter. Mol Cell 39:975–987. https://doi.org/10.1016/j.molcel.2010.08.027

Goley ED, Yeh Y-C, Hong S-H et al (2011) Assembly of the Caulobacter cell division machine. Mol Microbiol 80:1680–1698. https://doi.org/10.1111/j.1365-2958.2011.07677.x

Gonzalez D, Collier J (2013) DNA methylation by CcrM activates the transcription of two genes required for the division of Caulobacter crescentus. Mol Microbiol 88:203–218. https://doi.org/10.1111/mmi.12180

Grainge I (2010) FtsK–a bacterial cell division checkpoint? Mol Microbiol 78:1055–1057

Haakonsen DL, Yuan AH, Laub MT (2015) The bacterial cell cycle regulator GcrA is a σ70 cofactor that drives gene expression from a subset of methylated promoters. Genes Dev 29:2272–2286. https://doi.org/10.1101/gad.270660.115

Hayhurst EJ, Kailas L, Hobbs JK, Foster SJ (2008) Cell wall peptidoglycan architecture in Bacillus subtilis. Proc Natl Acad Sci U S A 105:14603–14608. https://doi.org/10.1073/pnas.0804138105

Hocking J, Priyadarshini R, Takacs CN et al (2012) Osmolality-dependent relocation of penicillin-binding protein PBP2 to the division site in Caulobacter crescentus. J Bacteriol 194:3116–3127. https://doi.org/10.1128/JB.00260-12

Holden SJ, Pengo T, Meibom KL et al (2014) High throughput 3D super-resolution microscopy reveals Caulobacter crescentus in vivo Z-ring organization. Proc Natl Acad Sci U S A 111:4566–4571. https://doi.org/10.1073/pnas.1313368111

Hottes AK, Shapiro L, McAdams HH (2005) DnaA coordinates replication initiation and cell cycle transcription in Caulobacter crescentus. Mol Microbiol 58:1340–1353. https://doi.org/10.1111/j.1365-2958.2005.04912.x

Huang KC, Mukhopadhyay R, Wen B et al (2008) Cell shape and cell-wall organization in gram-negative bacteria. Proc Natl Acad Sci U S A. https://doi.org/10.1073/pnas.0805309105

Huitema E, Pritchard S, Matteson D et al (2006) Bacterial birth scar proteins mark future flagellum assembly site. Cell 124:1025–1037. https://doi.org/10.1016/j.cell.2006.01.019

Karimova G, Dautin N, Ladant D (2005) Interaction network among Escherichia coli membrane proteins involved in cell division as revealed by bacterial two-hybrid analysis. J Bacteriol 187:2233–2243. https://doi.org/10.1128/JB.187.7.2233-2243.2005

Kelly AJ, Sackett MJ, Din N et al (1998) Cell cycle-dependent transcriptional and proteolytic regulation of FtsZ in Caulobacter. Genes Dev 12:880–893

Kiekebusch D, Michie KA, Essen L-O et al (2012) Localized dimerization and nucleoid binding drive gradient formation by the bacterial cell division inhibitor MipZ. Mol Cell 46:245–259. https://doi.org/10.1016/j.molcel.2012.03.004

Krupka M, Cabré EJ, Jiménez M et al (2014) Role of the FtsA C terminus as a switch for polymerization and membrane association. mBio 5:e02221-14. https://doi.org/10.1128/mBio.02221-14

Lam H, Schofield WB, Jacobs-Wagner C (2006) A landmark protein essential for establishing and perpetuating the polarity of a bacterial cell. Cell 124:1011–1023. https://doi.org/10.1016/j.cell.2005.12.040

Lambert A, Vanhecke A, Archetti A et al (2018) Constriction rate modulation can drive cell size control and homeostasis in C. crescentus. bioRxiv. https://doi.org/10.1101/266817

Lan G, Wolgemuth CW, Sun SX (2007) Z-ring force and cell shape during division in rod-like bacteria. Proc Natl Acad Sci U S A 104:16110–16115. https://doi.org/10.1073/pnas.0702925104

Lan G, Daniels BR, Dobrowsky TM et al (2009) Condensation of FtsZ filaments can drive bacterial cell division. Proc Natl Acad Sci U S A 106:121–126. https://doi.org/10.1073/pnas.0807963106

Lariviere PJ, Szwedziak P, Mahone CR et al (2018) FzlA, an essential regulator of FtsZ filament curvature, controls constriction rate during Caulobacter division. Mol Microbiol 107:180–197. https://doi.org/10.1111/mmi.13876

Laub MT, McAdams HH, Feldblyum T et al (2000) Global analysis of the genetic network controlling a bacterial cell cycle. Science 290:2144–2148

Laub MT, Chen SL, Shapiro L, McAdams HH (2002) Genes directly controlled by CtrA, a master regulator of the Caulobacter cell cycle. Proc Natl Acad Sci U S A 99:4632–4637. https://doi.org/10.1073/pnas.062065699

Li Z, Trimble MJ, Brun YV, Jensen GJ (2007) The structure of FtsZ filaments in vivo suggests a force-generating role in cell division. EMBO J 26:4694–4708. https://doi.org/10.1038/sj.emboj.7601895

Liu B, Persons L, Lee L, de Boer PAJ (2015) Roles for both FtsA and the FtsBLQ subcomplex in FtsN-stimulated cell constriction in Escherichia coli. Mol Microbiol 95:945–970. https://doi.org/10.1111/mmi.12906

Liu X, Meiresonne NY, Bouhss A, den Blaauwen T (2017) The Combined activities of FtsW and MurJ are required for lipid II translocation during cell division in Escherichia coli. bioRxiv. https://doi.org/10.1101/230680

Loose M, Mitchison TJ (2014) The bacterial cell division proteins FtsA and FtsZ self-organize into dynamic cytoskeletal patterns. Nat Cell Biol 16:38–46. https://doi.org/10.1038/ncb2885

Löwe J (1998) Crystal structure determination of FtsZ from Methanococcus jannaschii. J Struct Biol 124:235–243. https://doi.org/10.1006/jsbi.1998.4041

Löwe J, Amos LA (1998) Crystal structure of the bacterial cell-division protein FtsZ. Nature 391:203–206. https://doi.org/10.1038/34472

Löwe J, Amos LA (1999) Tubulin-like protofilaments in Ca2+–induced FtsZ sheets. EMBO J 18:2364–2371. https://doi.org/10.1093/emboj/18.9.2364

Lutkenhaus J, Sanjanwala B, Lowe M (1986) Overproduction of FtsZ suppresses sensitivity of lon mutants to division inhibition. J Bacteriol 166:756–762

Lyu Z, Coltharp C, Yang X, Xiao J (2016) Influence of FtsZ GTPase activity and concentration on nanoscale Z-ring structure in vivo revealed by three-dimensional Superresolution imaging. Biopolymers 105:725–734. https://doi.org/10.1002/bip.22895

Ma X, Ehrhardt DW, Margolin W (1996) Colocalization of cell division proteins FtsZ and FtsA to cytoskeletal structures in living Escherichia coli cells by using green fluorescent protein. Proc Natl Acad Sci U S A 93:12998–13003

Martin ME, Trimble MJ, Brun YV (2004) Cell cycle-dependent abundance, stability and localization of FtsA and FtsQ in Caulobacter crescentus. Mol Microbiol 54:60–74. https://doi.org/10.1111/j.1365-2958.2004.04251.x

McAdams HH, Shapiro L (2009) System-level design of bacterial cell cycle control. FEBS Lett 583:3984–3991. https://doi.org/10.1016/j.febslet.2009.09.030

Meeske AJ, Riley EP, Robins WP et al (2016) SEDS proteins are a widespread family of bacterial cell wall polymerases. Nature 537:634–638. https://doi.org/10.1038/nature19331

Meier EL, Razavi S, Inoue T, Goley ED (2016) A novel membrane anchor for FtsZ is linked to cell wall hydrolysis in Caulobacter crescentus. Mol Microbiol 101:265–280. https://doi.org/10.1111/mmi.13388

Meier EL, Daitch AK, Yao Q et al (2017) FtsEX-mediated regulation of the final stages of cell division reveals morphogenetic plasticity in Caulobacter crescentus. PLoS Genet 13:e1006999. https://doi.org/10.1371/journal.pgen.1006999

Milam SL, Erickson HP (2013) Rapid in vitro assembly of Caulobacter crescentus FtsZ protein at pH 6.5 and 7.2. J Biol Chem 288:23675–23679. https://doi.org/10.1074/jbc.M113.491845

Modell JW, Hopkins AC, Laub MT (2011) A DNA damage checkpoint in Caulobacter crescentus inhibits cell division through a direct interaction with FtsW. Genes Dev 25:1328–1343. https://doi.org/10.1101/gad.2038911

Modell JW, Kambara TK, Perchuk BS, Laub MT (2014) A DNA damage-induced, SOS-independent checkpoint regulates cell division in Caulobacter crescentus. PLoS Biol 12:e1001977. https://doi.org/10.1371/journal.pbio.1001977

Mohammadi T, van Dam V, Sijbrandi R et al (2011) Identification of FtsW as a transporter of lipid-linked cell wall precursors across the membrane. EMBO J 30:1425–1432. https://doi.org/10.1038/emboj.2011.61

Mohammadi T, Sijbrandi R, Lutters M et al (2014) Specificity of the transport of lipid II by FtsW in Escherichia coli. J Biol Chem 289:14707–14718. https://doi.org/10.1074/jbc.M114.557371

Möll A, Thanbichler M (2009) FtsN-like proteins are conserved components of the cell division machinery in proteobacteria. Mol Microbiol 72:1037–1053. https://doi.org/10.1111/j.1365-2958.2009.06706.x

Möll A, Schlimpert S, Briegel A et al (2010) DipM, a new factor required for peptidoglycan remodelling during cell division in Caulobacter crescentus. Mol Microbiol 77:90–107. https://doi.org/10.1111/j.1365-2958.2010.07224.x

Monteiro JM, Pereira AR, Reichmann NT et al (2018) Peptidoglycan synthesis drives an FtsZ-treadmilling-independent step of cytokinesis. Nature. https://doi.org/10.1038/nature25506

Morgenstein RM, Bratton BP, Nguyen JP et al (2015) RodZ links MreB to cell wall synthesis to mediate MreB rotation and robust morphogenesis. Proc Natl Acad Sci U S A 112:12510–12515. https://doi.org/10.1073/pnas.1509610112

Mukherjee A, Lutkenhaus J (1994) Guanine nucleotide-dependent assembly of FtsZ into filaments. J Bacteriol 176:2754–2758

Mukherjee A, Lutkenhaus J (1998) Dynamic assembly of FtsZ regulated by GTP hydrolysis. EMBO J 17:462–469. https://doi.org/10.1093/emboj/17.2.462

Mukherjee A, Lutkenhaus J (1999) Analysis of FtsZ assembly by light scattering and determination of the role of divalent metal cations. J Bacteriol 181:823–832

Mukherjee A, Dai K, Lutkenhaus J (1993) Escherichia coli cell division protein FtsZ is a guanine nucleotide binding protein. Proc Natl Acad Sci U S A 90:1053–1057

Oliva MA, Cordell SC, Löwe J (2004) Structural insights into FtsZ protofilament formation. Nat Struct Mol Biol 11:1243. https://doi.org/10.1038/nsmb855

Osawa M, Erickson HP (2013) Liposome division by a simple bacterial division machinery. Proc Natl Acad Sci U S A 110:11000–11004. https://doi.org/10.1073/pnas.1222254110

Osawa M, Erickson HP (2018) Turgor pressure and possible constriction mechanisms in bacterial division. Front Microbiol 9. https://doi.org/10.3389/fmicb.2018.00111

Osawa M, Anderson DE, Erickson HP (2008) Reconstitution of contractile FtsZ rings in liposomes. Science 320:792–794. https://doi.org/10.1126/science.1154520

Osawa M, Anderson DE, Erickson HP (2009) Curved FtsZ protofilaments generate bending forces on liposome membranes. EMBO J 28:3476–3484. https://doi.org/10.1038/emboj.2009.277

Osorio A, Camarena L, Cevallos MA, Poggio S (2017) A new essential cell division protein in Caulobacter crescentus. J Bacteriol 199:e00811–e00816. https://doi.org/10.1128/JB.00811-16

Paez A, Mateos-Gil P, Hörger I et al (2009) Simple modeling of FtsZ polymers on flat and curved surfaces: correlation with experimental in vitro observations. PMC Biophys 2:8. https://doi.org/10.1186/1757-5036-2-8

Pichoff S, Lutkenhaus J (2005) Tethering the Z ring to the membrane through a conserved membrane targeting sequence in FtsA. Mol Microbiol 55:1722–1734. https://doi.org/10.1111/j.1365-2958.2005.04522.x

Pichoff S, Lutkenhaus J (2007) Identification of a region of FtsA required for interaction with FtsZ. Mol Microbiol 64:1129–1138. https://doi.org/10.1111/j.1365-2958.2007.05735.x

Poggio S, Takacs CN, Vollmer W, Jacobs-Wagner C (2010) A protein critical for cell constriction in the gram-negative bacterium Caulobacter crescentus localizes at the division site through its peptidoglycan-binding LysM domains. Mol Microbiol 77:74–89. https://doi.org/10.1111/j.1365-2958.2010.07223.x

Pogliano J, Pogliano K, Weiss DS et al (1997) Inactivation of FtsI inhibits constriction of the FtsZ cytokinetic ring and delays the assembly of FtsZ rings at potential division sites. Proc Natl Acad Sci U S A 94:559–564

Ptacin JL, Lee SF, Garner EC et al (2010) A spindle-like apparatus guides bacterial chromosome segregation. Nat Cell Biol 12:791–798. https://doi.org/10.1038/ncb2083

Quardokus EM, Brun YV (2002) DNA replication initiation is required for mid-cell positioning of FtsZ rings in Caulobacter crescentus. Mol Microbiol 45:605–616

Radhakrishnan SK, Pritchard S, Viollier PH (2010) Coupling prokaryotic cell fate and division control with a bifunctional and oscillating oxidoreductase homolog. Dev Cell 18:90–101. https://doi.org/10.1016/j.devcel.2009.10.024

RayChaudhuri D, Park JT (1992) Escherichia coli cell-division gene ftsZ encodes a novel GTP-binding protein. Nature 359:251–254. https://doi.org/10.1038/359251a0

Rothfield L, Justice S, García-Lara J (1999) Bacterial cell division. Annu Rev Genet 33:423–448. https://doi.org/10.1146/annurev.genet.33.1.423

Sackett MJ, Kelly AJ, Brun YV (1998) Ordered expression of ftsQA and ftsZ during the Caulobacter crescentus cell cycle. Mol Microbiol 28:421–434

Sauvage E, Derouaux A, Fraipont C et al (2014) Crystal structure of penicillin-binding protein 3 (PBP3) from Escherichia coli. PLoS One 9:e98042. https://doi.org/10.1371/journal.pone.0098042

Scheffers D-J, Driessen AJM (2002) Immediate GTP hydrolysis upon FtsZ polymerization. Mol Microbiol 43:1517–1521

Schofield WB, Lim HC, Jacobs-Wagner C (2010) Cell cycle coordination and regulation of bacterial chromosome segregation dynamics by polarly localized proteins. EMBO J 29:3068–3081. https://doi.org/10.1038/emboj.2010.207

Sham L-T, Butler EK, Lebar MD et al (2014) Bacterial cell wall. MurJ is the flippase of lipid-linked precursors for peptidoglycan biogenesis. Science 345:220–222. https://doi.org/10.1126/science.1254522

Shapland EB, Reisinger SJ, Bajwa AK, Ryan KR (2011) An essential tyrosine phosphatase homolog regulates cell separation, outer membrane integrity, and morphology in Caulobacter crescentus. J Bacteriol 193:4361–4370. https://doi.org/10.1128/JB.00185-11

Shtylla B (2017) Mathematical modeling of spatiotemporal protein localization patterns in C. crescentus bacteria: a mechanism for asymmetric FtsZ ring positioning. J Theor Biol 433:8–20. https://doi.org/10.1016/j.jtbi.2017.08.011

Söderström B, Skoog K, Blom H et al (2014) Disassembly of the divisome in Escherichia coli: evidence that FtsZ dissociates before compartmentalization. Mol Microbiol 92:1–9. https://doi.org/10.1111/mmi.12534

Söderström B, Mirzadeh K, Toddo S et al (2016) Coordinated disassembly of the divisome complex in Escherichia coli. Mol Microbiol 101:425–438. https://doi.org/10.1111/mmi.13400

Spratt BG (1977) Temperature-sensitive cell division mutants of Escherichia coli with Thermolabile penicillin-binding proteins. J Bacteriol 131:293–305

Strauss MP, Liew ATF, Turnbull L et al (2012) 3D-SIM super resolution microscopy reveals a bead-like arrangement for FtsZ and the division machinery: implications for triggering cytokinesis. PLoS Biol 10:e1001389. https://doi.org/10.1371/journal.pbio.1001389

Stricker J, Maddox P, Salmon ED, Erickson HP (2002) Rapid assembly dynamics of the Escherichia coli FtsZ-ring demonstrated by fluorescence recovery after photobleaching. Proc Natl Acad Sci U S A 99:3171–3175. https://doi.org/10.1073/pnas.052595099

Strobel W, Möll A, Kiekebusch D et al (2014) Function and localization dynamics of bifunctional penicillin-binding proteins in Caulobacter crescentus. J Bacteriol 196:1627–1639. https://doi.org/10.1128/JB.01194-13

Sundararajan K, Miguel A, Desmarais SM et al (2015) The bacterial tubulin FtsZ requires its intrinsically disordered linker to direct robust cell wall construction. Nat Commun 6:7281. https://doi.org/10.1038/ncomms8281

Szwedziak P, Wang Q, Freund SM, Löwe J (2012) FtsA forms actin-like protofilaments. EMBO J 31:2249–2260. https://doi.org/10.1038/emboj.2012.76

Szwedziak P, Wang Q, Bharat TAM et al (2014) Architecture of the ring formed by the tubulin homologue FtsZ in bacterial cell division. eLife 3:e04601. https://doi.org/10.7554/eLife.04601

Taschner PE, Ypenburg N, Spratt BG, Woldringh CL (1988) An amino acid substitution in penicillin-binding protein 3 creates pointed polar caps in Escherichia coli. J Bacteriol 170:4828–4837. https://doi.org/10.1128/jb.170.10.4828-4837.1988

Terrana B, Newton A (1975) Pattern of unequal cell division and development in Caulobacter crescentus. Dev Biol 44:380–385. https://doi.org/10.1016/0012-1606(75)90409-1

Thanbichler M, Shapiro L (2006) MipZ, a spatial regulator coordinating chromosome segregation with cell division in Caulobacter. Cell 126:147–162. https://doi.org/10.1016/j.cell.2006.05.038

Tsang M-J, Bernhardt TG (2015) A role for the FtsQLB complex in cytokinetic ring activation revealed by an ftsL allele that accelerates division. Mol Microbiol 95:925–944. https://doi.org/10.1111/mmi.12905

Typas A, Banzhaf M, Gross CA, Vollmer W (2012) From the regulation of peptidoglycan synthesis to bacterial growth and morphology. Nat Rev Microbiol 10:123–136. https://doi.org/10.1038/nrmicro2677

Uehara T, Parzych KR, Dinh T, Bernhardt TG (2010) Daughter cell separation is controlled by cytokinetic ring-activated cell wall hydrolysis. EMBO J 29:1412–1422. https://doi.org/10.1038/emboj.2010.36

Vaughan S, Wickstead B, Gull K, Addinall SG (2004) Molecular evolution of FtsZ protein sequences encoded within the genomes of archaea, bacteria, and Eukaryota. J Mol Evol 58:19–29. https://doi.org/10.1007/s00239-003-2523-5

Wang X, Lutkenhaus J (1993) The FtsZ protein of Bacillus subtilis is localized at the division site and has GTPase activity that is dependent upon FtsZ concentration. Mol Microbiol 9:435–442

Wang SCE, West L, Shapiro L (2006) The bifunctional FtsK protein mediates chromosome partitioning and cell division in Caulobacter. J Bacteriol 188:1497–1508. https://doi.org/10.1128/JB.188.4.1497-1508.2006

White CL, Kitich A, Gober JW (2010) Positioning cell wall synthetic complexes by the bacterial morphogenetic proteins MreB and MreD. Mol Microbiol 76:616–633. https://doi.org/10.1111/j.1365-2958.2010.07108.x

Williams B, Bhat N, Chien P, Shapiro L (2014) ClpXP and ClpAP proteolytic activity on divisome substrates is differentially regulated following the Caulobacter asymmetric cell division. Mol Microbiol 93:853–866. https://doi.org/10.1111/mmi.12698

Wissel MC, Weiss DS (2004) Genetic analysis of the cell division protein FtsI (PBP3): amino acid substitutions that impair septal localization of FtsI and recruitment of FtsN. J Bacteriol 186:490–502

Woldemeskel SA, Goley ED (2017) Shapeshifting to survive: shape determination and regulation in Caulobacter crescentus. Trends Microbiol 25:673–687. https://doi.org/10.1016/j.tim.2017.03.006

Woldemeskel SA, McQuillen R, Hessel AM et al (2017) A conserved coiled-coil protein pair focuses the cytokinetic Z-ring in Caulobacter crescentus. Mol Microbiol 105:721–740. https://doi.org/10.1111/mmi.13731

Wolfe A, Phipps K, Weitao T (2014) Viral and cellular SOS-regulated motor proteins: dsDNA translocation mechanisms with divergent functions. Cell Biosci 4:31. https://doi.org/10.1186/2045-3701-4-31

Xiao J, Goley ED (2016) Redefining the roles of the FtsZ-ring in bacterial cytokinesis. Curr Opin Microbiol 34:90–96. https://doi.org/10.1016/j.mib.2016.08.008

Yakhnina AA, Gitai Z (2013) Diverse functions for six glycosyltransferases in Caulobacter crescentus cell wall assembly. J Bacteriol 195:4527–4535. https://doi.org/10.1128/JB.00600-13

Yang X, Lyu Z, Miguel A et al (2017) GTPase activity–coupled treadmilling of the bacterial tubulin FtsZ organizes septal cell wall synthesis. Science 355:744–747. https://doi.org/10.1126/science.aak9995

Yao Q, Jewett AI, Chang Y-W et al (2017) Short FtsZ filaments can drive asymmetric cell envelope constriction at the onset of bacterial cytokinesis. EMBO J 36:1577–1589. https://doi.org/10.15252/embj.201696235

Yeh Y-C, Comolli LR, Downing KH et al (2010) The Caulobacter Tol-pal complex is essential for outer membrane integrity and the positioning of a polar localization factor. J Bacteriol 192:4847–4858. https://doi.org/10.1128/JB.00607-10

Yu XC, Margolin W (1997) Ca2+−mediated GTP-dependent dynamic assembly of bacterial cell division protein FtsZ into asters and polymer networks in vitro. EMBO J 16:5455–5463. https://doi.org/10.1093/emboj/16.17.5455

Zielińska A, Billini M, Möll A et al (2017) LytM factors affect the recruitment of autolysins to the cell division site in Caulobacter crescentus. Mol Microbiol 106:419–438. https://doi.org/10.1111/mmi.13775

Chromosome Segregation in *Alphaproteobacteria*

Haibi Wang, Anna I. Bowman, and Grant R. Bowman

Abstract

In *Alphaproteobacteria*, the initiation of chromosome replication is closely followed by chromosome segregation. In the early phases of chromosome segregation, the two newly replicated copies of the chromosomal centromere are separated and then directed toward opposite cell poles. Centromere translocation is an energy-dependent process that is carried out by the parABS system, the components of which are widely conserved through bacteriophage and bacterial kingdom, and are adapted for pole-directed chromosome segregation in this clade. The centromeres are lead elements in the process of chromosome segregation, and after they are tethered to the cell poles via a polar scaffolding protein called PopZ, the remaining parts of the chromosome fill in behind, in an ordered procession that is contemporaneous with ongoing DNA replication. The latter phases of segregation are mediated by nucleoid condensation proteins that are structurally and functionally analogous to chromosome organization factors that operate in all kingdoms of life.

H. Wang
Department of Plant Pathology, University of Georgia, Athens, GA, USA

A. I. Bowman
Department of Neurology, University of Colorado Anschutz Medical Campus, Denver, CO, USA

G. R. Bowman (✉)
Department of Molecular Biology, University of Wyoming, Laramie, WY, USA
e-mail: grant.bowman@uwyo.edu

E. Biondi (ed.), *Cell Cycle Regulation and Development in Alphaproteobacteria*,
https://doi.org/10.1007/978-3-030-90621-4_5

1 Overview of Chromosome Segregation

A cell that replicates its genome is also faced with the challenge of segregating the DNA as two equal halves and ensuring that each daughter cell receives a complete copy. In a broad sense, all cells have similar means to accomplish this task. That is, they possess mechanisms for recognizing a defined location within each chromosome, called a centromere, and for directing force on those centromeres such that sister chromosomes are split apart and driven into separate daughter cells. Unlike eukaryotic cells, which separate DNA replication and chromosome segregation into discrete phases, bacteria begin to segregate chromosomal DNA soon after the initiation of DNA replication. This is possible because of the relative proximity of the origin of replication to the chromosome centromere, which is usually separated by just a few thousand bases, or 3% of the total size of the replicon (Livny et al. 2007). Thus, the centromere is present in two copies shortly after replication initiation, and the process of segregating these loci can begin. Subsequent to centromere segregation, the remaining sections of the chromosome follow in temporal and spatial order as the progress of the replication fork continues (Hong and McAdams 2011). The mechanisms for maintaining order during this bulk segregation phase are distinct from those involved in centromere segregation.

This chapter will focus on our knowledge of the molecular mechanisms of chromosome segregation in *Alphaproteobacteria*. An important factor that limits the breadth of this review is that experimental investigations on this subject have thus far been limited to a small number of species in this clade. *Caulobacter crescentus*, which is best known as a model organism for investigations in cell cycle regulation and bacterial cell biology, has been studied intensively with respect to chromosome segregation (Bowman et al. 2010; Easter and Gober 2002; Ebersbach et al. 2006; Figge et al. 2003; Laloux and Jacobs-Wagner 2013; Lim et al. 2014; Ptacin et al. 2010; Schwartz and Shapiro 2011; Shebelut et al. 2010; Surovtsev et al. 2016a; Taylor et al. 2017; Toro et al. 2008; Tran et al. 2017a; Viollier et al. 2004), and *Agrobacterium tumefaciens* (Ehrle et al. 2017; Howell et al. 2017; Kahng and Shapiro 2003), *Sinorhizobium meliloti* (Kahng and Shapiro 2003; Frage et al. 2016), and *Brucella abortus* (De Bolle et al. 2015; Deghelt et al. 2014) have been the subject of additional studies. In all of these species, centromeres that are not undergoing segregation are localized to a cell pole (Figge et al. 2003; Kahng and Shapiro 2003), and bacteria that place their centromeres in this manner are known to package the rest of the chromosome in an organized fashion, with the chromosome arms extending outward toward mid-cell (Viollier et al. 2004; Umbarger et al. 2011). In considering mechanisms for chromosome segregation, it is important to note that the *Caulobacter crescentus* genome consists of only one large circular chromosome, whereas *S. meliloti* and *A. tumefaciens* have two and four separate replicons, respectively. While the genomic sequences of these species suggest that each independent replicon has its own distinct set of chromosome segregation proteins, the extent to which their mechanisms are functionally and/or temporally interrelated is unknown.

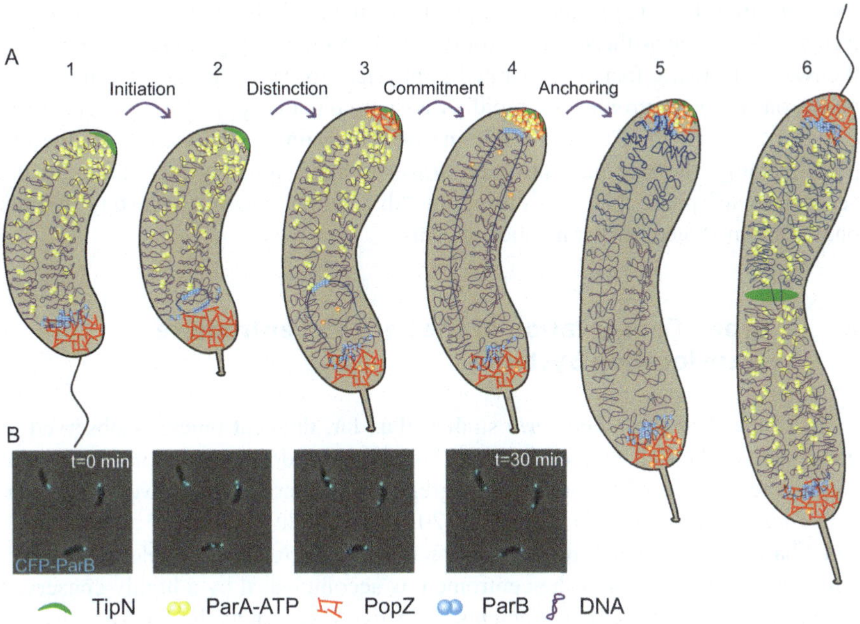

Fig. 1 Chromosome segregation in *Caulobacter crescentus*. (**a**) In new-born flagellated cells (sketch #1), the circular chromosome is arranged such that the origin of replication and nearby *parS* region are located at the flagellar pole, and the two chromosome arms extend across the cell to the opposite pole, where the replication terminus is located. Prior to the initiation of DNA replication, the cell establishes a gradient of DNA-bound ParA-ATP that emanates from the pole opposite the flagellum, which will become the destination for one of the newly replicated centromeres. Replication occurs during the "initiation" phase (sketch #1–#2), which is associated with a developmental transition that replaces the polar flagellum with a stalk and a different set of cell cycle regulatory proteins, and releases the *parS* region from the cell pole. After the *parS* region is duplicated, one of the two centromeres is chosen during the "distinction phase" (sketch #2–#3) as the substrate for rapid translocation to the opposite cell pole. During the subsequent "commitment phase" (sketch #3–#4), the translocating centromere is directed by the concentration gradient of ParA-ATP. The contributions of the polar proteins TipN and PopZ in setting up and maintaining the ParA gradient are discussed in the main text. When the centromere reaches the opposite pole (sketch #4–#5), it is anchored in place through interactions with PopZ. The initiation, distinction, and commitment phases finish long before the completion of S-phase, and the anchoring of the centromere to the pole is a geometric cue that establishes the global arrangement chromosomal DNA as replication continues (sketch #5–#6). (B) Fluorescence images of live *Caulobacter crescentus* cells, which express CFP-tagged ParB (in cyan) as a way of showing the location of the chromosome centromeres. The image panels show centromere localization through the initiation, distinction, and commitment phases. The whole cell cycle is around 90–120 min under these experimental conditions

Close observation and quantitative assessment of time-lapse movies of chromosome segregation in *Caulobacter* suggest that the process occurs in three distinct phases (Shebelut et al. 2010) (Fig. 1): (1) During "initiation," the centromere is detached from the "old" cell pole shortly before it is duplicated by passage of the replication fork. (2) In the subsequent "distinction" phase, one of the two centromeres is chosen as the substrate for the chromosome segregation machinery, and the two centromeric foci are separated in physical space. (3) In the "commitment" phase, one of the centromeres is rapidly transported to the opposite pole (henceforth called the "destination" pole), while the other remains in the vicinity of the old pole. Notably, the rate of centromere travel is approximately 50% faster in the "commitment" phase than it is during the "distinction" phase, suggesting that the phases are mediated by distinct mechanisms.

2 Broad Conservation of the ParABS Centromere Translocation System

In all of the *Alphaproteobacteria* studied thus far, the centromere is observed to travel across the cell and become anchored to the destination pole during the commitment phase of chromosome segregation. Generally, the travel time is on the order of 10–60 min (Ehrle et al. 2017; De Bolle et al. 2015; Thanbichler and Shapiro 2006) and represents a fraction of the total cell cycle. The rapid, directional translocation of the centromere is accomplished by a highly conserved co-functioning set of genetic elements collectively known as the ParABS system, which will be covered in detail in this chapter.

ParABS genes are by no means unique to *Alphaproteobacterial* chromosomes. Close homologs are present in the chromosomes of nearly all bacterial genera (Livny et al. 2007), and a highly homologous set of genetic elements (often called RepABC) is responsible for the partitioning of a broad range of single-copy megaplasmids and related replicons across the phylogenetic spectrum (Austin and Abeles 1983; Castillo-Ramírez et al. 2009; Cevallos et al. 2008; Chai and Winans 2005; Koper et al. 2016; MacLellan et al. 2006; Petersen et al. 2009; Pinto et al. 2012). Interestingly, some or all of the chromosomally encoded ParABS components are lost in some bacterial lineages, including members of the *Alphaproteobacteria* (Livny et al. 2007). This scattered pattern of loss suggests that the ParABS system originated in a common ancestor in ancient evolutionary time, and that alternative mechanisms for chromosome segregation have arisen in multiple lineages at later times in evolution. One of the best characterized examples of chromosome segregation without ParABS occurs in *Escherichia coli*, a member of the *Gammaproteobacteria* clade. The mechanisms associated with this form of chromosome partitioning appear to be more closely related to the segregation of bulk DNA that follows ParABS-dependent segregation in *Caulobacter crescentus*, and are discussed elsewhere (Reyes-Lamothe et al. 2012; Woldringh et al. 2015).

Typical chromosomally encoded ParABS systems include three components. *parA* and *parB* are protein-encoding genes that work together to drive centromere segregation, and are nearly always present as adjoining loci on the chromosome. *parS* refers to the cis-acting DNA component of the ParABS chromosome translocation system.

3 Centromeres Are Defined by *parS* Nucleotide Sequences

In bacteria, the term "centromere" refers to the region of DNA that is acted upon directly by the ParABS chromosome partitioning mechanism, and it is the first part of the chromosome to move toward the destination pole. Centromeres are defined by the presence and distribution of one or more discrete nucleotide sequences called *parS* sites, which can be referred to collectively as a *parS* region. Each individual *parS* site is a 16 nucleotide inverted repeat consensus sequence, which is recognized and bound by a ParB dimer (Livny et al. 2007; Lin and Grossman 1998; Sanchez et al. 2015). Most *parS* regions include multiple *parS* sites (usually less than 10) that are arranged in a loosely defined cluster with spacing that ranges from tens of base pairs to several kb (Livny et al. 2007; Tran et al. 2017a; Jecz et al. 2015). Because DNA replication initiation is often in close temporal association with centromere duplication, *parS* regions are most often found within a few kb of the origin of replication, or 3% of the total size of the replicon (Livny et al. 2007). In *Caulobacter crescentus*, experimental evidence indicates that there are seven *parS* sites in a 10 kb cluster that is centered 8 kb from the origin of replication. *parS* sites with the highest affinity for ParB are situated 5 kb apart in the middle of the cluster, and lower affinity sites that are divergent in sequence by 6–8 base pairs lie at greater distances (Tran et al. 2017b).

Experimental evidence in *Caulobacter crescentus* indicates that the insertion of an additional *parS* site within the *parS* region is well tolerated, but that chromosome segregation defects occur when a *parS* site is inserted outside of this region (Tran et al. 2017b). *parS* insertion sites that are more than 500 kb from the natural *parS* region are not viable. This lethal phenotype has been used as the basis of a screen to identify *parS* sites in *Caulobacter* and other species. In these experiments, suspected sequences of chromosomal DNA are cloned into multicopy plasmids, and those that include the *parS* region are not stably maintained (Toro et al. 2008). Together, these studies suggest that the ParABS mechanism cannot properly segregate DNA if it is confused by the presence of an additional *parS* region at an ectopic locus.

By inserting recombination sites at specific locations in the chromosome, it is possible to invert sections DNA such that the origin of replication and the *parS* region are separated by a longer distance. Under laboratory conditions, *Caulobacter crescentus* maintains viability when this distance is increased from the normal 8 kb to 400 kb (Toro et al. 2008). This recombination has the expected effect of increasing the amount of time between replication initiation and chromosome segregation and of re-orienting the relative locations of chromosomal loci such that the centromeres remain at the poles whereas the origin is located farther toward the mid-cell region. Despite the viability of these recombinant strains, the information obtained from genomic sequencing of wildtype *Alphaproteobacterial* species makes it clear that large distances between replication origin and *parS* region are not favored in evolution.

The benefits that come from holding *parS* regions and origins of replication in proximity are unclear, but may be related to the ability to temporally coordinate replication initiation and chromosome segregation. Separating the chromosomes may be simpler when there is less DNA to separate, and an early start on chromosome segregation might allow more rapid cell division. Additionally, there is evidence that tethering of the centromere to the cell pole brings the replication origin close to polar regulatory proteins that control the timing of replication initiation (Chen et al. 2011; Lasker et al. 2016). Thus, the proximity of these two elements may be a device that physically connects mechanisms for cytoplasmic organization (i.e., polar tethering of centromeres and localization of polar signaling proteins) to those that regulate the timing of replication initiation.

4 ParB Assembles into a Complex Superstructure at the *parS* Region

In all ParABS systems, *parS* sequences are palindromes or near-palindromes that are recognized and bound directly by the protein ParB in its dimeric state. In cytoplasm, nearly all of the ParB is thought to be in dimeric form, and it binds to *parS* sites at relatively high affinity and also to non-specific DNA sequences at somewhat lower affinity (Sanchez et al. 2015; Song et al. 2017; Taylor et al. 2015). The assembly of ParB on bacterial DNA is analogous to the formation of kinetochores on eukaryotic chromosomes, in that these structures define the centromere and act as an interface between DNA and the segregation machinery that moves it. Chromosomally encoded ParB proteins in *Alphaproteobacteria* are essential for cell viability, and are members of a broad ParB sub-type called Type 1A, which is common in several other bacterial clades as well as phage and plasmids (Oliva 2016). This variant of ParB has three domains: a C-terminal dimerization domain, a central helix–turn–helix domain that binds to DNA, and an N-terminal domain that allows association between ParB dimers (Chen et al. 2015) and interaction with ParA (Scholefield et al. 2011).

Visualization of ParB in cells using fluorescence microscopy reveals one, two, or more tightly localized puncta, each corresponding to a chromosomal centromere (Ehrle et al. 2017; Bowman et al. 2008; Gruber and Errington 2009; Iniesta 2014). These observations must somehow be reconciled with the fact that there are far more ParB proteins in the cell [experimentally measured to be about 360 dimers/cell in *Caulobacter* (Lim et al. 2014)] than *parS* binding sites (7 sites, Tran et al. 2017b), and suggests that the centromeres are comprised of higher-order associations among many ParB proteins. A quantitative assessment of this question in a single-copy plasmid-based ParABS system indicates that more than 90% of ParB, or several hundreds of molecules, are localized in partitioning complexes (Sanchez et al. 2015).

Older models suggested that clusters are formed by "spreading," where an initiating ParB dimer binds to a *parS* site, and subsequent binding of other ParB molecules is stabilized through lateral interactions (Murray et al. 2006). Based on newer evidence, current models propose that clusters are held together by ParB's N-terminal domain, which facilitates interactions between ParB dimers bound stably at *parS* sites and other dimers bound non-specifically to random sites on the chromosome, which are often nearby in linear sequence but could be several kb distant (Sanchez et al. 2015; Broedersz et al. 2014). The models are supported by a crystal structure of the ParB–DNA complex, which shows that each N-terminal domain in a ParB dimer can separately form tetrameric complexes with three other ParB dimers (Chen et al. 2015). Further, ParB is able to compact *parS*-containing DNA sequences in vitro, but this activity is blocked by mutations in ParB's N-terminal dimer interaction domain (Song et al. 2017).

Because ParB-DNA networks are formed from a multivalent network of interactions, large centromeric assemblies can form even when the affinities between individual components (whether it is ParB binding non-specifically to DNA or an interaction between ParB dimers) are relatively weak (Fig. 2a). The result, which is known as the "nucleation and caging model," is a dynamic cluster of ParB–DNA complexes (the cage) that centers on the *parS* region (the nucleation site), because the *parS*–ParB interaction is stable relative to other connections within the network (Sanchez et al. 2015). Although single *parS* sites are sufficient for the formation of a ParB cluster and centromere segregation in vivo (Jecz et al. 2015), genomic sequences of large replicons suggest that there are advantages in retaining several *parS* sites, perhaps as a mechanism for strengthening the ParB cluster. As discussed earlier in this chapter, placing an additional *parS* site outside of the cluster causes lethal defects (Toro et al. 2008; Tran et al. 2017b), probably because this creates a multicentric replicon.

5 The ParA ATPase Is Essential for Chromosome Segregation

The ParA component of the ParABS chromosome segregation system is a Walker A-type ATPase protein. In *Alphaproteobacteria*, the *parA* gene is essential for viability (Christen et al. 2011; Curtis and Brun 2014), and expressing ATPase deficient mutants has a dominant negative effect on the translocation of centromeres during chromosome segregation (Toro et al. 2008; Schofield et al. 2010). Chromosomally encoded ParA proteins are highly conserved throughout the bacterial kingdom (even more so than ParB), thus it is likely that all homologs share the same basic set of biochemical activities, including ATPase activity and ParB interaction. Our knowledge of the general mechanism of ParA-ATPase cycling is primarily derived from research in *Bacillus* (Scholefield et al. 2011; Hester and Lutkenhaus 2007; Leonard et al. 2005). ParA dimerizes after binding ATP, and in this state it binds to DNA non-specifically. When ParB interacts with DNA-bound ParA-ATP, the ATP is hydrolyzed, and ParA becomes monomeric and loses its affinity for DNA (Lim et al. 2014; Scholefield et al. 2011).

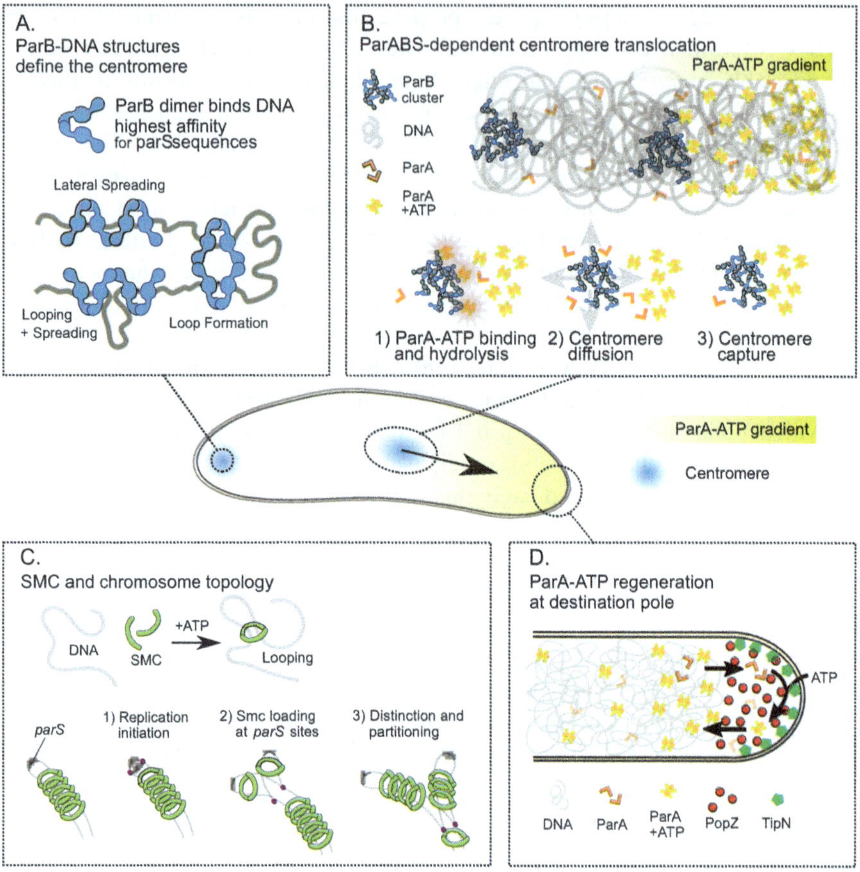

Fig. 2 Molecular mechanisms associated with chromosome segregation. (**a**) ParB-DNA structures define the centromere. ParB dimers bind DNA non-specifically, but bind to centromere-defining *parS* sites at significantly higher affinity. The N-terminal domain of ParB facilitates interactions between ParB dimers, both in cis- and trans-orientations. The combined effects of high affinity *parS* binding, multimerization, and lower affinity binding to non-specific DNA sequences enable ParB to form a dynamic cluster of molecules that cage *parS* and nearby DNA sequences. (**b**) ParABs-dependent centromere translocation. Contact between ParB and DNA-bound ParA-ATP stimulates ATP hydrolysis (step 1), which releases ParA from DNA and breaks the interaction. Upon release (step 2), the ParB cluster will continue to diffuse, though subsequent interactions with other ParA-ATP molecules (step 3) act like a ratchet in biasing net movement up the ParA-ATP concentration gradient. Continued directional translocation of the centromere requires a mechanism for controlling the distribution of ParA-ATP such that it is always highest in the direction of the destination pole. (**c**) ParA-ATP regeneration at destination pole. Monomeric ParA molecules do not bind DNA and diffuse rapidly within the cytoplasm. A current model holds that when ParA molecules encounter proteins that are specifically localized to the destination pole, such as the matrix-forming scaffold protein PopZ or the polar transmembrane protein TipN, they are induced to dimerize and bind ATP. As ParA-ATP molecules diffuse away from the PopZ matrix, they encounter DNA, to which they bind non-specifically. A gradient is formed because the most ParA-ATP molecules encounter and bind DNA soon after exiting the recycling point at the destination pole. (**d**) Smc and chromosome topology. Smc-ATP dimerizes and forms a ring that

6 ParA Dynamics and ATP-Cycling Drive Centromere Translocation

A major question in the field is how ParB-mediated stimulation of ParA-ATP relates to the mechanical forces that drive chromosome segregation. In recent years, studies on ParABS systems that function on low-copy number plasmids (Hu et al. 2017) and on chromosome centromeres in *Caulobacter* (Lim et al. 2014; Surovtsev et al. 2016a) have converged on the same basic mechanism. Interestingly, the energy from ATP hydrolysis does not produce a mechanical force as it does in myosin motors and the like, but is instead used as a symmetry-breaking switch that provides directional guidance to what would otherwise be random motion. This phenomenon is generally described as a "burnt-bridge Brownian ratchet," and is most clearly demonstrated in a reconstituted in vitro system that uses a minimal set of components: *parS*, ParB, and ParA from a plasmid system (Hwang et al. 2013; Vecchiarelli et al. 2013). In these systems, centromeres are created by coating a polystyrene bead with *parS* DNA and ParB. ParA-ATP is then laid across a carpet of non-specific DNA sequence. When the beads are placed on the carpet, they move across the carpet of ParA-ATP. Because contact with ParB stimulates ParA-ATPase activity, the areas where the bead has passed are devoid of ParA (the burnt-bridge component of the model). The consequence of this is that the polystyrene bead, which explores movement in all directions through Brownian motion, fails to get trapped by new ParA–ATP interactions when it diffuses in the retrograde direction, but is briefly held in place when it interacts with a part of the carpet that has ParA-ATP (the ratchet component of the model).

A similar set of mechanochemical interactions are thought to be at play during the segregation of chromosomal centromeres in *Caulobacter* cells, though there are some interesting differences in the proposed mechanisms compared to the in vitro reconstituted systems. ParA-ATP is distributed in a concentration gradient that increases in the direction of the destination pole (Schofield et al. 2010). This may facilitate directional guidance simply because the Brownian ratchet is even more strongly biased toward trapping the centromere when it happens to move toward an area of relatively high ParA-ATP concentration. It has also been proposed that the non-uniform distribution of ParA-ATP is a mechanism for providing a mechanical force that draws the centromere up the concentration gradient, as described by the "DNA-relay" model (Lim et al. 2014; Surovtsev et al. 2016a). According to this model, the elastic nature of DNA implies that some of the contacts between ParB

Fig. 2 (continued) encloses two strands of DNA. In *Caulobacter* and other *Alphaproteobacteria*, this process is thought to facilitate the early stages of centromere partitioning. Shortly after the initiation of DNA replication (step 1), the replisomes (colored purple) duplicate the *parS* site. The *parS* sites on the newly replicated DNA are loading sites for new Smc loops (step 2). As replication continues, more Smc loops are added at each *parS* site (step 3), and the complexes of Smc loops hold the growing chromosome arms in topologically separated structures

and ParA-ATP are stretched out of equilibrium. As the tension is relaxed (and before ParA-ATP hydrolysis), the centromere is pulled in the direction of the most ParA-ATP contacts, which is up the concentration gradient. Because tethers can be formed in any direction, the Brownian nature of the system remains, and it is the energy put into the creation of the ParA-ATP concentration gradient that imparts directionality to the movement of the chromosome centromere. The energy boost that comes from DNA-relay type forces may be important for achieving a sufficient rate of travel as the centromere moves through the densely crowded environment (Le Gall et al. 2016) of the bacterial nucleoid.

7 Chromosome Segregation Is Under Control of Polar Guidance Cues

The models that consider the segregation of low-copy plasmids suggest that the nucleoid and ParABS components of the DNA-relay model are sufficient to explain all aspects of their movement within cells, including the oscillatory movement of single ParB foci between poles and the separation of newly replicated centromeres (Hu et al. 2017; Surovtsev et al. 2016b; Ebersbach and Gerdes 2004; Ringgaard et al. 2009). However, it is important to note that the chromosomal centromeres in *Alphaproteobacteria* (and also in other clades) display a significantly different pattern of movement. In *Caulobacter, Brucella, and Agrobacterium*, the centromere is located at the old cell when present as a single copy, and after chromosome replication, one of the centromeres remains in place while the other travels across the cell and becomes stably localized to the opposite pole (Fig. 1) (Ehrle et al. 2017; Deghelt et al. 2014; Thanbichler and Shapiro 2006). By contrast, the *parS* regions of low-copy plasmids are rarely found at the cell poles, even when present in multiple copies (Ebersbach and Gerdes 2004; Derman et al. 2008). This discrepancy suggests that the localization and movement of chromosomal centromeres are being subject to additional guidance cues outside of ParABS.

In *Caulobacter*, chromosome segregation is affected by two polar guidance cues, named PopZ and TipN. PopZ is localized to the new pole and the old pole. In vitro, it is capable of interacting with both ParA and ParB, and in *popZ* knockout cells the chromosomal centromeres drift in the cytoplasm instead of becoming anchored to cell poles (Bowman et al. 2008; Ebersbach et al. 2008; Ptacin et al. 2014). The centromere localization defect has also been observed in *Agrobacterium* (Ehrle et al. 2017). Together, these observations suggest that PopZ's role is to tether the centromeres to the cell poles. Direct interaction between centromeric ParB and polar PopZ may account for a part of this mechanism. The significance of PopZ's interaction with ParA is less clear, even though experiments in which wildtype *popZ* was replaced with protein-specific binding mutants suggest that the ParA–PopZ interaction is more important for chromosome segregation and polar localization than the ParB–PopZ interaction in vivo (Ptacin et al. 2014).

One compelling model for PopZ as a polar guidance factor is that it provides a specific location for regenerating ParA-ATP (Fig. 2d). In this model, ParA molecules that have fallen off of DNA in the wake of the moving centromere diffuse as monomers in cytoplasm until coming into contact with PopZ at the cell pole. The interaction between ParA monomers and polar PopZ may stimulate ATP binding and dimerization of ParA, or simply provide a place for ParA accumulation as ParA-ATP is regenerated at an intrinsic rate. As the regenerated population of ParA-ATP molecules diffuse out of the polar zone they will bind to DNA, thereby creating a gradient of DNA-bound ParA-ATP molecules that emanates from the pole.

This model is supported by time-lapse fluorescence microscope experiments, wherein the localizations of ParB/centromeres and ParA have been directly observed in live cells by expressing genetically encoded fluorescent proteins in fusion with ParA and ParB (as summarized in Fig. 1 and 2b). Similar observations have been made in *Caulobacter crescentus* (Schofield et al. 2010) and the *Gammaproteobacteria Vibrio cholerae* (Fogel and Waldor 2006). In each study, the centromeres are observed to move up a visible concentration gradient of nucleoid-bound ParA that emanates from the destination pole. As the centromeres move across the nucleoid, they are observed to leave an area of no ParA in their wake, while the concentration of ParA emanating from the destination pole becomes progressively higher.

The *parS* centromeres on low-copy number plasmids have also been observed to follow ParA concentration gradients across the nucleoid, with the notable difference that the centromeres do not move all of the way out to the cell poles, and the concentration of ParA re-forms in the area between the segregated centromeres (Ringgaard et al. 2009). Thus, it would appear that the chromosomal ParABS systems use the cell poles as locations for regenerating ParA-ATP, whereas this occurs spontaneously in the case of plasmid-based systems, and this distinction explains why these different systems have different patterns of centromere movement.

The other polar guidance cue that has been identified in *Caulobacter* is TipN, which is localized exclusively at the destination pole (Huitema et al. 2006; Lam et al. 2006; Yeh et al. 2010). TipN is a large transmembrane protein that interacts directly with ParA (Ptacin et al. 2010; Lam et al. 2006). In the absence of TipN, time-lapse fluorescence microscopy experiments show that the concentration of ParA at the destination pole is far less robust than in wildtype cells. Instead, ParA often accumulates behind the traveling ParB focus. This has the effect of increasing the frequency at which the traveling centromere reverses direction, thereby slowing the overall rate of chromosome segregation and frequently resulting in incomplete segregation (Schofield et al. 2010).

Based on these observations, it appears that TipN and PopZ play overlapping roles in directing chromosome segregation, and this view is further supported by evidence of genetic interactions between *tipN* and *popZ* in *Caulobacter* (Schofield et al. 2010). First, the centromere localization defect observed in a *popZ* knockout strain can be rescued by overproducing TipN, suggesting the overabundance of one polar cue can compensate for the loss of the other. Second, the combination of defects in *tipN* and *popZ* single mutants becomes inviable when they are combined as a *tipN popZ* double mutant, suggesting a lethal defect in chromosome segregation when both polar cues are absent. The lethal phenotype is consistent with the fact that *parA* and *parB* are both essential genes in *Caulobacter* (Christen et al. 2011; Mohl et al. 2001). Surprisingly, TipN is not as well conserved through *Alphaproteobacteria* as PopZ, suggesting that these two proteins are not co-evolving, and that TipN may only be important in special contexts. In *Caulobacter*, PopZ is localized to both cell poles, and the focus of TipN at the destination pole may help to provide an unambiguous directional cue. In contrast, PopZ is found only at the destination pole during centromere translocation in *Agrobacterium* and *Brucella* (Ehrle et al. 2017; Deghelt et al. 2014), and these species do not carry a *tipN* homolog, perhaps because an additional directional cue is not necessary.

8 Segregation Activities Outside of ParABS

Despite the broad conservation of the ParABS mechanism across the prokaryotic kingdom (Livny et al. 2007) end even into archaea (Schumacher et al. 2015), these components are not essential for successful chromosome partitioning or survival in many bacterial species. In the *Gammaproteobacteria Vibrio cholerae*, for example, deletions in either *parA* or *parB* produce viable strains. In these mutants, the chromosomes are correctly partitioned between daughter cells even though the centromeres fail to complete the journey to their normal anchoring point at the destination pole (Fogel and Waldor 2006; Kadoya et al. 2011). In Gram positive *Bacillus subtilis*, deletion of the *parA* homolog (called *soj*) has no detectable effect on chromosome partitioning, whereas deletion of the *parB* homolog (*spo0J*) results in a 100-fold increase in the production of anucleate cells (Ireton et al. 1994). Notably, *E. coli* and related *Enterobacteriales* species manage chromosome segregation entirely without the ParABS system. Together, these and other examples suggest that the bacterial kingdom includes additional mechanisms for chromosome partitioning that operate along with or in place of ParABS.

An increasing body of literature is revealing ParABS-independent chromosome partitioning activities in *Caulobacter*. Part of the evidence for these mechanisms has come from quantitative assessment of chromosome segregation, which suggests that segregation occurs in multiple distinct phases (Shebelut et al. 2010) (Fig. 1). Notably, the rate of centromere travel is approximately 50% faster in the ParABS-mediated commitment phase than it is during the distinction phase, and neither centromere detachment from the old pole nor the separation of centromeres is blocked by expression of a dominant-negative ParA mutant protein that fully inhibits the commitment step (Shebelut et al. 2010). This suggests that the early phases of chromosome partitioning occur by ParABS-independent mechanisms, but these are not well understood. It is possible that the short-distance movements of the centromeres at early stages occur by the release of tension within the nucleoid, as may occur after breaking the PopZ-dependent polar anchor, and by the creation of new tension during the replication of DNA. In both cases, the source of the tension could be proteins that mediate the compaction of bulk DNA.

9 SMC/Condensin and Topologically Associated Domains

SMC is a DNA-associated protein that can mediate DNA compaction and may play a role in the early phases of chromosome partitioning. SMC proteins (also known as condensins) have been a subject of significant interest in bacterial cell biology because they are found in nearly all prokaryotes, and are strikingly homologous in structure and function to a broad class of eukaryotic proteins with the same name. Eukaryotic condensins organize chromatin by binding ATP and subsequently forming dimers that hold DNA strands together in a closed loop (Haering et al. 2002).

The strongest evidence that SMC is involved in bacterial chromosome segregation comes from *smc* mutant phenotypes in species outside of *Alphaproteobacteria*. In *B. subtilis*, for example, *Δsmc* strains produce a large fraction of anucleate cells, indicating failure in chromosome segregation (Gruber et al. 2014; Wang et al. 2014). *B. subtilis* SMC is loaded onto origin-proximal regions of the chromosome by a ParB-dependent process (Gruber and Errington 2009). Recent chromosome capture studies have compared chromosomal architecture in wild type and *smc* mutants, and found that the contacts between chromosome arms are fewer in number and more widely distributed in *smc* mutants, suggesting that the arms are misaligned when SMC is not loaded onto the chromosome (Wang et al. 2015). Restoring SMC loading restores the alignment of chromosome arms, initially at the *parS* site where SMC is loaded and later through the rest of the chromosome as the area of SMC coverage spreads from its loading site (Wang et al. 2017). A model that describes the role of SMC in chromosome partitioning, called "loop extrusion," holds that SMC is loaded onto DNA at *parS* regions shortly after the passage of the replisome, and that the collective looping and DNA condensation activities of multiple SMC molecules on each of the newly replicated DNA segments result in the formation of two separate topologically associated domains (TADs). Progressive addition of SMC molecules on the two separating *parS* sites drives further extrusion of the TADs, thus driving the initial phase of chromosome partitioning (Wang et al. 2014).

E. coli also requires a nucleoid-associated SMC-like ATPase called MukB for chromosome partitioning. *mukB* deletion mutants are defective not only in chromosome segregation but also in DNA compaction (Danilova et al. 2007; Lioy et al. 2018), which is consistent with the idea that an SMC-like protein also establishes chromosomal TAD's in this organism. Unlike SMC, MukB will load onto DNA non-specifically, and instead of being loaded onto DNA at *parS* sites, it is restricted to the area around the origin of replication(Danilova et al. 2007) by the exclusionary activity of a different TAD that forms at the terminal region (Nolivos et al. 2016). Instead of segregating newly replicated DNA by loop extrusion, the TAD's established by MukB may drive partitioning through entropy-driven demixing (discussed below).

In *Caulobacter*, SMC is loaded onto the chromosome at *parS* sites through interaction with ParB, where it holds the chromosome arms in close proximity (Tran et al. 2017a) (Fig. 2c). Consistent with this, chromosome arms are misaligned in

Δsmc mutants (Le et al. 2013), suggesting that *Caulobacter* SMC forms a TAD in the region of the chromosomal centromere, much as it does in *B. subtilis*. Currently, the most direct evidence that SMC affects chromosome partitioning in *Caulobacter* is that the expression of a dominant-negative ATPase deficient form of SMC causes replicated centromeres to be held in close proximity (Schwartz and Shapiro 2011). Presumably, either the distinction or the commitment step in centromere segregation is blocked because DNA strands are linked by SMC loops that are too numerous or too strong to allow partitioning to move forward. However, efforts to gain a deeper understanding of SMC's role have been hampered by the lack of an obvious chromosome partitioning phenotype in *Δsmc* knockout strains, which suggests that the TAD formed by SMC is not necessary for segregation in *Caulobacter*. Instead, the *parABS* elements, which are essential in *Caulobacter* but not in *B. subtilis*, appear to be the dominant players in chromosome segregation in this species.

10 Potential Roles for Other Nucleoid-Associated Proteins

SMC is just one of several nucleoid-associated proteins that may be involved in establishing TADs or otherwise participating in chromosome segregation in *Alphaproteobacteria*, but at the current time, the roles of these other factors are even less clear. The *Caulobacter* genome includes two genes that encode the histone-like nucleoid-associated protein HU, but the double knockout strain exhibits only a minor degree of chromosome de-compaction (Le et al. 2013) and has not been reported to affect chromosome segregation. Three recent publications have described the function of another nucleoid-associated protein called GapR (Taylor et al. 2017; Arias-Cartin et al. 2017; Ricci et al. 2016). GapR is highly conserved among *Alphaproteobacteria* and is essential in *Caulobacter* under normal growth conditions. A significant fraction of cells in *gapR* mutant cultures are anucleate or otherwise lacking DNA in polar regions, suggesting defects in DNA compaction or chromosome partitioning (Arias-Cartin et al. 2017). However, the phenotype of *gapR* mutants is highly pleiotropic, and also includes delayed chromosome replication and differences in gene expression across the chromosome, which has made it difficult to assess the direct role of GapR in chromosome segregation. Quantitative fluorescence microscopy has shown that the passage of the replisome displaces GapR from DNA, leaving a broad section of GapA-free nucleoid in its wake (Arias-Cartin et al. 2017), in a manner that is analogous to the clearance of ParA after passage of the centromeres. Although it is tempting to speculate that GapA is associated with a replication-dependent DNA compaction activity that facilitates chromosome segregation, current evidence is not sufficient to support this notion.

11 Partitioning Origin-Distal DNA

In *Caulobacter*, as well as in other species, the origin of replication or the closely associated *parS* region is the first part of the chromosome to be partitioned during chromosome segregation. How is the rest of the chromosome partitioned after the ParABS-dependent movement has completed? Theoretical simulations predict that this aspect of chromosome partitioning is an entropy-driven process, in which the demixing of the intertwined DNA polymers is the lowest energy outcome (Jun and Mulder 2006). While complete partitioning is not spontaneous and probably requires energy input (Minina and Arnold 2014), there is disagreement over where energy is added to the system. Modeling studies that are based on the *E. coli* system suggest that differential compaction of TADs leads to a non-uniform distribution of forces within the nucleoid that drives partitioning (Junier et al. 2014). Alternatively, another model suggests that supercoiling imposes a certain plectonemic structure on chromosomal DNA, and that in vivo crowding prevents the lowest energy conformation. Here, initial centromere displacement by ParABS, in conjunction with formation of relatively open, nucleoid-free space due to cell elongation and local DNA compaction in the vicinity of the replisome, provides sufficient entropic force for the partitioning of the chromosome toward the destination pole (Hong et al. 2013). An interesting demonstration of partitioning forces in the origin-distal parts of the chromosome comes from the observation of double-strand break repair in *Caulobacter* cells (Badrinarayanan et al. 2015). Here, a single double-strand break causes the surrounding ~300 kb section of the chromosome to become highly mobile within the cytoplasm, and after it pairs with its homologous partner for recombination-repair, the traveling section of DNA rapidly returns to its normal position in the cell, via a ParA-independent mechanism that presumably involves the same entropic forces that are at play during chromosome segregation.

12 Concluding Remark

In this chapter, we discuss the ParABS system as the central mechanism for generating the force to drive chromosome segregation in *Alphaproteobacteria*. Although the vast majority of information on chromosome segregation in this clade has come from studies that focused exclusively on *Caulobacter crescentus*, the high degree of conservation among ParABS and the limited number of studies that have considered chromosome in other *Alphaproteobacteria* species suggest that, for the most part, the lessons from *Caulobacter* can be broadly applied. There are, however, some notable exceptions to this idea. Some *Alphaproteobacteria* lack one or more ParABS components, and the mechanisms that compensate for such deficiencies are not known. Similarly, the polar factors that facilitate ParABS function in *Caulobacter* are not as well conserved as ParABS itself, and the question of how chromosome segregation is supported in the absence of these factors has not been explored. Another important difference between *Caulobacter* and many other *Alphaproteobacteria* species is the number of chromosomes being segregated. For those that harbor two or more replicons, each usually carries its own complete ParABS mechanism. It will be interesting to learn how these species' physiology has changed to accommodate these additions.

Owing to the fact that ParABS systems have been characterized in detail in a wide range of systems across the bacterial kingdom, much is known about this core mechanism. However, there are gaps in our understanding of the interfacing between core ParABS operations and additional factors, which seem to have evolved in a clade-specific manner. Consequently, current models are incomplete. For example, the establishment and maintenance of a cytoplasmic gradient of DNA-bound ParA would seem to require a pole-localized regenerator of ParA-ATP, but the molecular mechanism is not known. Additionally, if there is a pole-directed directional cue, the mechanism by which this machinery reliably chooses one, but not both centromeres has not been adequately described. In a broader sense, the fact that segregation is limited to one centromere and one time per cell cycle suggests that the system is also controlled by negative regulation. However, this and other intriguing possibilities await further investigation.

References

Arias-Cartin R et al (2017) Replication fork passage drives asymmetric dynamics of a critical nucleoid-associated protein in Caulobacter. EMBO J 36:301–318

Austin S, Abeles A (1983) Partition of unit-copy miniplasmids to daughter cells. I. P1 and F miniplasmids contain discrete, interchangeable sequences sufficient to promote equipartition. J Mol Biol 169:353–372

Badrinarayanan A, Le TBK, Laub MT (2015) Rapid pairing and resegregation of distant homologous loci enables double-strand break repair in bacteria. J Cell Biol 210:385–400

Bowman GR et al (2008) A polymeric protein anchors the chromosomal origin/ParB complex at a bacterial cell pole. Cell 134:945–955

Bowman GR et al (2010) Caulobacter PopZ forms a polar subdomain dictating sequential changes in pole composition and function. Mol Microbiol 76:173–189

Broedersz CP et al (2014) Condensation and localization of the partitioning protein ParB on the bacterial chromosome. Proc Natl Acad Sci U S A 111:8809–8814

Castillo-Ramírez S, Vázquez-Castellanos JF, González V, Cevallos MA (2009) Horizontal gene transfer and diverse functional constrains within a common replication-partitioning system in Alphaproteobacteria: the repABC operon. BMC Genomics 10:536

Cevallos MA, Cervantes-Rivera R, Gutiérrez-Ríos RM (2008) The repABC plasmid family. Plasmid 60:19–37

Chai Y, Winans SC (2005) RepB protein of an agrobacterium tumefaciens Ti plasmid binds to two adjacent sites between repA and repB for plasmid partitioning and autorepression. Mol Microbiol 58:1114–1129

Chen YE et al (2011) Spatial gradient of protein phosphorylation underlies replicative asymmetry in a bacterium. Proc Natl Acad Sci U S A 108:1052–1057

Chen B-W, Lin M-H, Chu C-H, Hsu C-E, Sun Y-J (2015) Insights into ParB spreading from the complex structure of Spo0J and parS. Proc Natl Acad Sci U S A 112:6613–6618

Christen B et al (2011) The essential genome of a bacterium. Mol Syst Biol 7:528

Curtis PD, Brun YV (2014) Identification of essential alphaproteobacterial genes reveals operational variability in conserved developmental and cell cycle systems. Mol Microbiol 93:713–735

Danilova O, Reyes-Lamothe R, Pinskaya M, Sherratt D, Possoz C (2007) MukB colocalizes with the oriC region and is required for organization of the two Escherichia coli chromosome arms into separate cell halves. Mol Microbiol 65:1485–1492
De Bolle X, Crosson S, Matroule J-Y, Letesson J-J (2015) Brucella abortus cell cycle and infection are coordinated. Trends Microbiol 23:812–821
Deghelt M et al (2014) G1-arrested newborn cells are the predominant infectious form of the pathogen Brucella abortus. Nat Commun 5:4366
Derman AI, Lim-Fong G, Pogliano J (2008) Intracellular mobility of plasmid DNA is limited by the ParA family of partitioning systems. Mol Microbiol 67:935–946
Easter J, Gober JW (2002) ParB-stimulated nucleotide exchange regulates a switch in functionally distinct ParA activities. Mol Cell 10:427–434
Ebersbach G, Gerdes K (2004) Bacterial mitosis: partitioning protein ParA oscillates in spiral-shaped structures and positions plasmids at mid-cell. Mol Microbiol 52:385–398
Ebersbach G et al (2006) Regular cellular distribution of plasmids by oscillating and filament-forming ParA ATPase of plasmid pB171. Mol Microbiol 61:1428–1442
Ebersbach G, Briegel A, Jensen GJ, Jacobs-Wagner C (2008) A self-associating protein critical for chromosome attachment, division, and polar organization in caulobacter. Cell 134:956–968
Ehrle HM et al (2017) Polar organizing protein PopZ is required for chromosome segregation in agrobacterium tumefaciens. J Bacteriol 199
Figge RM, Easter J, Gober JW (2003) Productive interaction between the chromosome partitioning proteins, ParA and ParB, is required for the progression of the cell cycle in Caulobacter crescentus. Mol Microbiol 47:1225–1237
Fogel MA, Waldor MK (2006) A dynamic, mitotic-like mechanism for bacterial chromosome segregation. Genes Dev 20:3269–3282
Frage B et al (2016) Spatiotemporal choreography of chromosome and megaplasmids in the Sinorhizobium meliloti cell cycle. Mol Microbiol 100:808–823
Gruber S, Errington J (2009) Recruitment of condensin to replication origin regions by ParB/SpoOJ promotes chromosome segregation in B. subtilis. Cell 137:685–696
Gruber S et al (2014) Interlinked sister chromosomes arise in the absence of condensin during fast replication in B. subtilis. Curr Biol 24:293–298
Haering CH, Löwe J, Hochwagen A, Nasmyth K (2002) Molecular architecture of SMC proteins and the yeast cohesin complex. Mol Cell 9:773–788
Hester CM, Lutkenhaus J (2007) Soj (ParA) DNA binding is mediated by conserved arginines and is essential for plasmid segregation. Proc Natl Acad Sci U S A 104:20326–20331
Hong S-H, McAdams HH (2011) Compaction and transport properties of newly replicated Caulobacter crescentus DNA. Mol Microbiol 82:1349–1358
Hong S-H et al (2013) Caulobacter chromosome in vivo configuration matches model predictions for a supercoiled polymer in a cell-like confinement. Proc Natl Acad Sci U S A 110:1674–1679
Howell M et al (2017) Absence of the polar organizing protein PopZ results in reduced and asymmetric cell division in agrobacterium tumefaciens. J Bacteriol 199
Hu L, Vecchiarelli AG, Mizuuchi K, Neuman KC, Liu J (2017) Brownian ratchet mechanism for faithful segregation of low-copy-number plasmids. Biophys J 112:1489–1502
Huitema E, Pritchard S, Matteson D, Radhakrishnan SK, Viollier PH (2006) Bacterial birth scar proteins mark future flagellum assembly site. Cell 124:1025–1037
Hwang LC et al (2013) ParA-mediated plasmid partition driven by protein pattern self-organization. EMBO J 32:1238–1249
Iniesta AA (2014) ParABS system in chromosome partitioning in the bacterium Myxococcus xanthus. PLoS One 9:e86897
Ireton K, Gunther NW, Grossman AD (1994) spo0J is required for normal chromosome segregation as well as the initiation of sporulation in Bacillus subtilis. J Bacteriol 176:5320–5329
Jecz P, Bartosik AA, Glabski K, Jagura-Burdzy G (2015) A single parS sequence from the cluster of four sites closest to oriC is necessary and sufficient for proper chromosome segregation in Pseudomonas aeruginosa. PLoS One 10:e0120867

Jun S, Mulder B (2006) Entropy-driven spatial organization of highly confined polymers: lessons for the bacterial chromosome. Proc Natl Acad Sci U S A 103:12388–12393

Junier I, Boccard F, Espéli O (2014) Polymer modeling of the E coli genome reveals the involvement of locus positioning and macrodomain structuring for the control of chromosome conformation and segregation. *Nucleic Acids Res* 42:1461–1473

Kadoya R, Baek JH, Sarker A, Chattoraj DK (2011) Participation of chromosome segregation protein ParAI of Vibrio cholerae in chromosome replication. J Bacteriol 193:1504–1514

Kahng LS, Shapiro L (2003) Polar localization of replicon origins in the multipartite genomes of agrobacterium tumefaciens and Sinorhizobium meliloti. J Bacteriol 185:3384–3391

Koper P, Żebracki K, Marczak M, Skorupska A, Mazur A (2016) RepB proteins of the multipartite Rhizobium leguminosarum bv trifolii genome discriminate between centromere-like parS sequences for plasmid segregational stability. *Mol Microbiol* 102:446–466

Laloux G, Jacobs-Wagner C (2013) Spatiotemporal control of PopZ localization through cell cycle-coupled multimerization. J Cell Biol 201:827–841

Lam H, Schofield WB, Jacobs-Wagner C (2006) A landmark protein essential for establishing and perpetuating the polarity of a bacterial cell. Cell 124:1011–1023

Lasker K, Mann TH, Shapiro L (2016) An intracellular compass spatially coordinates cell cycle modules in Caulobacter crescentus. Curr Opin Microbiol 33:131–139

Le Gall A et al (2016) Bacterial partition complexes segregate within the volume of the nucleoid. Nat Commun 7:12107

Le TBK, Imakaev MV, Mirny LA, Laub MT (2013) High-resolution mapping of the spatial organization of a bacterial chromosome. Science 342:731–734

Leonard TA, Butler PJ, Löwe J (2005) Bacterial chromosome segregation: structure and DNA binding of the Soj dimer–a conserved biological switch. EMBO J 24:270–282

Lim HC et al (2014) Evidence for a DNA-relay mechanism in ParABS-mediated chromosome segregation. elife 3:e02758

Lin DC, Grossman AD (1998) Identification and characterization of a bacterial chromosome partitioning site. Cell 92:675–685

Lioy VS et al (2018) Multiscale structuring of the E. coli chromosome by nucleoid-associated and condensin proteins. Cell 172:771–783.e18

Livny J, Yamaichi Y, Waldor MK (2007) Distribution of centromere-like parS sites in bacteria: insights from comparative genomics. J Bacteriol 189:8693–8703

MacLellan SR, Zaheer R, Sartor AL, MacLean AM, Finan TM (2006) Identification of a megaplasmid centromere reveals genetic structural diversity within the repABC family of basic replicons. Mol Microbiol 59:1559–1575

Minina E, Arnold A (2014) Induction of entropic segregation: the first step is the hardest. Soft Matter 10:5836–5841

Mohl DA, Easter J, Gober JW (2001) The chromosome partitioning protein, ParB, is required for cytokinesis in Caulobacter crescentus. Mol Microbiol 42:741–755

Murray H, Ferreira H, Errington J (2006) The bacterial chromosome segregation protein Spo0J spreads along DNA from parS nucleation sites. Mol Microbiol 61:1352–1361

Nolivos S et al (2016) MatP regulates the coordinated action of topoisomerase IV and MukBEF in chromosome segregation. Nat Commun 7:10466

Oliva MA (2016) Segrosome complex formation during DNA trafficking in bacterial cell division. Front Mol Biosci 3:51

Petersen J, Brinkmann H, Pradella S (2009) Diversity and evolution of repABC type plasmids in Rhodobacterales. Environ Microbiol 11:2627–2638

Pinto UM, Pappas KM, Winans SC (2012) The ABCs of plasmid replication and segregation. Nat Rev Microbiol 10:755–765

Ptacin JL et al (2010) A spindle-like apparatus guides bacterial chromosome segregation. Nat Cell Biol 12:791–798

Ptacin JL et al (2014) Bacterial scaffold directs pole-specific centromere segregation. Proc Natl Acad Sci U S A 111:E2046–E2055

Reyes-Lamothe R, Nicolas E, Sherratt DJ (2012) Chromosome replication and segregation in bacteria. Annu Rev Genet 46:121–143
Ricci DP et al (2016) Cell cycle progression in Caulobacter requires a nucleoid-associated protein with high AT sequence recognition. Proc Natl Acad Sci U S A 113:E5952–E5961
Ringgaard S, van Zon J, Howard M, Gerdes K (2009) Movement and equipositioning of plasmids by ParA filament disassembly. Proc Natl Acad Sci U S A 106:19369–19374
Sanchez A et al (2015) Stochastic self-assembly of ParB proteins builds the bacterial DNA segregation apparatus. Cell Syst 1:163–173
Schofield WB, Lim HC, Jacobs-Wagner C (2010) Cell cycle coordination and regulation of bacterial chromosome segregation dynamics by polarly localized proteins. EMBO J 29:3068–3081
Scholefield G, Whiting R, Errington J, Murray H (2011) Spo0J regulates the oligomeric state of Soj to trigger its switch from an activator to an inhibitor of DNA replication initiation. Mol Microbiol 79:1089–1100
Schumacher MA et al (2015) Structures of archaeal DNA segregation machinery reveal bacterial and eukaryotic linkages. Science 349:1120–1124
Schwartz MA, Shapiro L (2011) An SMC ATPase mutant disrupts chromosome segregation in Caulobacter. Mol Microbiol 82:1359–1374
Shebelut CW, Guberman JM, van Teeffelen S, Yakhnina AA, Gitai Z (2010) Caulobacter chromosome segregation is an ordered multistep process. Proc Natl Acad Sci U S A 107:14194–14198
Song D, Rodrigues K, Graham TGW, Loparo JJ (2017) A network of cis and trans interactions is required for ParB spreading. Nucleic Acids Res 45:7106–7117
Surovtsev IV, Lim HC, Jacobs-Wagner C (2016a) The slow mobility of the ParA partitioning protein underlies its steady-state patterning in Caulobacter. Biophys J 110:2790–2799
Surovtsev IV, Campos M, Jacobs-Wagner C (2016b) DNA-relay mechanism is sufficient to explain ParA-dependent intracellular transport and patterning of single and multiple cargos. Proc Natl Acad Sci U S A 113:E7268–E7276
Taylor JA et al (2015) Specific and non-specific interactions of ParB with DNA: implications for chromosome segregation. Nucleic Acids Res 43:719–731
Taylor JA, Panis G, Viollier PH, Marczynski GT (2017) A novel nucleoid-associated protein coordinates chromosome replication and chromosome partition. Nucleic Acids Res 45:8916–8929
Thanbichler M, Shapiro L (2006) MipZ, a spatial regulator coordinating chromosome segregation with cell division in Caulobacter. Cell 126:147–162
Toro E, Hong S-H, McAdams HH, Shapiro L (2008) Caulobacter requires a dedicated mechanism to initiate chromosome segregation. Proc Natl Acad Sci U S A 105:15435–15440
Tran NT, Laub MT, Le TBK (2017a) SMC progressively aligns chromosomal arms in Caulobacter crescentus but is antagonized by convergent transcription. Cell Rep 20:2057–2071
Tran NT et al (2017b) Permissive zones for the centromere-binding protein ParB on the Caulobacter crescentus chromosome. Nucleic Acids Res. https://doi.org/10.1093/nar/gkx1192
Umbarger MA et al (2011) The three-dimensional architecture of a bacterial genome and its alteration by genetic perturbation. Mol Cell 44:252–264
Vecchiarelli AG, Hwang LC, Mizuuchi K (2013) Cell-free study of F plasmid partition provides evidence for cargo transport by a diffusion-ratchet mechanism. Proc Natl Acad Sci U S A 110:E1390–E1397
Viollier PH et al (2004) Rapid and sequential movement of individual chromosomal loci to specific subcellular locations during bacterial DNA replication. Proc Natl Acad Sci U S A 101:9257–9262
Wang X, Tang OW, Riley EP, Rudner DZ (2014) The SMC condensin complex is required for origin segregation in Bacillus subtilis. Curr Biol 24:287–292
Wang X et al (2015) Condensin promotes the juxtaposition of DNA flanking its loading site in Bacillus subtilis. Genes Dev 29:1661–1675

Wang X, Brandão HB, Le TBK, Laub MT, Rudner DZ (2017) Bacillus subtilis SMC complexes juxtapose chromosome arms as they travel from origin to terminus. Science 355:524–527
Woldringh CL, Hansen FG, Vischer NOE, Atlung T (2015) Segregation of chromosome arms in growing and non-growing Escherichia coli cells. Front Microbiol 6:448
Yeh Y-C, Comolli LR, Downing KH, Shapiro L, McAdams HH (2010) The caulobacter Tol-pal complex is essential for outer membrane integrity and the positioning of a polar localization factor. J Bacteriol 192:4847–4858

Modeling the Cell Cycle of *Caulobacter crescentus*

Ralf Blossey

Abstract

Given the accumulated detailed knowledge about the molecular regulators in the cell cycle of *Caulobacter crescentus*, the latter has become a topic of several attempts of mathematical modeling, following earlier such work on the eukaryotic cell cycle. In this review, I highlight some particular aspects of modeling attempts of the *Caulobacter crescentus* cell cycle. Firstly, I briefly discuss the use of ordinary differential equations and the methods of nonlinear dynamics in such models. Secondly, I introduce and discuss three examples of 'minimal' or 'small' models of cell cycle aspects, one of them in some detail. Finally, I comment on the opposite case of 'large' models, *i.e.* models which try to give a more global view of cell cycle regulation. The aim of this text is to convey what and how one can learn from modeling approaches about the workings of the cell cycle that experiment alone cannot teach us.

1 Introduction

In its most general form the notion of the cell cycle covers the full set of molecular and structural events that take place in a cell resulting in its division and duplication into two daughter cells (Morgan 2007). For both eukaryotes and prokaryotes, our advances in the understanding of the molecular interactions driving these transitions increasingly allow us to formulate mathematical models which help in a quantitative understanding of these processes (Lenz and Søgaard-Andersen 2011; Tyson and Novak 2014). Here I am concerned with the approaches to model the

R. Blossey (✉)
Univ. Lille, CNRS, UMR8576 - UGSF, Unité de Glycobiologie Structurale et Fonctionnelle, Lille, France
e-mail: ralf.blossey@univ-lille.fr

E. Biondi (ed.), *Cell Cycle Regulation and Development in Alphaproteobacteria*,
https://doi.org/10.1007/978-3-030-90621-4_6

cell cycle of the alphaproteobacterium *Caulobacter crescentus*, *Caulobacter* in the following. *Caulobacter* differentiates between a mobile swarmer cell which moves with a flagellum in its aqueous environment and a stationary stalker cell which is immobilized at a surface. The bacterium therefore has an asymmetric cell division: it is composed of a sequence of swarmer to stalked differentiation (G1/S transition), the DNA replication phase (S), together with the development of the early pre-divisional cell, and finally the phase of cellular division, *i.e.* the splitting of the late pre-divisional cell into the daughter stalked and swarmer cell (G2). This progression is illustrated in Fig. 1.

The asymmetry of the cell cycle—the differentiation into stalked and swarmer cells—has developed into an attractive topic for researchers interested in mathematical models of the cell cycle. One motivating aspect has been the question of the *universality* of the cell cycle in at least two respects. The first is how a bacterial cell cycle relates to the cell cycle in eukaryotes. The second relates to an intrinsic

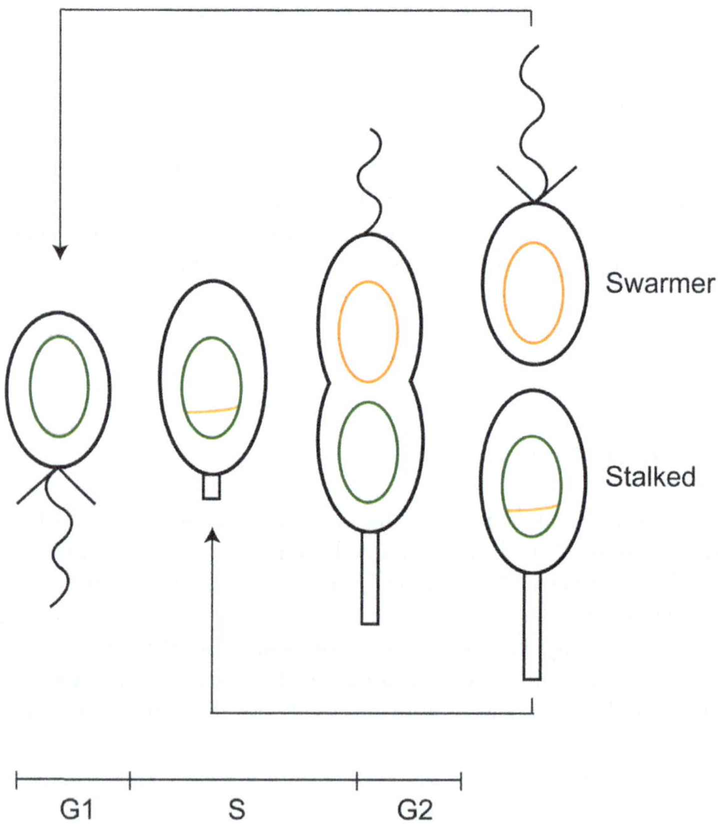

Fig. 1 Cell cycle progression in *Caulobacter*: the swarmer cell first transitions to an anchored cell (G1 → S); the DNA replicates in S-phase, cell division leads to daughter stalked and swarmer cells (G2). Figure taken from Vandecan et al. (2016), with permission

evolutionary question, since *Caulobacter* is a member of a large family of bacteria which all have evolved their specific lifestyles. Both comparisons across domains of life, and among different members of the same family, therefore offer unique opportunities to understand what makes the cell cycle work at a molecular level, and what in it is *'essential'*, or even universal.

In reviewing models of the *Caulobacter* cell cycle in this article, I have made the choice to mainly focus on a discussion of three different 'small' models, whereby 'small' refers to the number of molecular components that is considered. The intention of this text is to highlight how such choices can be justified, and what conclusions the resulting models allow to generate. In my personal impression it does not make much sense to attempt to formulate a mathematical model that contains all known molecular components; there is no good reason to try to build an artificial or virtual *Caulobacter*. Building a model must serve a scientific purpose, and completeness by itself is no good argument. A model should be formulated to test hypotheses and get answers to specific questions that can be posed; as such it should be able to rationalize experimental findings and to make testable predictions. In comparing experiment and the model, mutant strains of the bacteria also play a crucial role.

In order to gather a 'hands-on' understanding of these aspects, the present text is organized as follows. Section 2 briefly discusses the mathematical basis that is used for the models I introduce, which are ordinary differential equations and the methods of nonlinear dynamics. Section 3 introduces three specific 'small' models of aspects of the *Caulobacter* cell cycle and their discussion. Section 4 then discusses some of the 'large' models of the cell cycle, as well as giving an outlook for further work in this direction. Section 5 concludes with a discussion. Finally, I have tried as much as possible to a give an up-to-date bibliography on the chosen topics; I apologize right away to authors whose work I may have missed.

2 Cell Cycle Models: ODE's and Nonlinear Dynamics

The mathematical method of choice for models of the cell cycle is ordinary differential equations, and for their analysis the tools of nonlinear dynamics. (I will briefly comment on stochastic approaches in Sect. 4.) These methods build on a deterministic description of molecular constituents by their concentrations. Typically, the underlying interaction networks are built from proteins that are generated by the transcription of genes, which are under control by the proteins. The simplest reactions one can have in such a gene-based protein interaction network for the case of two proteins is repression or activation, described by changes of the concentrations

$$[\dot{A}] = \mathbf{Rep}([A], [B]) \tag{1}$$

and

$$[\dot{B}] = \mathbf{Act}([A], [B]), \tag{2}$$

where the bracketed symbol $[A]$ denotes the concentration of protein $[A]$ and where the $[\dot{A}]$ stands for the time derivative $d[A]/dt$. The functions on the RHS refer to nonlinear expressions for activation and repression we define and explain below.

It should be noted that concentration is not generally a good quantity: the numbers of molecules in a cell can be very small, notably in a transcription process that just starts. One should thus always think of concentrations in these models as average quantities over synchronized cell ensembles. This also removes spatial dependences and obviously is quite a drastic simplification of matters, but a highly successful one.

If, in our simple example, protein A activates B, and B represses A, and we further allow the change in concentration of each species by protein degradation, we can e.g. have for the two functions on the right-hand sides

$$\mathbf{Rep}(A, B) = \frac{k_1}{k_2 + B^2} - k_3 A \tag{3}$$

and

$$\mathbf{Act}(A, B) = \frac{k_4 A^2}{k_5 + A^2} - k_6 B\,. \tag{4}$$

Both functions are nonlinear; protein B represses A in an indirect way: by binding to an operator site on the gene a, thereby impeding the binding of RNA-polymerase and thus transcription. In the same way, A can bind to a promoter site of gene b, and thereby help to recruit RNA-polymerase to transcribe the gene. We describe this process in a simple graph, a *wiring diagram*, shown in Fig. 2. Note that the concentrations of A and B appear with an exponent 2. This is the *Hill-exponent* which describes the degree of cooperativity of the proteins; they mostly act as dimers (hence the value of 2). The degradation terms add as linear contributions, but also these can in fact be nonlinear.

Although the equations depend on two variables only, A and B, they contain already six parameters, that in this simple case are related to the reaction rates of the underlying molecular reactions. (Counting the Hill-exponent separately, there are even seven.) We will see that the resulting equations built from the above functions can lead to rather unwieldy expressions. One can simplify the equations by using the

Fig. 2 The wiring diagram of the interaction system underlying equations (1), (2)

assumption that the denominators (also more general ones than what we introduced above) are roughly constant during the dynamics. This then yields a general form of *all* dynamic equations for protein concentrations X_i as

$$\tau_i \frac{dX_i}{dt} = (s_i + \sum_j a_{ij} X_j^2 (1 - X_i) - (d_i + \sum_j b_{ij} X_j^2) X_i \,, \tag{5}$$

where the first entries in the two terms on the RHS correspond to activating and inactivating (degradation) rates and the exponents refer, as before, to the Hill exponents, here set with a value of 2 again. The a_{ij} and b_{ij} are symmetric matrices that can be deduced from the wiring diagram of the chosen interaction system; their dimension corresponds to the number of interacting partners X_i. The parameter τ_i in front of the time derivative defines a time-scale (a relaxation rate). The model defined by Eq. (5) can be derived from basic models of gene regulation, as detailed in the Appendices of Pfeuty and Kaneko (2009); Pfeuty (2012).

The analysis of ordinary differential equations is mostly numerical; it involves solving them in a specific software using standard algorithms for the numerical solution of ODEs, as e.g. implemented in MATHEMATICA. Another step is to look at stationary solutions of the equations: this involves setting the time derivatives to zero and solving the resulting algebraic equations (again, in general numerical). The stability of the fixed-points can be tested by performing a linear stability analysis around these stationary points. We will see this applied in the following in our third example.

3 'Small' Models of the Cell Cycle Dynamics of *Caulobacter crescentus*

In order to construct dynamical models of the cell cycle of *Caulobacter*, we need first to determine a wiring diagram, for which in turn we need to identify the key molecular regulators. They are listed as follows:

- **CtrA** is the *Caulobacter's* master regulator. It is essentially linked to all upstream and downstream processes of the cell cycle (Quon et al 1996, 1998; Laub et al 2000, 2002; Biondi et al 2006). It is under control of two promoters: a weak one, P1, is induced by GcrA. Growing concentrations of CtrA repress P1 and turn on the strong promoter P2 which leads to a rapid increase of CtrA. Its phosphorylated form CtrA-P is involved in the formation of the Z-Ring, a polymer formed by the protein FtsZ which acts in the division of the cell. In the context of DNA replication, its task is to repress the origin of replication by blocking the action of its antagonist, DnaA (Collier et al 2006).
- **GcrA** is a secondary regulator involving many different genes, including *ctrA*, as just mentioned. It is antagonistic to DnA, so our basic *A*-*B* circuit discussed before applies here.

- **DivK**. Its phosphorylated form promotes the degradation of CtrA in the stalked cell after cell division. It inactivates CtrA while it is in turn activated by CtrA. Both CtrA and DivK are the endpoints of phosphotransfer modules: DivL-CckA-CtrA and DivJ-PleC-DivK (Hecht et al 1995; Jacobs et al 1999; Jacobs et al. 2001; Chen et al 2009). Activation and inactivation of these molecules are modulated by switching between phosphorylated and dephosphorylated forms and their spatial localization in the predivisional cell.
- **DnaA** is the replication initiation protein. It binds DNA at a position upstream of a specific origin of replication (*oriC*), causing DNA strand separation, preparing for further separation by a helicase motor protein. It is also involved in the activation of transcription of several cell cycle genes (Hottes et al 2005). DnaA also activates the transcription of the *ctrA* gene.
- **CcrM.** Induced (activated) by CtrA it methylates the new DNA of strand in order to shut down the expression of several genes, among them *ctrA*.

We can now turn to the first model which was developed in order to compare the prokaryotic and eukaryotic cell cycle.

3.1 Model 1: Brazhnik and Tyson

The first model I discuss has been developed by Brazhnik and Tyson (2006). The motivation of this early paper on the topic has been to highlight similarities and differences in the cell cycle of bacteria and a unicellular eukaryotic organism, yeast. The Brazhnik–Tyson model contains six variables, four of which correspond to proteins: CtrA, GcrA, DivK, and its phosphorylated component, DivK-P. The two remaining components are considered as 'phenomenological variables' in the sense as they correspond to behaviours rather than molecular identities: Z describes the closing of the Z-ring which completes cell division, and I is a delay between CtrA activation and the closing of the Z ring. The full set of equations is given by, using the definitions CtrA $\equiv X_1$, GcrA $\equiv X_2$, DivK $\equiv X_3$, DivK-P $\equiv X_4$, Z and I,

$$[\dot{X}_1] = \frac{k_1 J_1^2 [X_2]}{J_1^2 + [X_1]^2} + \frac{k_2 [X_1]^2}{L_1^2 + [X_1]^2} - \left(k_3 + \frac{k_4 [X_4]^2}{J_2^2 + [X_4]^2} \right) [X_1] \tag{6}$$

$$[\dot{X}_2] = \frac{k_5 J_3^2}{J_3^2 + [X_1]^2} - k_6 [X_2] \tag{7}$$

$$[\dot{X}_3] = k_7 [X_1] + k_8 [X_4] - k_9 (1 - [Z]) [X_3] - k_{10} [X_3] \tag{8}$$

$$[\dot{X}_4] = -k_8 [X_4] + k_9 (1 - [Z]) [X_3] - k_{10} [X_4] \tag{9}$$

$$[\dot{I}] = k_{11} [X_1] - k_{12} [I] \tag{10}$$

$$[\dot{Z}] = \frac{k_{13} (1 - [Z])}{J_7 + (1 - [Z])} - \left(k_{14} + k_{15} \left(\frac{[I]}{L_8} \right)^4 \right) \frac{[Z]}{J_8 + [Z]} . \tag{11}$$

We have used here a slightly different ordering scheme of the equations than in Brazhnik and Tyson (2006), since we prefer to put the four equations for the proteins first. The rate constants k_i have been numbered chronologically in these equations which makes them different from the original definitions in Brazhnik and Tyson (2006); these differences should, however, be clear from a comparison between the two sets of equations here and in the original paper. The numerical values of the constants are listed in the original paper, as well as how they have been obtained.

A first look at the equations reveals the following. The first two equations involve activation and repression terms, as we discussed in the simple model of Sect. 2. The next two equations describe the phosphorylation and dephosphorylation of DivK. The last two equations describe the cell division process in a phenomenological way; in order to have the transition occur abruptly, note that a high value of the exponent on $[I]$ has been chosen.

Figure 3 displays the dynamical curves obtained for the variables from the numerical solution of the equations, obtained from programming them in MATHEMATICA, and using the parameters determined in Brazhnik and Tyson (2006). One sees that CtrA concentration grows until cell division, marked by rapid decay of Z. GcrA fulfils a perfectly symmetric oscillation in one cycle, while DivK and DivK-P show antagonistic variations. As shown in the bottom graph, the sum of their concentrations varies only little over one cycle.

Since the Brazhnik–Tyson model contains also phenomenological variables and not only proteins, it is not evident how to represent the underlying wiring diagram. This task becomes much easier by looking at the proteins only. Further, since the experimental data used to compare the model with (see Brazhnik and Tyson (2006)) in fact do not distinguish between the two forms of DivK, we look at the dynamics of DivK_{total} = DivK + DivK-P which is obtained from summing up the third and fourth equations of the Brazhnik–Tyson model, as shown in Fig. 3 (bottom). The wiring diagram corresponding to this reduced model is shown in Fig. 4.

3.2 Model 2: Murray et al.

Recent work by Murray et al (2013), combining experimental and theoretical analysis, considered the regulation of CtrA via the two different promoters as discussed in the summary of protein regulators at the beginning of this section.

Their model contains four equations for the variables GcrA $= X_1$, CtrA $= X_2$, CtrA-P $= X_3$ und CckA-P $= X_4$. They read as

$$[\dot{X}_1] = \frac{k_1 J_{CG}}{J_{CG} + [X_3]} - k_2[X_1] \tag{12}$$

$$\begin{aligned}[\dot{X}_2] = (k_3[X_1] + k_4)\frac{J_{C1}}{J_{C1} + [X_3]}((1-m)S + m) + k_5\frac{[X_2]}{J_{C2} + [X_2]} \\ + k_6[X_3] - k_7[X_1][X_4] - k_8([X_{4,T}] - [X_4])[X_2]\end{aligned} \tag{13}$$

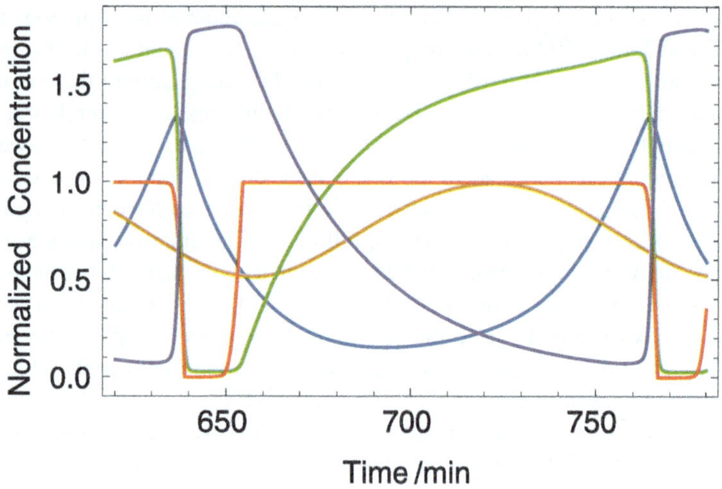

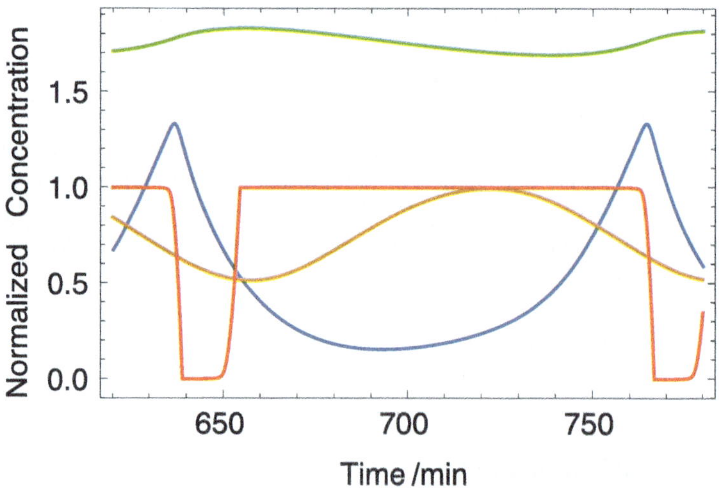

Fig. 3 Change of protein concentrations during the *Caulobacter* cell cycle. Top: Time courses of scaled concentrations of CtrA (blue), GcrA (brown), DivK (green) and DivK-P (purple) for one division cycle, marked by the end of the red curve for Z. Bottom: the graph shows the same profile with $DivK_{total}$ instead of the both components

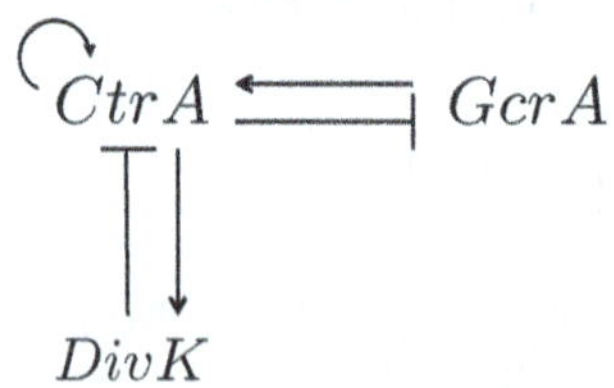

Fig. 4 Brazhnik–Tyson model reduced to its protein components. DivK in this graph refers to total DivK, i.e. both the phosphorylated and dephosphorylated species

$$[\dot{X}_3] = k_7[X_1][X_4] - k_8[X_3] - k_9([X_{4,T}] - [X_4])[X_3] \tag{14}$$

$$[\dot{X}_4] = (k_{10} + k_{11}S)([X_{4,T}] - [X_4]) - k_{12}[X_4], \tag{15}$$

where $[X_{4,T}]$ is the total concentration of CckA and CckA-P. In this model there also appear phenomenological parameters, m and S, whereby the latter changes between the two values 0 and 1. These parameters describe the switch involving the change from the weak promoter P_1 to the strong promoter P_2 and involve methylation states of DNA which would originate via the kinetics of CcrM, which the model does not contain explicitly.

The work by Murray et al (2013) adds an important aspect to the modeling discussion, as it is indeed concerned with the necessity of specific proteins for the progression of the cell cycle - a discussion alluded to before in Li et al (2008). The authors built a minimal model in which they explicitly included the two-promoter structure controlling the *ctrA* gene and furthermore modeled the regulation of the phosphorylation of CtrA by CckA and its phosphorylated form, CckA-P. Replication initiation is used as a 'kick-off' for the dynamics. The most interesting result of this work is that GcrA turns out to be dispensable, which could be verified experimentally. Figure 5 shows the reduced wiring diagram of this network.

In addition, also CcrM was found not generically necessary for cell cycle progression. On the basis of an evolutionary analysis, it at appears that both CcrM and GcrA are often either both present or absent in the family members of *Alphaproteobacteria*, and that they are absent in the tree root of this family. This paper shows the whole beauty of small models: by considering a key subset of regulators and providing a detailed experimental analysis, the case for the dispensability of previously believed key regulators could be convincingly argued - but also since beyond the observed dynamical properties of the model, evolutionary insights have been used in combination.

We now turn to the last 'small' model, which will be discussed in some more technical detail.

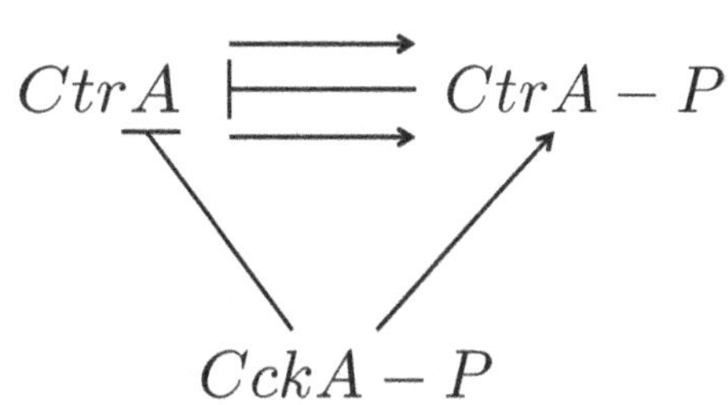

Fig. 5 (Reduced) wiring diagram of the model by Murray et al.. The complex regulation of CtrA and CtrA-P is key in this model

3.3 Model 3: Vandecan, Biondi and Blossey

Given the non-essentiality of GcrA discussed before it seems natural to replace it by the essential protein DnaA, which is done in the third model we discuss. The main architecture of the interactions retained in this model is summarized in Fig. 6, following Vandecan et al. (2016). In the view of the authors, this three-tier protein network describes the core module of cell cycle regulation in *Caulobacter*, hence its 'minimal' regulatory circuit.

Turning to the mathematical formulation, this time we make use of the approach developed in Pfeuty and Kaneko (2009); Pfeuty (2012) which is best suited for the analysis of core networks. Denoting the proteins now by $X_i, i = 1, .., 3$ we have (CtrA, DivK, DnaA) $\equiv (X_1, X_2, X_3)$. Each of the proteins fulfils the dynamic equation (5) in which, as a reminder, the τ_i are time constants, the s_i the basal activation constants for each protein, and the a_{ij} and b_{ij} are activation and inactivation matrices that govern the interactions shown in Fig. 6.

The diagonal elements a_{ii} are the self-interactions of the proteins, shown as curved arrows in Fig. 6. Since dimerization is experimentally verified in all three proteins, CtrA (Reisenauer et al 1999), DivK (Guillet et al 2002; Hung and Shapiro 2002; Brazhnik and Tyson 2006) and DnaA (Erzberger et al. 2002) the value of the Hill coefficient has been fixed to the value of 2.

The cross-diagonal parameters in both the activation and inactivation matrices describe the couplings between the different proteins. The remaining parameters d_i are the basal degradation rates of each of the proteins; in absence of the constant terms and the coupling terms, protein concentrations decay exponentially like $x_i \sim \exp -(d_i/\tau_i)t$.

According to the wiring of Fig. 6, for *Caulobacter* one has the following activation and inactivation matrices a_{ij} and b_{ij} which contain five activating and two inactivating interactions as their entries:

$$a_{ij} = \begin{pmatrix} a_{11} & 0 & a_{13} \\ a_{21} & a_{22} & 0 \\ 0 & 0 & a_{33} \end{pmatrix} \quad b_{ij} = \begin{pmatrix} 0 & b_{12} & 0 \\ 0 & 0 & 0 \\ b_{31} & 0 & 0 \end{pmatrix} \tag{16}$$

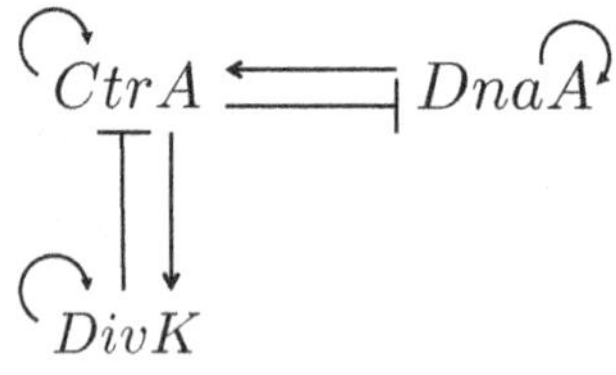

Fig. 6 The minimal regulatory circuit of *Caulobacter crescentus* is a system of three proteins CtrA, DivK and DnaA, connected by synergistic and opposing interactions in such a manner as to form a *coupled limit-cycle oscillator*, i.e. CtrA-DivK and CtrA-DnaA. Arrows correspond to activation, turnstiles to inactivation: e.g. CtrA activates DivK, while DivK inactivates CtrA. Following Vandecan et al. (2016)

Table 1 Experimental data extracted from Collier et al (2006) for CtrA and DnaA, and from Jacobs et al. (2001) for DivK. From Vandecan et al. (2016), with permission

Time/min	CtrA	DivK	DnaA
0	1.0	1.0	1.0
20	0.0	1.0	2.22
40	0.0	1.05	2.0
60	0.0	1.05	0.67
80	0.63	1.25	0.33
100	1.22	1.25	0.22
120	1.06	1.30	0.89

such that the model can be explicitly written as

$$\tau_1 \frac{\mathrm{d}[X_1]}{\mathrm{d}t} = (s_1 + a_{11}[X_1]^2 + a_{13}[X_3]^2)(1 - [X_1]) - (d_1 + b_{12}[X_2]^2)[X_1] \tag{17}$$

$$\tau_2 \frac{\mathrm{d}[X_2}{\mathrm{d}t} = (s_2 + a_{21}[X_1]^2 + a_{22}[X_2]^2)(1 - [X_2]) - d_2 X_2 \tag{18}$$

$$\tau_3 \frac{\mathrm{d}[X_3]}{\mathrm{d}t} = (s_3 + a_{33}[X_3]^2)(1 - X_3) - (d_3 + b_{31}[X_1]^2)[X_3]\,. \tag{19}$$

In order to determine the parameter values of the model, we need to obtain experimental information. Qualitatively, DnaA and CtrA change during the cell cycle progression in an antagonistic fashion, with the DnaA levels increasing when CtrA is degraded and disappearing in predivisional cells when CtrA is synthesized again by the GcrA-dependent transcriptional activation (Collier et al 2006). The total DivK level does not vary much over the cell cycle (see our first model) with the concentration of the protein proportional to the cellular volume (Jacobs et al. 2001). The protein concentrations can thus be estimated as relative amounts (%) with respect to the time 0 concentration, and CtrA, DnaA and DivK data were extracted from earlier works (Collier et al 2006; Jacobs et al. 2001).

These data are assembled in Table 1. We note that the extraction of the experimental data for DivK requires reading from a gel, with obvious difficulties due to the low variation of their concentrations.

We now turn to the parametrization of our model, taking the experimental results as our point of reference. Rather than presenting one unique data-set, optimized for agreement with experimental data, we select two. Figure 7(top) and (bottom) show the time traces for these two different parametrizations of the system (17); the parameters are summarized in Table 2. We call these parametrizations *weak* and *strong coupling*, respectively, whereby these terms are motivated by the two distinct types of dynamics one observes. Specifically, the notions 'weak' and 'strong' were chosen with regards to the amplitude variations of DivK, reflecting its coupling to the other two regulators.

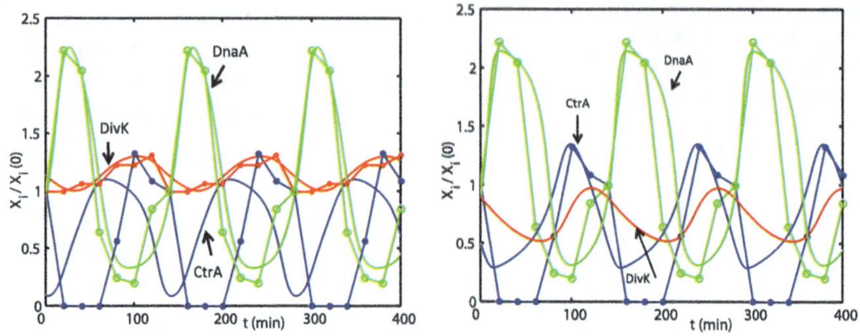

Fig. 7 Time traces of the cell cycle in *Caulobacter*, for CtrA (blue), DnaA (green) and DivK (red) for the 'weakly' coupled case (left) and the 'strongly' coupled case (right). The smooth lines have been computed; the straight lines connect the experimental data points from Table 1. From Vandecan et al. (2016), with permission

Table 2 Model parametrization for the 'weak coupling' and 'strong coupling' cases. The basal degradation constants were put equal to 1 in the comparisons to the experimental data. The value of a_{22} (being either a small number or zero) has no effect on the dynamics. From Vandecan et al. (2016), with permission

	a_{11}	a_{13}	a_{21}	a_{22}	a_{33}	b_{12}	b_{31}	τ_1	τ_2	τ_3	s_1	s_2	s_3
Weak	4	0.3	0.5	0	5	1	3.6	0.2	3	0.42	0.006	0.1	0.14
Strong	6	0.3	2	0	4	10	3	0.44	4	0.1	0.009	0	0.17

In Fig. 7(top), DivK varies only weakly, and the fit to the experimental data is quite good with respect to both DnaA and DivK, but fairly poor with respect to CtrA. In particular, we note that the phasing of protein concentrations is not perfectly respected, since the maximum of CtrA appears before the minimum in DnaA. In the strong coupling case of Fig. 7(bottom) we see that, on the contrary, DnaA and CtrA are well phased with the dynamics of DivK being considerably more pronounced than the experimental data show. One can therefore conclude that it if one insists to remain with the simple model presented here, it is better to interpret DivK not just as the total component, but rather as an active fraction of its total concentration which is involved in controlling the dynamics of CtrA. In particular, we note that it is due to this effect that the proper phasing of the two motifs is achieved. Thus, the dynamic model already teaches us something about the behaviour of its components and indicates a direction for improvement.

Our two different parametrizations of the model show different qualitative and quantitative behaviours which also have consequences beyond the normal cell behaviour on which these parametrizations are usually based. This can be shown by considering cell cycle arrest. There is experimental evidence that proteotoxic stress arises via a depletion of the Hsp70 chaperone, DnaK, either through genetic manipulation or heat shock to which *Caulobacter* is exposed. This acts on DnaA by promoting its rapid degradation (Jonas et al 2013).

This mechanism can be qualitatively captured in our model by an increase of the degradation parameter of DnaA, d_3^*. We have therefore studied the behaviour of the model that emerges when d_3^* is taken as a so-called *bifurcation parameter* which is the one that controls the stability of the dynamics. For this variation we have determined the fixed-points of the system (17) and studied the local and global stability of the solutions.

The two resulting one-parameter bifurcation diagrams are shown in Fig. 8. Figure 8(top) corresponds to the weak coupling case. The limit cycle shrinks upon approach to the critical value of the degradation constant of DnaA and the transition arises in a continuous fashion. We find that at a critical value of $d_{3,crit}^* < 1.15$ the system displays what is called a *supercritical Hopf bifurcation*, in which the change of behaviour occurs smoothly.

The situation we encounter in the strong coupling case of Fig. 8b is more complex. We identify four regimes which are again more easy to explain when considering the lowering of d_3^*. For $d_3^* > 1.163$, the system has two unstable and one stable fixed-point which is the lowest of these branches. At a critical value $d_{3,crit}^* = 1.163$, it becomes unstable with respect to a limit cycle again via a supercritical Hopf bifurcation. This limit cycle has a very small width and remains stable only within a small interval until $d_3^* = 1.158$ when it collapses with respect to the above-discussed *Caulobacter* limit cycle via a *saddle-node bifurcation of limit cycles* (SNLC). This case is akin to the scenario SNLC_1 discussed in Pfeuty (2012). Until $d_3^* = 1.163$, the dynamics is still influenced by the former stable fixed-point around which the trajectory continues to spiral.

We like to be more specific about our procedure. We followed the approach developed in Pfeuty (2012), where the starting point of the bifurcation analysis is the study of the fixed-points, determined by the intersection of the *nullclines* defined by the stationary equations

$$(s_1 + a_{11}[X_1]^2 + a_{13}[X_3]^2)(1 - [X_1]) - (d_1 + b_{12}[X_2]^2)[X_1] = 0 \tag{20}$$

$$(s_2 + a_{21}[X_1]^2 + a_{22}[X_2]^2)(1 - [X_2]) - d_2[X_2] = 0 \tag{21}$$

$$(s_3 + a_{33}[X_3]^2)(1 - [X_3]) - (d_3 + b_{31}[X_1]^2)[X_3] = 0. \tag{22}$$

The stability of the numerically obtained fixed-points is obtained by studying the *Jacobian* $\mathbf{J}(\underline{\mathbf{x}}^*)$ with $\underline{\mathbf{x}} = (x_1^*, x_2^*, x_3^*)$. Upon decreasing the bifurcation parameter d_3^* from large values, whenever appears a stable fixed-point for both the weak and strong coupling cases, there occurs a supercritical Hopf bifurcation to a limit cycle. For the strong coupling case, the resulting cycle remains very small upon further decrease in d_3^*.

Upon expansion at a decrease of d_3^*, the limit cycle undergoes a *saddle-node bifurcation of limit cycles* akin to the SNLC_1 scenario described in Pfeuty (2012). The following Figures illustrate the progression from the limit cycle behaviour to cell cycle arrest in the strong coupling parametrization, as summarized in the bifurcation diagram of Fig. 8(bottom). Figure 9 shows the concentrations of CtrA,

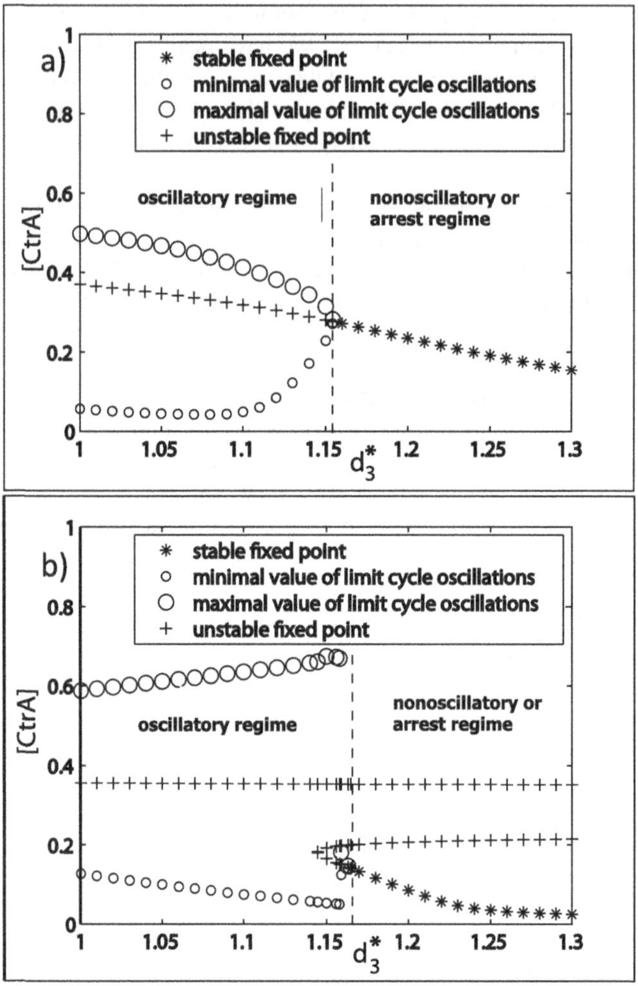

Fig. 8 The two cell cycle arrest scenarios in the 'weak' coupling (top, **a**)) and 'strong' coupling (bottom, **b**)) cases. From Vandecan et al. (2016), with permission. For the explanation, see main text

DivK and DnaA upon an increase of the degradation parameter d_3^*. The appearance of a second limit cycle of smaller width is signaled by additional oscillations in the concentrations, seen when going from Fig. 10(top) to (bottom). The system then settles into a second, smaller limit cycle, further illustrated by Fig. 10(bottom) which shows the exchange of stability of the limit cycles.

Pfeuty (2012) rejects the above observed SNLC scenarios on general grounds for biological systems as there is no general biological mechanism to rationalize a transition between two different limit cycles with different amplitudes over an extended range of parameters. In our case the transition sequence occurs over a

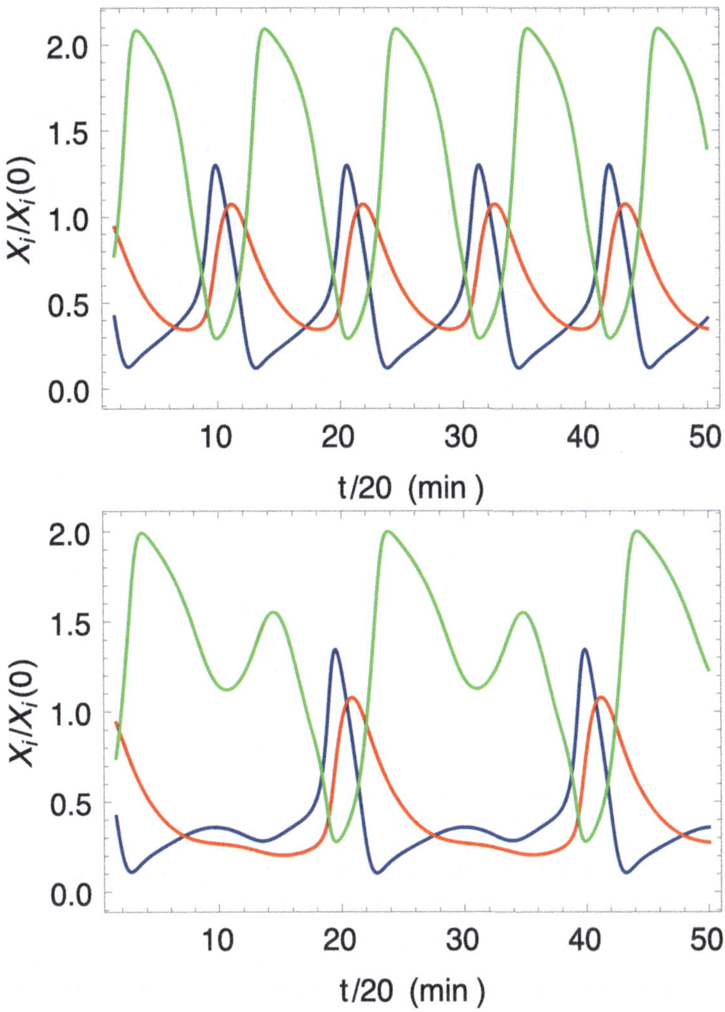

Fig. 9 Two steps of the progression of the 'strong'-coupling model through the bifurcation scenario. Top, for $d_3^* = 1.13$ one has the limit cycle behaviour; bottom, for $d_3^* = 1.15$ the system is in the large limit cycle but spirals around an additional unstable fixed-point between two periodic orbits. From Vandecan et al. (2016), Supplementary Material

very small interval of the parameter d_3^* so that, if such a transition exists, it might be easily missed in experiment. A *global* difference between the two scenarios should, however, be visible in corresponding experiments: in the weak coupling case the amplitude of the limit cycle oscillations shrinks continuously when the critical values of cell cycle arrest are approached, while in the strong coupling case, a limit cycle of finite width for $d_3^* << d_{3,crit}^*$ appears to break down abruptly to the arrest state for $d_3^* >> d_{3,crit}^*$. In terms of the width of the limit cycle, even the trend

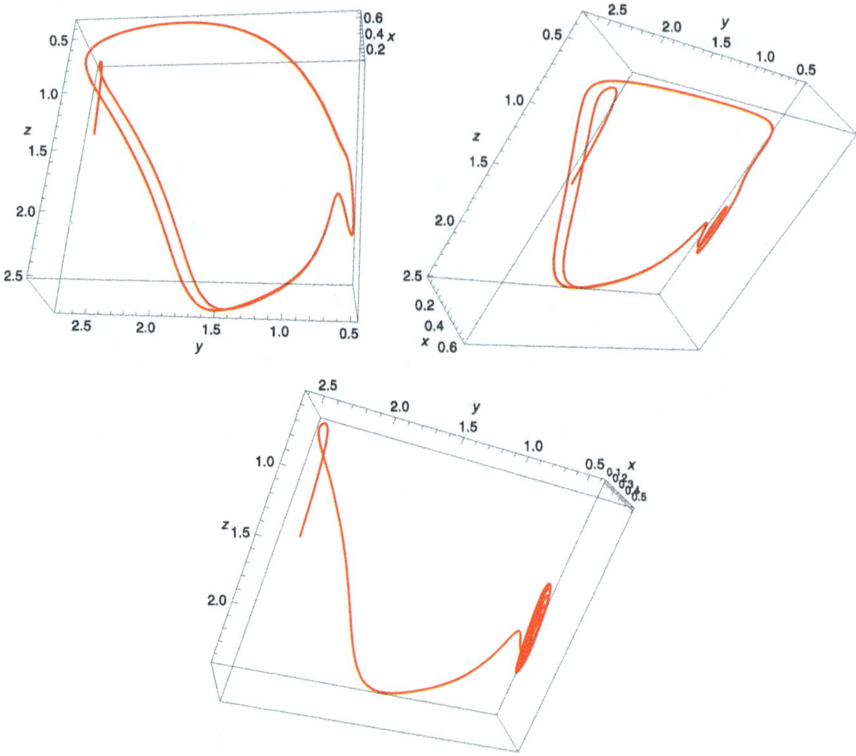

Fig. 10 Progression of the limit cycles of strong-coupling model. Top left: for $d_3^* = 1.154$, the second limit cycle announces its appearance as the trajectory circles around an additional unstable fixed-point. Top right: for $d_3^* = 1.156$ the system spirals around the still unstable fixed-point while still being set in the large limit cycle. Bottom: for $d_3^* = 1.159$ the system is now set in the smaller limit cycle. From Vandecan et al. (2016), Supplementary Material

upon the approach of $d_{3,crit}^*$ between the two cases is opposite: a shrinking limit cycle in the weak coupling case vs an expanding limit cycle in the strong coupling case. In this latter case the transition would thus not look much different than in the SNIC/SH-transitions described in Pfeuty (2012).

It is time to make a quick stop here before continuing. What have we learnt so far? First of all, the key regulators of *Caulobacter* are quite well-established; it can be no doubt that CtrA, DnaA and DivK are key, without them, the cycle cannot function.

We have also seen how modeling can help in understanding. First of all, a modeler has a lot of freedom in selection of the components retained in the model: the confrontation to experiment allows clues as to what component is essential, see model 2. Parametrization is a generally touchy issue: linking model parameters to experimental ones can be tricky. While each parametrization of parameters yields a unique dynamics, not every one is supported by experiments: most will need to be rejected. However, different parametrizations may also uncover different systems behaviours, as we saw in the perturbation of model 3: originally parametrized by a functioning cell cycle, the model was then perturbed to study cell cycle arrest. For the two parametrizations, very different behaviours are found that can be confronted to experiment. This is the general scheme under which mathematical models of the cell cycle should be understood.

We now turn to a review of other works on *Caulobacter's* cell cycle that have appeared in recent years and discuss their results in the light of the foregoing insights. The focus is now on 'large' models, i.e. models that attempt to cover a large set of molecular actors.

4 'Large' Models of *Caulobacter's* Cell Cycle

In this section, we review a number of papers of outstanding interest in the modeling of the cell cycle of *Caulobacter*. The work by Brazhnik and Tyson (2006) constitutes the first attempt of the Tyson group to work on the bacterial cell cycle. John J. Tyson is well-known from his development of large-scale models for the eukaryotic cell cycle, often together with Bela Novak. Subsequent to his first work, Tyson developed a number of complex models for different aspects of the *Caulobacter* cell cycle.

The first works in this direction are the papers (Li et al 2008, 2009) which cover describe 'large-scale' models of the *Caulobacter* cell cycle, adopting the same philosophy as used for the eukaryotic cell cycle before. They can be seen as extensions of the first model discussed in this paper, in which four proteins (CtrA, GcrA, Div-K, DivK-P) are discussed together with two phenomenological variables (I and Z). In the subsequent paper (Li et al 2008), this set is augmented by two more proteins (DnaA and CcrM), a protein entity englobing a set of different regulators involved in Z-ring formation (FtsZ) and a set of additional state variables in the same vein as I and Z: four covering the different methylation states of the DNA sequence of *oriC*, and of the genes *ftsZ*, *ctrA* and *ccrM*. Two further state variables cover the concentrations of replication forks and their progression (elongation). Finally, the amount of DNA is monitored as well. Altogether, this makes 16 variables, and a correspondingly large parameter set (see Table 2 in that paper): truly a large-scale model.

In the paper, the mathematical model is confronted with behaviours from several mutant strains which help constrain the model. On the other hand, the model then allows to make predictions on the expected behaviour of other mutants. The authors also already suggest that since the key regulators are also present in other family members of *Caulobacter*, they could provide interesting systems for comparative study.

The paper by Li et al (2009) further expands this model to 28 variables with 96 parameters and continues the same philosophy of using mutants to constrain and predicting new mutants. They key new feature of this model is the inclusion of the components of the phosphorelay module controlling the decision of a cell to switch between a swarmer and a stalked cell. Remembering from the lifestyle of the bacteria, swarmer cells must *decide* to settle and grow a stalk in order to begin cell division. Further aspects of this bistable dynamics are discussed in the later papers of the Tyson group (Subramanian et al 2013, 2015). Finally, the most recent work has been oriented towards the discussion of spatial effects (protein diffusion) (Subramanian et al 2015; Subramanian and Tyson 2017) and stochastic models (Li et al 2016).

These two points merit special attention. As said in the beginning, the mathematical discussion presented here in some detail has focused on the use of ordinary differential equations. These assume essentially well synchronized cell ensembles. However, they entirely neglect any spatial dependence: there is also the underlying assumption that the proteins are well-mixed. This is obviously not correct in general: already the formation of the Z-ring requires the localization of proteins. In the context of continuous variables, the correction for spatial behaviour is principally simple: it is reasonable to assume the proteins as diffusing in the cell. Thus, the ordinary differential equations are turned into nonlinear diffusion equations by adding a diffusion term for each protein, $\sim \Delta X_i$, where Δ is the Laplace operator (i.e. the second order spatial derivatives).

While technically simple, this entirely changes the behaviour of the mathematical system of equations. They now have become partial differential equations, and need much more effort to solve them numerically. Obviously spatial effects become relevant in the process of Z-ring formation which requires the spatial recruitment of molecules; a modeling study using such equations is by Shtylla (2017). This is one reason why cell cycle models often rather use fictitious or phenomenological parameters to cut a given system into different compartments - like the stalked and swarmer cells are. The dynamics can be considered again as well-mixed in the respective sub-compartments, and the spatial derivatives can again be neglected.

The second aspect, stochastic models, is even more subtle. Stochasticity is present in reality in the form of 'intrinsic' noise which describes the fact that protein numbers are small in the initial stages of the transcription process, and they may fluctuate in both time and space. These effects are beyond any continuum model, and they refer to a model of a single cell, not of a synchronized cell ensemble. They only make sense to apply when the model can be compared to experiment on a single molecule level inside the cell. An example of such a study is the work by Lin et al (2010). We will here not go any further in this direction and stick with the simpler (but well founded) continuum models.

In closing this section it is worthwhile to comment on a series of recent works which continue with a global view of *Caulobacter* regulation and use modern computational techniques. The first of these is Shen et al (2015), which applies a hybrid simulation technique and model checking methodology to model various possible regulatory circuits. This approach uses both continuous and discrete variables, and the formalized control mechanism allows to use a clean approach to tackle the parametrization problem. An approach employing a similar philosophy is by Quiñones-Valles et al (2014) which uses a Boolean description together with an exhaustive literature search for inputting data.

It is perceivable that high-throughput analyses genomic analyses as performed by Zhou et al. (2015) on the transcriptional regulatory network of *Caulobacter*, or the most recent work by Venetz et al. (2019) in which a reduced synthetic network was deduced from *Caulobacter* can pave the way, together with hybrid simulation techniques and 'big data'-type machine learning approaches will allow us to gain large-scale views of the function of the genomes of *Alphaproteobacteria*.

5 Conclusions

In this brief review, I have attempted to provide an overview of models of the cell cycle of *Caulobacter crescentus*. My focus has been to explain the use of ordinary differential equations to build 'small' models of the cell cycle which can be interrogated with the means of the methods of nonlinear dynamics. This approach has been illustrated by the discussion of three models describing key aspects of the dynamics of the key regulators underlying the cell cycle. The third model, based on the three regulators DnaA, CtrA and DivK, has been discussed in some mathematical detail, in order to illustrate the problem of parametrization of such models, as well as what kind of predictions can be obtained from them. In particular, the approach taken by Murray et al (2013) has shown the success of the combination of 'small' models, dedicated experiments and evolutionary analysis—a major advantage in the study of *Alphaproteobacteria*.

Where will the field go? I will close with my personal opinion. 'Small' models as discussed here will remain of interest if they can be combined with dedicated experiments and complementary analysis. The development of large-scale models will benefit from the wave of high-throughput approaches that appear on the scene. I can easily imagine the combination of synthetic gene circuits, interrogated with hybrid simulations and formalized model checking: big data in modeling complements and reinforces big data in experiment. *Caulobacter crescentus* could be an ideal target for such ambitions.

References

Biondi EG, Reisinger SJ, Skerker JM, Arif M, Perchuk BS, Ryan KR, Laub MT (2006) Regulation of the bacterial cell cycle by an integrated genetic circuit. Nature 444:899–904

Brazhnik P, Tyson JJ (2006) Cell cycle control in bacteria and yeast: a case of convergent evolution? Cell Cycle 5:522–529

Chen EY, Tsokos CG, Biondi EG, Perchuk BS, Laub MT (2009) Dynamics of two phosphorelays controlling cell cycle progression in *Caulobacter crescentus*. J Bacteriol. 191:7417–7429

Collier J, Murray SR, Shapiro L (2006) DnaA couples DNA replication and the expression of two cell cycle master regulators. EMBO J 25:346–356

Erzberger JP, Pirruccello MM, Berger JM (2002) The structure of bacterial DnaA: implications for general mechanisms underlying DNA replication initiation. EMBO J 21:4763

Guillet V, Ohta N, Cabantous S, Newton A, Samama JP (2002) Crystallographic and biochemical studies of DivK reveal novel features of an essential response regulator in *Caulobacter crescentus*. J Biol Chemistry 277:42003–42010

Hecht GB, Lane T, Ohta N, Sommer JM, Newton A (1995) An essential single-domain response regulator required for normal-cell division and differentiation in *Caulobacter crescentus*. EMBO J 14:3915–3924

Hottes AK, Shapiro L, McAdams HH (2005) DnaA coordinates replication initiation and cell cycle transcription in *Caulobacter crescentus*. Mol Microbiol. 58:1340–1353

Hung DY, Shapiro L (2002) A signal transduction protein cues proteolytic events critical to *Caulobacter* cell cycle progression. Proc Natl Acad Sci (USA) 99:13161

Jacobs C, Domian IJ, Maddock JR, Shapiro L (1999) Cell cycle-dependent polar localization of an essential bacterial histidine kinase that controls DNA replication and cell division. Cell 97:111–120

Jacobs C, Hung D, Shapiro L (2001) Dynamic localization of a cytoplasmic signal transduction response regulator controls morphogenesis during the *Caulobacter* cell cycle. Proc Natl Acad Sci (USA) 98:4095–4100

Jonas K, Liu J, Chien P, Laub MT (2013) Proteotoxic stress induces a cell cycle arrest by stimulating Lon to degrade the replication initiator DnaA. Cell 154:623–626

Laub MT, McAdams HH, Feldblyum T, Fraser CM, Shapiro L (2000) Global analysis of the genetic network controlling a bacterial cell cycle. Science 290:2144–2148

Laub MT, Chen SL, Shapiro L, McAdams HH (2002) Genes directly controlled by CtrA, a master regulator of the *Caulobacter* cell cycle. Proc Natl Acad Sci (USA) 99:4632–4637

Li S, Brazhnik P, Sobral B, Tyson JJ (2008) A quantitative study of the division cycle of *Caulobacter crescentus* stalked cells. PLoS Comp Biol 4:e9

Li S, Brazhnik P, Sobral B, Tyson JJ (2009) Temporal controls of the asymmetric cell division cycle in *Caulobacter crescentus*. PLoS Comp Biol 5:e10000463

Li F, Subramanian K, Chen M, Tyson JJ, Cao Y (2016) A stochastic spatiotemporal model of a response-regulator network in the *Caulobacter crescentus* life cycle. Phys Biol 13:035007

Lin Y, Crosson S, Scherer NF (2010) Single-gene tuning of *Caulobacter* cell cycle period and noise, swarming motility, and surface adhesion. Mol Syst Biol 6:445

Lenz P, Søgaard-Andersen L (2011) Temporal and spatial oscillations in bacteria. Nature Rev Microbiol 9:565–571

Morgan DO (2007) The cell cycle: Principles of control. New Science Press, London

Murray SM, Panis G, Fumeaux C, Viollier PH, Howard M (2013) Computational and genetic reduction of a cell cycle to its simplest, primordial components. PLoS Biology 11:e100179

Pfeuty B (2012) Dynamical principles of cell-cycle arrest: reversible, irreversible, and mixed strategies. Phys Rev E 86:021917

Pfeuty B, Kaneko K (2009) The combination of positive and negative feedback loops confers exquisite flexibility to biochemical switches. Phys Biol 6:046013

Quiñones-Valles C, Sánchez-Osorio I, Martinez-Antonio A (2014) Dynamical modeling of the cell cycle and cell fate emergence in *Caulobacter crescentus*. PLoS One 9:e111116

Quon KC, Marczynski GT, Shapiro L (1996) Cell cycle control by an essential bacterial two-component signal transduction protein. Cell 84:83–93

Quon KC, Yang B, Domian IJ, Shapiro L, Marczynski GT (1998) Negative control of bacterial DNA replication by a cell cycle regulatory protein that binds at the chromosome origin. Proc Natl Acad Sci (USA) 95:120–125

Reisenauer A, Quon K, Shapiro L (1999) Bacterial DNA methylation: a cell cycle regulator? J Bacteriol 181:2430–2439

Shen X, Collier J, Dill D, Shapiro L, Horowitz M, McAdams HH (2015) Architecture and inherent robustness of a bacterial cell-cycle control system. Proc Natl Acad Sci 15:11340–11345

Shtylla B (2017) Mathematical modeling of spatiotemporal protein localization patterns in *C. crescentus* bacteria: A mechanism for asymmetric FtsZ ring positioning. J Theor Biol 433:8–20

Subramanian K, Tyson JJ (2017) Spatiotemporal models of the asymmetric division cycle of *Caulobacter crescentus*. Res Probl Cell Differ 61:23–48

Subramanian K, Paul MR, Tyson JJ (2013) Potential role of a bistable histidine kinase switch in the asymmetric division cycle of *Caulobacter crescentus*. PLoS Comput Biol 9:e1003221

Subramanian K, Paul MR, Tyson JJ (2015) Dynamical localization of DivL and PleC in the asymmetric division cycle of *Caulobacter crescentus*: a theoretical investigation of alternative models. PLoS Comp Biol 11:e1004348

Tyson JJ, Novak B (2014) Control of cell growth, division and death. Interface Focus 4:20130070

Vandecan Y, Biondi E, Blossey R (2016) Core-oscillator model of *Caulobacter crescentus*. Phys Rev E 93:062413

Venetz JE, Medico LD, Wölfle A, Schächle P, Bucher Y, Appert D, Tschan F, Flores-Tinoco CE, van Kooten M, Guennoun R, Deutsch S, Christen M, Christen B (2019) Chemical synthesis rewriting of a bacterial genome to achieve design flexibility and biological functionality. PNAS 116:8070–8079

Zhou B, Schrader JM, Kalogeraki VS, Abeliuk E, Dinh CB, Pham JQ, Cui ZZ, Dill DL, McAdams HH, Shapiro L (2015) The global regulatory architecture of transcription during the *Caulobacter* cell cycle. PLoS Genetics 11:e1004831

Diversity of Growth Patterns in the Alphaproteobacteria

Michelle A. Williams, David T. Kysela, and Pamela J. B. Brown

Abstract

Bacterial cell shape is highly variable in the Alphaproteobacteria, suggesting that multiple molecular mechanisms contribute to the generation of distinct morphologies. Although many growth patterns including lateral elongation, pre-septal elongation, polar elongation, budding and mid-cell growth have been observed among Alphaproteobacteria, only a few of the underlying mechanisms are well understood. We summarize the variations in growth patterning that determine morphological diversity in each order of the Alphaproteobacteria, highlighting the current understanding of pre-septal elongation and prosthecate tip budding in the *Caulobacterales* and unipolar growth and prosthecate cell budding in the *Rhizobiales*. The Alphaproteobacteria have evolved novel growth patterns through the diversification and regulation of the cell elongation machinery. Ultimately, expanding studies of cell growth in the Alphaproteobacteria will deepen our understanding of the mechanisms that underlie morphological complexity in the domain Bacteria.

M. A. Williams
Division of Biological Sciences, University of Missouri, Columbia, MO, USA

Department of Biology, McMaster University, Hamilton, ON, Canada

D. T. Kysela
Department of Biology, Indiana University, Bloomington, IN, USA

Département de Microbiologie, Infectiologie et Immunologie, Université de Montréal, Montréal, Canada

P. J. B. Brown (✉)
Division of Biological Sciences, University of Missouri, Columbia, MO, USA
e-mail: brownpb@missouri.edu

E. Biondi (ed.), *Cell Cycle Regulation and Development in Alphaproteobacteria*,
https://doi.org/10.1007/978-3-030-90621-4_7

1 Introduction

1.1 Ecological Diversity of Alphaproteobacteria

The breadth of environmental niches occupied by Alphaproteobacteria ranges from free-living bacteria in salt water, freshwater, and soil environments, to hydrothermal vents, and Antarctic sea ice. In addition to a wide range of geographical niches, the Alphaproteobacteria can also adopt various lifestyles including: plant symbionts, plant pathogens, facultative intracellular pathogens, and obligate intracellular pathogens. Given the assortment of lifestyles and niches, it is not surprising that these bacteria have a high degree of phylogenetic, metabolic, and morphological diversity (Fig. 1a; Randich and Brun 2015; Batut et al. 2004; Venter et al. 2004). The class Alphaproteobacteria is comprised of six highly diverse orders: *Rhizobiales, Caulobacterales*, *Rhodospirillales*, *Rhodobacterales, Sphingomonadales*, and *Rickettsiales* (Fig. 1a) and contains bacteria with a striking array of morphologies including asymmetric rods, buds, spirals, stars, and prosthecae (Fig. 1b). In this chapter, the term *prosthecae* is defined as appendages that are true extensions of the cell envelope and includes the stalks of the *Caulobacterales* (Staley 1968). These orders are rich in morphological diversity, but we are just beginning to understand the mechanisms that generate such morphological complexity. Looking only at cell shape and microcolony formation, early investigators surmised that many Alphaproteobacteria must grow fundamentally different from other rod-shaped bacteria (Whittenbury and Dow 1977; Hirsch 1974). These initial observations provided foundational insights into the growth mechanisms required to faithfully reproduce unique bacterial shapes. Still, the underlying molecular mechanisms and evolution of bacterial shapes remains largely unknown for a majority of the Alphaproteobacteria.

1.2 Methods to Visualize Bacterial Growth Patterning

Technological innovations and improvements in microscopy have enabled more detailed studies of bacterial growth patterning. Growth patterning in bacteria refers to the localized insertion of new cell wall (referred to as peptidoglycan) into the preexisting wall material. The peptidoglycan is a stress-bearing meshwork of proteins and sugars that is part of the cell envelope. In Gram-negative bacteria, the cell wall is a thin layer sandwiched between the cytoplasmic and outer membrane, while in gram-positive bacteria, the cell wall is thicker and connected by a myriad of proteins and polysaccharides. The cell wall has three main functions: first, to provide protection from osmotically stressful environments, second, provide support for cell envelope spanning structures such as flagella, pili and secretion systems, and third, maintain and faithfully reproduce the cell shape (Vollmer et al. 2008; Weidel and Pelzer 1964). To function properly, the peptidoglycan has a dedicated structure comprised of polysaccharide chains that are made up of two alternating subunits

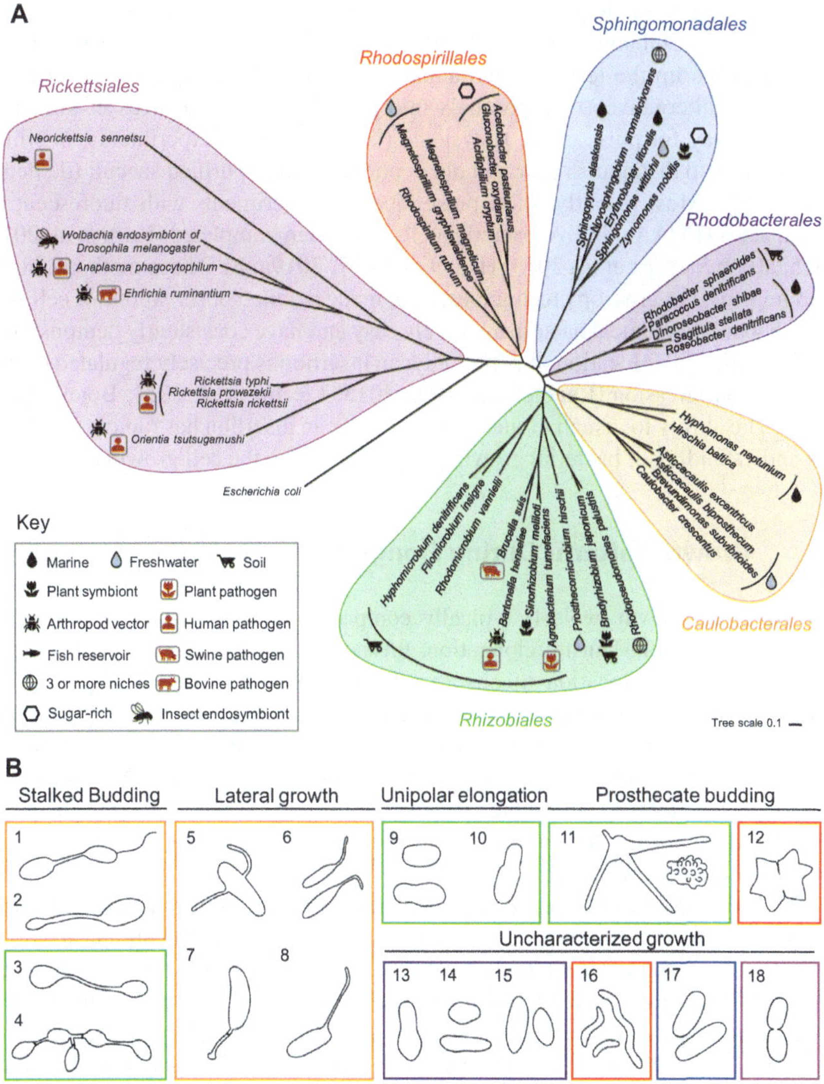

Fig. 1 Phylogeny and morphology of Alphaproteobacteria. (**a**) A concatenated alignment of 37 conserved protein sequences was obtained using Phylosift (Darling et al. 2014). The maximum likelihood phylogeny was estimated using RAxML (Stamatakis 2014) under the LG model of amino acid substitution with gamma-distributed rate variation and was formatted using iTOL (Letunic and Bork 2016). Tree scale represents the branch length with 0.1 substitutions per site. Clade colors indicate the phylogenetic order: *Caulobacterales* (orange), *Rhizobiales* (green), *Rhodobacterales* (purple), *Sphingomonadales* (blue), *Rhodospirillales* (red), and *Rickettsiales* (pink). *Escherichia coli* was used as an outgroup. Icons represent the primary niche for each species as listed in the key. (**b**) Schematics of morphological diversity found within the Alphaproteobacteria grouped by mode of reproduction. Colored boxes represent the phylogenetic order

of N-acetylglucosamine (Glc*N*Ac) and *N*-acetylmuramic acid (Mur*N*Ac). Attached to the Mur*N*Ac subunit is a pentapeptide chain with D-amino acids. Typically, D-alanine makes up the terminal two amino acids. The discovery that bacteria can incorporate alternative D-amino acids into peptidoglycan (Caparros et al. 1992) paved the way for pulse chase D-cysteine (D-cys) labeling experiments, in which it is possible to distinguish new and old peptidoglycan in purified sacculi (de Pedro et al. 1997). Most recently, short pulse labeling experiments with fluorescent D-amino acids (FDAAs) in live bacterial cells have been completed (Kuru et al. 2012, 2015, 2019; Siegrist et al. 2013; Hsu et al. 2017, 2019a, b). These methods, along with improved microscopy techniques, have made detailed observations of cell wall growth in non-canonical bacteria relatively easy and have consistently demonstrated that the subcellular location of peptidoglycan insertion is precisely regulated during cell cycle progression (Randich and Brun 2015; Liechti et al. 2014; Botella et al. 2017). The ability to visualize sites of peptidoglycan insertion has rapidly improved our understanding of bacterial growth patterning during the cell growth cycle.

1.3 Growth Patterns During Elongation

The bacterial growth cycle is typically comprised of two distinct phases: elongation and division. During elongation, the new peptidoglycan is synthesized at a bacterial-specific location to enable cells to increase in length (Fig. 2). The Alphaproteobacteria have evolved novel strategies for spatiotemporal regulation of peptidoglycan synthesis machinery during elongation, leading to the establishment of diverse growth patterns. Growth patterns during elongation include: insertion of peptidoglycan laterally along the sidewalls (Fig. 2d, e), budding (Fig. 2a, b, d), unipolar growth (Fig. 2c), insertion of peptidoglycan near mid-cell (Fig. 2e), and uncharacterized growth patterns (Fig. 1b).

The lateral insertion of peptidoglycan along the sidewalls of rod-shaped bacteria is best understood in *Escherichia coli* and *Bacillus subtilis*. Computational and high-resolution microscopy techniques have revealed that short filaments of the actin homolog MreB are targeted to areas of negative curvature and move circumferentially around the cell to maintain rod shape (Ursell et al. 2014; Dominguez-Escobar et al. 2011). Movement of MreB complexes is driven by cell wall synthesis (Garner et al. 2011; van Teeffelen et al. 2011). Cell wall synthesis during elongation requires enzymes including the monofunctional penicillin-binding protein (PBP)2 (DD-

Fig. 1 (continued) corresponding to color of the bacterial orders in A. 1. *Hyphomonas neptunium* 2. *Hirschia baltica* 3. *Hyphomicrobium denitrificans* 4. *Rhodomicrobium vannielii* 5. *Asticcacaulis biprosthecum* 6. *Brevundimonas halotolerans* 7. *Caulobacter crescentus* 8. *Asticcacaulis excentricus* 9. *Agrobacterium tumefaciens* 10. *Brucella abortus* 11. *Prosthecomicrobium hirschii* 12. *Stella humosa* 13. *Sagittula stellata* 14. *Roseovarius tolerans* 15. *Rhodobacter sphaeroides* 16. *Magnetospirillum magentotactum* 17. *Novosphingobium acidiphilum* 18. *Rickettsia typhi*

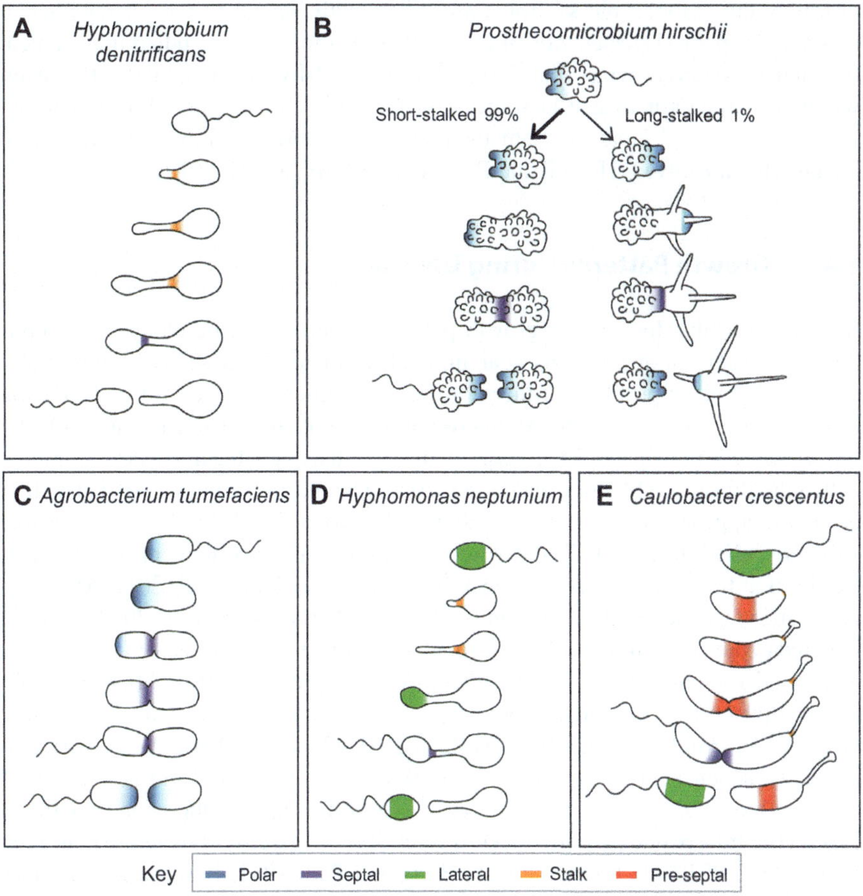

Fig. 2 Growth patterning in representative Alphaproteobacteria. Colors represent type of growth patterning: polar elongation (blue), septal elongation (purple), lateral elongation (green) stalk elongation (orange), pre-septal elongation (red). Typical Alphaproteobacterial cell cycle for the species shown is as follows: a motile, swarmer cell differentiates into a non-motile sessile cell, elongates, synthesizes a flagellum or flagella at the daughter cell pole, and divides producing a motile daughter cell and a non-motile mother cell that resumes growth

transpeptidase), the shape, elongation, division, and sporulation (SEDS) protein RodA (glycosyltransferase) and the bifunctional PBP1a (glycosyltransferase and DD-transpeptidase) (Spratt 1975; Banzhaf et al. 2012; Meeske et al. 2016; Cho et al. 2016) (see Sect. 7). The integral membrane protein RodZ acts as a linker between peptidoglycan biosynthesis machinery in the periplasm and the cytoplasmic scaffolding complex MreB (Morgenstein et al. 2015). Thus, peptidoglycan insertion and cytoplasmic scaffolds are intrinsically linked and work together to maintain proper rod shape. MreB-based cell elongation is well studied in canonical rod-shaped bacteria (Typas et al. 2011; Rohs et al. 2018; Shi et al. 2018). However, within the Alphaproteobacteria, some of the *Rhizobiales* maintain rod shape without MreB, and the molecular mechanisms underlying alternative modes of elongation are largely unresolved (Howell and Brown 2016; Brown et al. 2012).

1.4 Growth Patterns During Division

In contrast to the diversity of growth patterns observed during elongation, peptidoglycan insertion at positions near mid-cell prior to cell division is universally conserved (Fig. 2; purple). Furthermore, the machinery required for cell division is well-conserved among the Alphaproteobacteria. Cell division begins with the sequential recruitment of ~30 proteins to mid-cell that assemble into a large complex called the divisome (Mahone and Goley 2020; Du and Lutkenhaus 2019). The first protein to appear at mid-cell is the tubulin-like protein FtsZ that forms a ring-like structure called the Z-ring. The Z-ring is comprised of short filaments of FtsZ that treadmill at the division site (Bisson-Filho et al. 2017; Yang et al. 2017; Monteiro et al. 2018; Caldas et al. 2019; Perez et al. 2019; Baranova et al. 2020) and are tethered to the divisome complexes through C-terminal interactions with membrane-associated proteins such as FtsA and ZipA (Haney et al. 2001). After assembly of the divisome is complete, peptidoglycan biosynthesis is activated (Addinall et al. 1997; Muller et al. 2007; Moll and Thanbichler 2009). As constriction proceeds, septal peptidoglycan is synthesized inwards to build the new poles of the daughter cells. Septal peptidoglycan synthesis requires the monofunctional PBP3 (DD-transpeptidase), the SEDS protein FtsW (glycosyltransferase), and the bifunctional PBP1b (Cho et al. 2016; Botta and Park 1981; Bertsche et al. 2006; Taguchi et al. 2019). The order of recruitment and essentiality of divisome proteins may vary across species, but the overall mechanism of cell division appears to be broadly conserved (Lutkenhaus and Du 2017). In addition to peptidoglycan insertion for division and growth of the main cell body, some Alphaproteobacteria including *Caulobacte*r and *Hyphomonas* use a specialized form of zonal peptidoglycan insertion at the cell body-stalk junction to enable stalk elongation (Fig. 2a, d, e; orange).

Here, we will review the current models of growth patterning within each of the Alphaproteobacterial orders, with an emphasis on the orders *Rhizobiales* and *Caulobacterales*. We highlight more recent mechanistic studies of growth in a few model Alphaproteobacteria including: *Caulobacter*, *Hyphomonas*, and *Agrobacterium*, and consider outstanding questions and perspectives for future research on growth patterning within the Alphaproteobacteria.

2 Growth Patterns of Budding Bacteria in the *Rhizobiales*

One of the earliest indications of diversity in growth patterning of bacteria is evident by the abundance of budding bacteria within the order *Rhizobiales*. Classically defined budding bacteria insert new cell wall material into a small area of the mother cell to form a daughter cell that is typically smaller in size (Hirsch 1974). Here, we describe three types of budding: (1) Budding from the tip of a prosthecae (prosthecate tip budding), (2) budding from the cell body of a prosthecate cell (prosthecate cell budding), and (3) budding from one pole of a typical rod-shaped cell (unipolar elongation). All three types of budding are found within the *Rhizobiales* (Fig. 1b; green boxes), resulting in a high degree of morphological complexity in this bacterial order. The prevalence of prosthecate tip budding, prosthecate cell budding, and unipolar elongation suggests that budding is an ancestral trait in the *Rhizobiales* and may have a common molecular mechanism.

2.1 Growth Patterns of Prosthecate Tip Budding Bacteria

Hyphomicrobium (Figs. 1b:3; 2a) and *Rhodomicrobium* (Fig. 1b:4) are the best characterized genera of prosthecate tip budding bacteria (Moore 1981). The prosthecae was classically termed a hypha, defined as an appendage from the main cell body required for reproduction. These prosthecate tip budding bacteria have dimorphic life cycles and typically consist of an ovoid mother cell with a prosthecae extended from one pole that terminates in a daughter cell bud (Hirsch 1974). Based on observations of synchronized populations, four phases of growth have been described for *Hyphomicrobium* and *Rhodomicrobium*: swarmer cell maturation, prosthecae growth, bud formation, and bud detachment. Each phase occurs in a sequential manner such that growth from one phase will finish before growth of the next phase begins (Fig. 2a; Wali et al. 1980; Moore and Hirsch 1973). The swarmer cell has a single polar flagellum that is shed before prosthecae biogenesis is initiated. The prosthecae is synthesized at one pole of the cell, and the cell wall is contiguous between mother cell and prosthecae. Prosthecae growth appears to occur at the mother cell-prosthecae junction, suggesting that the machinery required for prosthecae growth is spatially restricted to this junction. The prosthecae is required to produce a bud, which develops as a terminal swelling of the prosthecae. Whether the growth machinery is spatially restricted in the bud or diffuse within the bud remains to be explored. Before the daughter cell is released, the septum forms at the junction between the prosthecae and the cell body. Thus, the cell division machinery must be specifically targeted to this junction near the end of the cell cycle to enable septum formation. Finally, the bud is released from the tip of the prosthecae producing two morphologically and functionally distinct cells; a motile swarmer cell and a prosthecate cell (Moore 1981; Moore and Hirsch 1973).

In addition to the typical prosthecae tip budding cell cycle outlined above (Fig. 2a), *Hyphomicrobium* species have been observed with more complex morphologies. For example, some species can adopt a branched and chained morphology, and helical prosthecae have been observed under different environmental conditions (Holm et al. 1996). The morphological complexity of the prosthecae tip budding bacteria is a product of spatial and temporal regulation of growth machinery across the cell cycle to produce distinct areas of growth within the cell, but how is this machinery localized and regulated to produce more complex prosthecate morphologies? More in-depth studies are required to fully understand this mode of growth and the impact of changing environmental conditions on growth patterning. Future studies of *Hyphomicrobium* and other prosthecae tip budding bacteria belonging to the *Rhizobiales* may be hampered by long generation times, which can range from 6.5 to 14.5 h (Wali et al. 1980), and the difficulty to culture these bacteria. Improved modeling and sequencing techniques may enable rapid prediction of growth requirements, and high-throughput culturing methods may help circumvent the challenge of studying cell wall biogenesis in slow-growing bacteria (Henry et al. 2010; Oberhardt et al. 2015; Connon and Giovannoni 2002).

Like *Hyphomicrobium*, *Rhodomicrobium* can exhibit complex morphologies (Fig. 1b:4) that are suggestive of a polymorphic lifestyle, in which it can adopt either a traditional prosthecate tip budding cell cycle or a more complex cell cycle. For example, *Rhodomicrobium* cells can form attached mother–daughter cell pairs. Even though the cells remain attached, a peptidoglycan crossband is formed in the prosthecae nearest to the daughter cell completing compartmentalization of the cells (Conti and Hirsch 1965). Attached mother–daughter cell pairs continue to grow from multiple prosthecae, leading to a ball and chain morphology (Fig. 1b:4). Remarkably, prosthecae may also terminate in an exospore rather than a daughter cell bud highlighting the complexity of *Rhodomicrobium* cell biology (Whittenbury and Dow 1977). The intricate morphologies of *Rhodomicrobium* species present an interesting opportunity to ask what environmental conditions determine the switch to the chained morphology? How are these phenotypes regulated in a cell cycle dependent manner? How is the growth machinery spatially and temporally regulated to produce a single growth pole between a mother–daughter pair that is part of a larger chain? Further studies of *Rhodomicrobium* cell cycle and growth are necessary to elucidate the regulation of more complex bacterial morphologies and lifestyles.

Morphological observations of other *Rhizobiales* species indicate that many have a dimorphic lifestyle and grow by prosthecae tip budding. For example, the cell cycle of *Rhodopseudomonas palustris* indicates that it grows by prosthecae tip budding (Whittenbury and Dow 1977; Whittenbury and McLee 1967) and this bacterium produces a unipolar polysaccharide (Fritts et al. 2017) and polar intracytoplasmic membranes (LaSarre et al. 2018). In addition, some *Pedomicrobium* species have multiple, lateral prosthecae rather than a single, polar prosthecae (Moore 1981; Cox and Sly 1997). Overall, the growth patterns of species within the genera *Rhodopseudomonas*, *Pedomicrobium*, and *Filomicrobium* are largely unstudied. Characterization of the essential genes in *R. palustris* predicted to function in peptidoglycan biosynthesis (Pechter et al. 2015), coupled with data mining from recent or ongoing genome sequencing projects of species in the genera

Pedomicrobium and *Filomicrobium* (Henriques and De Marco 2015), may enable additional studies about growth and development in these bacterial species.

2.2 Growth Patterns of Prosthecate Cell Budding Bacteria

Unlike the prosthecae tip budding bacteria, appendages of the prosthecate cell budding bacteria are non-reproductive (Fig. 2b, Staley 1968). All prosthecate cell budding bacteria studied thus far appear to reproduce by budding directly from one pole of a prosthecae mother cell. In contrast, several phenotypes are variable within these bacteria including the number and length of prosthecae, presence of gas vesicles, and utilization of a dimorphic lifestyle. For example, *Prosthecomicrobium litoralum* is non-motile with multiple, short prosthecae (Bauld et al. 1983), while *Ancalomicrobium adetum* is non-motile, contains gas vesicles, and has 1–20 long prosthecae that can be bifurcated at the ends (Staley 1968). Although the mechanisms of prosthecae biosynthesis in the prosthecate cell budding bacteria remain wholly unexplored, it's intriguing to speculate that the long, bifurcated prosthecae of *A. adetum* may be a product of both growth at the cell body-prosthecae junction to elongate the prosthecae and growth at the tip of the prosthecae to generate the bifurcated ends. Furthermore, how is the placement and arrangement of multiple long- or short-prosthecae determined? Studies exploring the synthesis and regulation of prosthecae in the prosthecate cell budding bacteria will contribute to our understanding of more specialized growth patterns.

Among the prosthecate cell budding bacteria, the cell cycle and growth pattern is best characterized in *Prosthecomicrobium hirschii* (Figs. 1b:11; 2b). *P. hirschii* can adopt a pleomorphic lifestyle, with both a non-motile, long-prosthecate form with 3–12 long prosthecae (Fig. 2b, right) and a motile, short-prosthecate form with numerous, regularly placed prosthecae (Fig. 2b, left). Both short-prosthecate and long-prosthecate forms persist in the same culture (Fig. 1b:11; Staley 1968, 1984), and short-prosthecate cells of *P. hirschii* have a dimorphic lifestyle (Fig. 2b; Williams et al. 2016). Swarmer cells have a single polar flagellum that is shed during the transition into a sessile cell that is capable of elongating. Labeling of short-prosthecate cells with FDAAs confirmed that short-prosthecate cells grow by prosthecae cell budding from a single pole. Furthermore, prior to cell division new cell wall material is synthesized at mid-cell. Under the conditions tested, a short-prosthecate mother cell gave rise to a short-prosthecate daughter cell 99% of the time. Rarely, a short-prosthecate mother cell would give rise to a long-prosthecate daughter cell. The long-prosthecate morphotype is non-motile and is equally likely to give rise to either another long-prosthecate cell or a short-prosthecate cell. The variations in the Alphaproteobacterial growth cycle that allow for such morphological plasticity to emerge remains to be determined, and some key questions regarding *P. hirschii* cell cycle remain to be addressed. At what level is the switch between the short-prosthecate and long-prosthecate morphotypes regulated? Perhaps the transition is regulated at the genetic level via DNA methylation or gene silencing. Alternatively, the switch may be determined at the cellular level via quorum sensing. *P. hirschii* produces an aryl-homoserine lactone signal that regulates biofilm formation and pigmentation but does not appear to impact the cell morphology (Liao et al. 2018). Ultimately, identifying the trigger for morphological changes will allow for a better understanding of how the cell elongation machinery is regulated during short and long-prosthecate growth of *P. hirschii*, and will likely inform studies of other prosthecate cell budding bacteria with pleomorphic lifestyles.

2.3 Unipolar Elongation of Rod-Shaped Bacteria

Bacteria with unusual morphologies such as prosthecae easily attracted scientific interest, and observations led to the hypothesis that these bacteria grow by budding; however, many genera in the *Rhizobiales* such as *Agrobacterium* (Fig. 1b:9) and *Brucella* (Fig. 1b:10) have a typical rod-shaped morphology. Despite their close relationship to other budding bacteria, it was assumed that the rod-shaped species in the *Rhizobiales* elongate using lateral insertion of peptidoglycan along the sidewall since this mechanism is shared by most rod-shaped bacteria such as *E. coli* (Kuykendall 2005).

An early indication that rod-shaped bacteria in the *Rhizobiales* may use an alternative mechanism of cell elongation is the atypical morphologies induced by blocks in cell division. When cell division is blocked, most rod-shaped bacteria form long, smooth filaments (Dai and Lutkenhaus 1991). In contrast, blocking cell division in *Rhizobium* and *Agrobacterium* species induces branching and bulging phenotypes (Fujiwara and Fukui 1972; Latch and Margolin 1997; Kaneshiro et al. 1983), indicating that these species may grow fundamentally differently from other rod-shaped bacteria. Careful examination of *A. tumefaciens* branching mutants concluded that branches were the result of a tip splitting event of the growing pole (Fujiwara and Fukui 1974). These observations along with genome sequences of several *Rhizobiales* confirm that these bacteria must use an alternative mechanism for cell elongation as the genes encoding the canonical elongation machinery including MreBCD, RodA, and PBP2 are absent from the genomes of all sequenced *Rhizobiales* (Brown et al. 2012; Margolin 2009; Daniel and Errington 2003).

A combination of time-lapse microscopy and labeling studies using Texas Red succinimidyl ester (TRSE), D-cys, or FDAAs led to the finding that *Agrobacterium tumefaciens* undergoes unipolar elongation (Brown et al. 2012; Kuru et al. 2012). Since then, *A. tumefaciens* has emerged as a model bacterial system for mechanistic studies of unipolar elongation, which consists of three phases (Fig. 2c). First, peptidoglycan is inserted at the new pole of the cell to allow cells to increase in length. Next, growth is terminated from the new pole and resumes near mid-cell. Finally, cell division occurs, producing two new poles which are primed for unipolar growth. While the growth pattern has been well-established, key questions remain. What enzymes are required for polar elongation? How is growth restricted to the pole during elongation? How is unipolar growth coordinated with other processes such as DNA replication, chromosome segregation, and cell division?

A bioinformatics approach to mine the *A. tumefaciens* genome for genes encoding enzymes capable of peptidoglycan biosynthesis revealed the presence of six genes encoding high molecular weight (HMW) PBPs (Cameron et al. 2014). A subset of these PBPs are likely to be recruited to the growth pole and function during cellular elongation. Labeling of *A. tumefaciens* cells with a fluorescent antibiotic that binds PBPs confirms that at least some PBPs are localized to the growth pole, and citrine-PBP1a is observed at the growth pole (Cameron et al. 2014). The chemical composition of *A. tumefaciens* peptidoglycan also suggests a potential role for LD-transpeptidases (LDTs) in peptidoglycan biosynthesis (Brown et al. 2012). LDTs are enriched in the genomes of *Rhizobiales*, and at least one LDT localizes to the growth pole of *A. tumefaciens*, supporting the hypothesis that these enzymes contribute to unipolar growth (Cameron et al. 2014). Additional investigations are needed to determine the contributions of PBPs and LDTs to peptidoglycan insertion at the pole during elongation, at mid-cell prior to cell division, or to both processes. In addition, it will be necessary to identify and characterize the proteins that are responsible for spatial and temporal regulation of the activity of the biosynthetic enzymes.

While the candidate enzymes for polar peptidoglycan insertion have been identified, regulatory mechanisms which restrict growth to the pole during elongation remain elusive. Polarity determinants with potential to serve as scaffolding proteins including FtsZ and FtsA persist at the growth pole (Cameron et al. 2014; Zupan et al. 2013); however, these proteins likely primarily function in cell division. Indeed, depletion of FtsZ does not block unipolar growth (Howell and Brown 2016). The polar organizing protein, PopZ also persists at the growth pole (Grangeon et al. 2015), but deletion or depletion of PopZ does not block unipolar growth (Grangeon et al. 2017; Howell et al. 2017). Instead, the loss of PopZ severely impairs chromosome segregation (Ehrle et al. 2017). Together these observations suggest that the growth pole is an important hub for the coordination of several important cell cycle events including elongation, DNA replication, chromosome segregation, and cell division. Recently, GPR (for Growth Pole Ring), a large apolipoprotein that forms a ring at the growth pole in *A. tumefaciens*, was shown to be essential for rod-shape maintenance, and depletion of this GPR led to rounded cells (Zupan et al. 2019). This phenotype implicates GPR as a likely candidate to scaffold PG enzymes during elongation. Future studies aimed at determining the role that GPR plays in recruiting proteins to the growth pole may shed light on the mechanisms that the Rhizobiales use to scaffold proteins during polar elongation.

While *A. tumefaciens* is emerging as a model for the study of unipolar growth in rod-shaped Alphaproteobacteria, bioinformatics searches illustrate that all *Rhizobiales* genomes lack the canonical machinery for dispersed lateral peptidoglycan insertion (Brown et al. 2012; Margolin 2009; Daniel and Errington 2003). This observation suggests other rod-shaped bacteria within the *Rhizobiales* may be likely to exhibit unipolar growth. TRSE or D-cys labeling experiments in *Brucella abortus*, *Sinorhizobium meliloti*, and *Ochrobactrum anthropic* have confirmed unipolar growth is the main mode of elongation in these rod-shaped species as well (Brown et al. 2012). In *S. meliloti*, a pair of essential proteins, RgsP and

RgsM, which contribute to unipolar elongation have been described (Schaper et al. 2018). RgsP is a 7-transmembrane receptor protein with no similarity to known proteins involved in the regulation of cell wall biosynthesis. Loss of RgsP causes growth inhibition, aberrant cell shape, and altered peptidoglycan composition and may serve as a scaffold for recruitment of growth pole proteins. RgsM is an endopeptidase that interacts with RgsP and may cleave peptidoglycan crosslinks to promote the insertion of new peptidoglycan strands. More recently, additional Rgs proteins as well as the Tol-Pal complex have been implicated in promoting unipolar growth and cell division of *S. meliloti*, though the function of most of these proteins in the regulation of growth remain unclear (Krol et al. 2020). The conservation of most of these proteins in Rhizobiales suggests that a growth pole complex may function in place of the canonical elongasome, effectively limiting peptidoglycan biosynthesis to one pole.

It is now clear that members of the order *Rhizobiales* elongate through different modes of budding that include prosthecate tip budding, prosthecate cell budding, and unipolar growth. The prevalence of these growth modes indicates that budding is widespread in the *Rhizobiales* and raises questions such as what is the advantage of unipolar growth and why is one mode of budding preferred over another? Ongoing studies in the *Rhizobiales*, using a combination of in silico, genetic, and imaging approaches may shed light on these questions.

3 Diverse Growth Patterns of Stalked Bacteria in the *Caulobacterales*

The Alphaproteobacterial order *Caulobacterales* is morphologically diverse with many species capable of producing stalks (within the *Caulobacterales* prosthecae are commonly referred to as stalks). Some of the stalked bacteria reproduce by budding using the stalks as reproductive structures (Fig. 1b:1–2). The stalked budding bacteria currently includes three genera: *Hyphomonas* (Fig. 1b:1; Fig. 2d), *Hirschia* (Fig. 1b:2), and *Hellea*. In contrast, the stalks are not used as reproductive structures in a distinct subset of the stalked bacteria (Fig. 1b:5–8), which includes the well-characterized *Caulobacter crescentus* (Fig. 1b:7; Fig. 2e). *C. crescentus* uses dispersed insertion of peptidoglycan along the lateral sidewalls during elongation of the main cell body and a specialized form of zonal growth to promote stalk elongation. Non-reproductive stalks are thought to function in nutrient uptake and starvation triggers stalk elongation (Schmidt and Stanier 1966). Finally, some members of the *Caulobacterales* do not produce stalks. It is generally inferred that non-stalked *Caulobacterales* elongate using the canonical MreB-dependent mechanism; however, few detailed studies of the growth pattern in these bacteria have been completed.

3.1 Stalked Budding Bacteria Within the *Caulobacterales*

Among the genera of stalked budding bacteria in the *Caulobacterales*, the growth pattern is best characterized in *Hyphomonas* (Fig. 2d). Based on its morphology and development, the stalked budding bacterium *Hyphomonas neptunium* was originally characterized as a member of the genus *Hyphomicrobium* (Leifson 1964), but was later reclassified into the order *Caulobacterales* based on metabolic analysis, DNA hybridization, and 16S rRNA sequencing (Weiner et al. 2000; Badger et al. 2006). Despite the reclassification, *Hyphomonas* species have a similar cell cycle and morphology to the stalked budding bacteria in the *Rhizobiales* (Compare Fig. 2a, d; Zerfas et al. 1997), the stalk cell membrane and cell wall is contiguous with the mother cell body, and a bud forms at the tip of the stalk (Zerfas et al. 1997). Stalk formation in *Hyphomonas* is essential for reproduction, and inhibition of DNA synthesis with nalidixic acid results in elongated stalks with no buds, suggesting bud formation is coupled to chromosome replication (Weiner and Blackman 1973). In *Hyphomonas neptunium* chromosome segregation requires a two-step process (Jung et al. 2019). First, the origins are partitioned to the mother cell poles. Next, chromosome is translocated across the stalk into the bud. The latter step is delayed until the bud is sufficiently large to accommodate the DNA. The recent development of genetic tools and image analysis software for the manipulation of *H. neptunium* (Jung et al. 2015; Hartmann et al. 2020) has opened doors to more in-depth studies of the growth and morphology of stalked budding bacteria.

Growth patterns during the *H. neptunium* cell cycle were revealed by FDAA labeling, and four distinct regions of peptidoglycan insertion were identified, which involves alternating rounds of dispersed and zonal peptidoglycan insertion (Fig. 2d; Cserti et al. 2017). First, dispersed peptidoglycan insertion occurs in the swarmer cell, followed by zonal peptidoglycan insertion at the base of the stalk to promote stalk growth. As the stalk forms, the cell becomes rounded near the old pole, and tapered at the stalked pole forming a tear drop shape. Upon completion of stalk elongation, another round of dispersed peptidoglycan insertion at the tip of the stalk occurs, which forms the daughter cell bud. Finally, zonal peptidoglycan insertion at the daughter cell bud neck drives cell division to release the newly formed daughter cell. The initial time for a mother cell to grow a stalk and produce a bud is approximately 4 h, but once the stalk is formed, new buds form at the tip of the stalk every 2.5 h (Cserti et al. 2017).

In contrast to the *Rhizobiales*, *H. neptunium* encodes genes for all the essential elongasome proteins including MreB and PBP2. Consistent with the zones of peptidoglycan insertion during growth, both MreB and PBP2 localize in a diffuse pattern in the mother cell, then to the base of the stalk during stalk elongation, and again in a diffuse pattern in the newly forming daughter cell (Cserti et al. 2017). Based on their localization patterns and the inability to construct deletion mutants, it appears that both MreB and PBP2 are essential members of the elongasome complex. Other peptidoglycan remodeling enzymes also likely contribute to elongation in *Hyphomonas*. For instance, *H. neptunium* contains three D-alanine carboxypeptidases, of which DacB is homologous to *E. coli* DacN, and DacH and DacL are specific to *H. neptunium*. While only the *dacB* mutant has morphological defects, DacL localizes to the base of the stalk during stalk elongation (Cserti et al. 2017). Identification and characterization of additional proteins unique to the stalked budding bacteria and that localize to sites of elongation will help identify key features of the stalked budding mode of growth. Given that *H. neptunium* and *C. crescentus* encode similar elongasome proteins and display similar peptidoglycan profiles, it is likely that the differences in growth patterning are a product of novel binding partners that specifically regulate stalk synthesis and the transition between stalk and dispersed growth. Therefore, dissecting the protein–protein interactions of elongasome machinery is an interesting avenue for future studies.

Other stalked budding genera in the *Caulobacterales* include *Hirschia* and *Hellea* (Schlesner et al. 1990; Alain et al. 2008). While detailed mechanistic studies of growth patterning have not been completed, both bacteria are reported to grow by budding from the end of a stalk. The completion of the genome of the type species *Hirschia baltica* (Chertkov et al. 2011) and *Hellea balneolensis* should enable comparative genomic approaches to uncover features that are common among the marine stalked budding bacteria. Since Alphaproteobacteria represent a majority of marine bacteria (Venter et al. 2004), these types of studies have the potential to identify proteins essential for stalked budding and may reveal the adaptations required for a dimorphic lifestyle in marine environments.

3.2 *Caulobacterales* with Non-reproductive Stalks

Bacteria with stalks are ubiquitous in water samples, and several genera of stalked bacteria have been identified including: *Caulobacter*, *Asticcacaulis*, and *Brevundimonas* (Fig. 1b:5–8). *Caulobacterales* stalks can vary in their length, placement, and number depending on species. Stalks in the order *Caulobacterales* are non-reproductive extensions of the main cell body that are devoid of cytoplasmic proteins and DNA, and are composed of a cell envelope that is separated by protein cross-bands (Poindexter 1964).

The stalk likely functions in nutrient uptake (Schlimpert et al. 2012; Hughes et al. 2010; Wagner et al. 2006; Ireland et al. 2002; Curtis 2017; Larson and Pate 1976; Tam and Pate 1985), and in phosphate-limiting conditions the *C. crescentus* stalk can elongate up to 10 times its normal length, which is thought to increase the rate of phosphate uptake (Schmidt and Stanier 1966; Gonin et al. 2000). In *C. crescentus*, stalk lengthening may not only be a mechanism to increase nutrient uptake. Instead, phosphate levels could be used as a proxy for the amount of available nutrients and stalk lengthening enables the cell body to move away from the surface and into a more favorable nutrient environment (Klein et al. 2013). This is a viable model for *C. crescentus* since the stalk is polarly localized (Fig. 1b:7), and the tip of the stalk produces an adhesin used for attachment to surfaces (Poindexter and Cohen-Bazire 1964; Berne et al. 2013; Merker and Smit 1988). In contrast, stalks in *Asticcacaulis excentricus* are typically sub-polar (Fig. 1b:8), and the adhesin is present on the cell pole (Poindexter and Cohen-Bazire 1964). *Asticcacaulis biprosthecum* produces two lateral stalks on opposite sides of the cell body and synthesizes a polar adhesin (Fig. 1b:5; Pate et al. 1973; Li et al. 2012). Thus, *Asticcacaulis* stalks are not capped with an adhesion and do not contribute to surface attachment, but the stalks still elongate in phosphate-limiting conditions. Finally, most *Brevundimonas* species produce short, polar stalks that are capped with a polar adhesin (Fig. 1b:6; Curtis 2017). A survey of 18 different *Brevundimonas* species showed that stalk formation is rare in nutrient rich conditions and that stalks may be produced on an as-needed basis (Curtis 2017). This illustrates that stalk formation is highly variable between species, and under different nutrient conditions.

Although the exact function of the stalk remains elusive, the mechanism of stalk biosynthesis has been examined. In *C. crescentus*, stalk elongation requires the localized insertion of peptidoglycan at the cell body-stalk junction, which is dependent on a specialized complex comprised of both elongation-specific (MreB, RodZ, RodA, PBP2) and division-specific (DipM, SdpA, SdpB, CrbA) components (Wagner et al. 2005; Divakaruni et al. 2007; Billini et al. 2019). MreB mediates the recruitment of both synthetic and lytic proteins to the stalked pole, enabling stalk elongation (Billini et al. 2019). Remarkably, peptidoglycan insertion occurs via an expansion of the polar cap with subsequent remodeling to form the new stalk segment. LD-transpeptidases make significant contributions to crosslinking the peptidoglycan in the stalk, though not as a core component of the polar complex (Billini et al. 2019; Stankeviciute et al. 2019). In addition, the cytoskeletal bactofilins BacA/B localize to the stalked pole and act as a scaffold to recruit the DD-transpeptidase PBPC to the base of the stalk (Kuhn et al. 2010). In turn, PBPC recruits the stalk-specific protein StpX (Hughes et al. 2010, 2013). The targeting of BacA to the stalked pole is independent of MreB, indicating that bactofillins may serve as a distinct morphogenetic module to promote stalk formation (Billini et al. 2019). Together, these findings suggest that repurposing of existing peptidoglycan synthesis machineries may enable the emergence of novel morphological features such as stalks. Stalk synthesis is coupled to cell cycle progression in *C. crescentus*. In fact, stalk synthesis is transcriptionally regulated by a cytoplasmic phosphorelay signaling pathway, suggesting stalk growth can be controlled from intracellular signals (Xu et al. 2009; Biondi et al. 2006). Regulation of stalk synthesis is likely multi-faceted, including external environmental signals such as phosphate levels and unknown internal signals.

Similar to *C. crescentus*, *A. biprothecum* stalks function in nutrient transport, and stalk synthesis is regulated by external signals such as phosphate starvation and also by intrinsic cell cycle signals (Pate et al. 1973; Porter and Pate 1975). Both BacA and SpmX have recently been implicated as key factors in promoting *A. biprosthecum* stalk biosynthesis (Caccamo et al. 2020). In contrast to *C. crescentus*, stalk synthesis is not cell cycle regulated in *Brevundimonas*, and there is little conservation of stalk-associated proteins between *Caulobacter* and *Brevundimonas* species (Curtis 2017). It will be necessary to explore the mechanisms of stalk biogenesis in multiple stalked bacteria to gain a comprehensive understanding of this process in the Caulobacterales.

In addition to studies detailing the mechanism of stalk biogenesis, labeling and localization studies in *C. crescentus* have shed light on the complex growth patterns that give rise to asymmetric growth and division. D-cys labeling of peptidoglycan and localization of the key elongation-specific proteins MreB and MurG showed that cell elongation in *C. crescentus* occurs by two distinct mechanisms: dispersed growth in swarmer cells, and pre-septal elongation in early and late-pre-divisional cells (Fig. 2e; Aaron et al. 2007; Takacs et al. 2013). The *Caulobacter* cell cycle begins with a motile swarmer cell that grows in a dispersed manner (Fig. 2e green) characterized by diffuse patterns of MreB and MurG (Aaron et al. 2007). Based on conservation of the core cell elongation proteins, dispersed growth of swarmer cells in *C. crescentus* likely occurs by a similar mechanism to *E. coli* and *B. subtilis*. Next, the swarmer cell sheds its flagellum and transitions into a sessile cell. In this early pre-septal cell, FtsZ is recruited to the mid-cell and in turn recruits MreB and MurG. Bands of FtsZ, MreB, and MurG at mid-cell initiate an FtsZ-dependent mode of pre-septal elongation that is responsible for the majority of peptidoglycan biosynthesis during elongation (Fig. 2e red). FtsZ is known to be a scaffold for peptidoglycan biosynthesis during cell division, but surprisingly it may also play a major role in cell elongation in many Alphaproteobacteria. Prior to cell division, MreB and MurG become dispersed again in the new pole daughter cell, while the stalked daughter cell resumes pre-septal elongation (Aaron et al. 2007; Takacs et al. 2013). Thus, growth patterning in *C. crescentus* has been well characterized, and shown to be regulated in part by specific scaffolding proteins. However, several questions on how the peptidoglycan biosynthesis machinery and cytoskeletal regulators interact and coordinate throughout the cell cycle remain to be explored. Since many aspects of the *Caulobacter* cell cycle are known to be under transcriptional, translational, and post-translational control, how is growth patterning also regulated at each level? *C. crescentus* is a well-established model system that is poised to answer some of the detailed mechanistic questions regarding regulation of cell growth patterning in the Alphaproteobacteria; however, the growth patterns of *Asticcacaulis* and *Brevundimonas* should also be experimentally determined as they may be distinct.

4 Cell Growth in the *Rhodospirillales*

Across the Alphaproteobacteria, many species have adapted strategies that allow them to occupy diverse niches (Fig. 1a). One of the most unique mechanisms is employed by magnetotactic bacteria in the order *Rhodospirillales*. Magnetotactic bacteria align themselves with the magnetic field of the earth to navigate toward microaerophilic environments mainly in fresh and salt water environments (Uebe and Schuler 2016). This type of motility is called magnetotaxis and requires magnetotactic bacteria to synthesize magnetosomes, which are comprised of crystals of magnetite found in inner membrane invaginations (Faivre and Schuler 2008). The cell cycle of magnetotactic bacteria must be highly coordinated to ensure that both daughter cells inherit magnetosomes, but cell growth and division of magnetotactic bacteria is just beginning to be resolved. Magnetotactic bacteria exhibit diversity in cell shape, as spiral, cocci, and vibriod morphologies have been described (Sakaguchi et al. 2002; Lefevre and Bazylinski 2013; Schleifer et al. 1991). Relatively little is known about the growth and development of coccus and vibrio-shaped magnetotactic bacteria, however *Magnetospirillum gryphiswaldense* (Fig. 1b:9) is emerging as model for mechanistic studies of cell cycle progression.

Studies in *M. gryphiswaldense* have primarily focused on the spatiotemporal regulation of magnetosomes during the cell cycle (Lin and Pan 2011; Katzmann et al. 2011; Staniland et al. 2010); however, these studies have also provided some indirect clues about the growth patterning in this bacterium. Tracking polyhydroxybutyrate (PHB) granules in elongating cells of *M. gryphiswaldense* revealed that PHB granules near the mid-cell separate as the cells grow in length, whereas PHB granules near the pole remain fixed (Katzmann et al. 2011). This suggests a mechanism of growth reminiscent of pre-septal elongation in *C. crescentus* (Aaron et al. 2007; Takacs et al. 2013). Treatment of *M. gryphiswaldense* with cephalexin causes cells to filament, similar to *C. crescentus*, consistent with the possibility that a combination of lateral growth and pre-septal elongation may be used during elongation. Following cell division, the two daughter cells often have different sizes: one daughter cell is typically ~15% smaller than the other daughter cell. Uneven daughter cell length suggests the constriction sites are placed asymmetrically and is consistent with the presence of an Alphaproteobacterial cell cycle in Magnetospirilla (Katzmann et al. 2011). Recent studies of cell division in *M. gryphiswaldense* indicate that MipZ1 contributes to spatiotemporal regulation of FtsZ-ring formation (Toro-Nahuelpan et al. 2019) and PopZ contributes to the establishment of polarity, cell division, and chromosome segregation (Pfeiffer et al. 2019). The specific mechanisms of elongation and division in *M. gryphiswaldense* are just beginning to be revealed, and it remains to be determined if these mechanisms are conserved in non-magnetic *Rhodospirillales* such as *Rhodospirillum rubrum*.

Another group within the *Rhodospirillales* contains prosthecate bacteria with a unique, star-shaped morphology. *Stella humosa* is the type strain for this group and resembles a flat, six-pointed star with radial symmetry (Fig. 1b, 12; Vasilyeva 1985). Microscopic observations of *S. humosa* illustrate that this bacterium divides by forming a cross-wall between its narrowest sides. After division, each daughter cell has three prosthecae and must regenerate three new prosthecae to maintain its six-pronged morphology (Vasilyeva 1985). The star-shaped morphology only persists in low-nutrient conditions, and these bacteria can be asymmetrical or sphere-shaped under different environmental conditions (Hirsch and Schlesner 1981). *S. humosa* and other members of the prosthecate *Rhodospirillales* are a fascinating example of the breadth of bacterial morphologies that await further characterization. The recent completion of the genome sequences for three *Stella* species may provide new insights into these captivating microbes (Shibai et al. 2019).

The *Rhodospirillales* also includes the acetic acid bacteria, which are able to grow in ethanol and sugar-rich environments and are often insect symbionts (Crotti et al. 2010). Although little is known about the mechanisms of elongation in the acetic acid bacteria, a recent analysis of 20 species representing the five major orders (*Rickettsiales* was not included) revealed that the peptidoglycan of the Alphaproteobacteria clusters into three groups based on similar peptidoglycan composition (Espaillat et al. 2016). Group 1 was exclusive to the acetic acid bacteria of the *Rhodospirillales*, and peptidoglycan from group 1 was surprisingly distinct from the other two groups, containing novel, L,D-(1-3) crosslinks and acylated muropeptides. These novel modifications were shown to confer resistance to endopeptidases of co-habiting bacteria and evade recognition by host immune responses, respectively. Therefore, the acetic acid bacteria may have evolved novel mechanisms for the synthesis or modification of peptidoglycan. Characterization of growth patterning in the acetic acid bacteria remains an interesting area for future investigations.

5 Cell Growth in the *Rhodobacterales*

Like many Alphaproteobacteria, species in the order *Rhodobacterales* demonstrate features of asymmetry, including rosette formation, polar holdfast production, and motile and non-motile cells in the same culture (Slightom and Buchan 2009; Bruhn et al. 2005; Labrenz et al. 1998). Since budding typically results in a narrower, pointed, or tapered cell pole prior to bud development these morphologies are used to identify species that likely elongate by budding. Many *Rhodobacterales* have these characteristics (Fig. 1b:13–15; Gonzalez et al. 2003; Labrenz et al. 1999). For example, both *Phaeobacter inhibens* and *Sagittula stellata* are clearly narrower at one pole and produce a unipolar polysaccharide at the opposite cell pole which enables rosette formation (Segev et al. 2015; Gonzalez et al. 1997). Together, these observations suggest that many *Rhodobacterales* may utilize budding as a mode of cell growth although their growth patterning is largely unexplored. More studies, including time-lapse microscopy, FDAA, D-cys, or TRSE labeling, are needed to better characterize the mode of growth utilized by *Rhodobacterales*. Two candidates: *Rhodobacter sphaeroides* and members of the *Roseobacter* clade are emerging as candidate model systems for the study of growth patterning in the *Rhodobacterales*.

R. sphaeroides is emerging as a model for studies of the bacterial cell cycle and its regulation. In particular, studies of motility (Wilkinson et al. 2011), quorum sensing (Puskas et al. 1997), and chemotaxis (Chiu et al. 2013) have been conducted. Morphological changes induced by antibiotic treatment suggest *R. sphaeroides* uses an MreB-dependent mode of pre-septal elongation (Slovak et al. 2005, 2006). Insertion of new peptidoglycan occurs along the side walls as cells elongate, and is focused at the mid-cell during cell division. After the completion of division, peptidoglycan biosynthesis returns to the sidewalls. Treatment with A22 (targets MreB) or mecillinam (targets PBP2) inhibits proper insertion of peptidoglycan, suggesting *R. sphaeroides* uses the canonical MreB-dependent pathway for elongation (Lin et al. 2019). Accordingly, PBP2, MreB, and MreC fused to fluorescent proteins form bands or rings at mid-cell in elongating cells. Once constrictions form, PBP2 transits from mid-cell to the three-quarter site of each daughter cell. PBP2 localization to the future division site in cells with a septum may indicate that both elongation and division machinery are active in pre-divisional cells concurrently (Slovak et al. 2005, 2006). Notably, the *R. sphaeroides* genome also contains a gene predicted to encode a PBP3 homolog, which presumably functions during division. It remains to be determined if both PBP2 and PBP3 function in discrete complexes for elongation and division simultaneously. In addition to the impact of antibiotic treatments, the environment can influence cell shape. For example, during aerobic growth, R. *sphaeroides* cells have a rod-shaped morphology with tapered ends, but adopt an ovococcous morphology with extensive membrane invaginations when growing photoheterotrophically (Woronowicz and Niederman 2010; Tucker et al. 2010). Thus, *R. sphaeroides* is a promising model system to study shifts in cell growth patterning that lead to rearrangements of cell shape in response to environmental conditions.

Marine niches, including open ocean, coastal waters, and sea ice are largely colonized by species of the *Roseobacter* clade (Brinkhoff et al. 2008; Buchan et al. 2005). Many marine *Roseobacters* are found in phytoplankton communities, in symbiosis with marine eukaryotes, and some can cause disease (Buchan et al. 2005). Interestingly, several *Prionitis* species were shown to form galls on marine algae (Ashen and Goff 2000), reminiscent of crown galls induced by *A. tumefaciens* infection. The marine algae symbiont *Dinoroseobacter shibae* displays remarkable diversity in cell shape (Patzelt et al. 2013). In this species, small ovoid or rod-shaped cells coexist with highly elongated and swollen cells in culture. Tracking growth of individual bacteria demonstrated that many *D. shibae* cells elongate by polar budding, and some cells alternate growth from pole-to-pole. In some cases, extreme asymmetric cell division events lead to the production of a large, swollen cell and a smaller, pleomorphic cell. Remarkably, in the same culture some ovoid cells undergo symmetric cell division events producing two identical ovoid-shaped cells. This phenotypic variation is controlled by quorum sensing, and quorum sensing mutants grow as a population of homogenous ovoid cells (Patzelt et al. 2013). The control of morphological variation by QS is an exciting and novel discovery that requires further investigation to determine its prevalence within the Alphaproteobacteria. In addition, mutation of the cell cycle control genes *ctrA*, *chpT,* and *cckA* each results in a loss of phenotypic variation in *D. shibae*. Together, these results may indicate that the CtrA phosphorelay is integrated into the QS signaling pathway and contributes to the regulation of morphological heterogeneity (Wang et al. 2014). Since many *Roseobacter* species display morphological heterogeneity and may grow by polar elongation, additional members of this bacterial family are prime candidates for the study of growth patterning in marine Alphaproteobacteria.

6 Cell Growth in *Rickettsiales*

Species in the order *Rickettsiales* have lost the ability to replicate outside of eukaryotic host cells and comprise many species that cause disease in humans. There are two families in the order *Rickettsiales*: *Rickettsiaceae* and *Anaplasmataceae*. The *Rickettsiaceae* are maintained through various animal reservoirs (Fig. 1a) and can be spread to humans through arthropod vectors such as flea, tick, louse, or mite (Eremeeva et al. 2005; Merhej et al. 2014). Two well-known species include *Rickettsia typhi,* which causes typhus, and *Rickettsia rickettsii*, which causes Rocky Mountain spotted fever. Both bacteria primarily replicate in the cytoplasm of vascular endothelial cells during human infections (Merhej et al. 2014). The *Anaplasmataceae* are generally transferred to humans through the bite of a tick and can replicate in the vacuole of a variety of immune cells including erythrocytes, neutrophils, monocytes, macrophages, and endothelial cells (Battilani et al. 2017).

Since the Rickettsia have adopted an obligate intracellular lifestyle, they have reduced genome sizes compared to closely related free-living bacteria. In addition, many *Rickettsiales* require less peptidoglycan since they replicate inside cells with an osmotically stable environment (Sallstrom and Andersson 2005). Given their obligate intracellular lifestyle, *Rickettsiales* are challenging to culture, which has hampered the study of cell growth patterning; however, comparative genomic analyses have provided some insight into the mechanisms of elongation. Species that do not encode genes for lipid II biosynthesis (MurA-G, MurY), SEDS proteins (FtsW, RodA), or class A (PBP1a/b) and class B PBPs (PBP2/3) do not synthesize a detectable peptidoglycan structure; this includes: *Anaplasma phagocytophilum*, *Ehrlichia ruminantium*, and *Neorickettsia sennetsu* (Otten et al. 2018). Based on the absence of these signature peptidoglycan proteins, several *Anaplasma*, *Ehrlichia*, and *Neorickettsia* species may lack peptidoglycan.

Interestingly, a group of *Rickettsiales* species that were classified as having low levels or incomplete peptidoglycan were shown to encode a subset of proteins involved in peptidoglycan biosynthesis (Otten et al. 2018). The genomes of peptidoglycan intermediate species encode the full set of lipid II biosynthesis proteins, at least one of the SEDS proteins, and at least one class B PBP, but no class A PBPs. For example, *Anaplasma marginale*, *Wolbachia* species, and *Orientia tsutsugamushi* may all synthesize an intermediate peptidoglycan. Indeed, lipid II biosynthesis genes and the gene encoding PBP2 are expressed in *O. tsutsugamushi*, and cells label with FDAAs, suggesting that peptidoglycan is indeed synthesized in this bacterium (Atwal et al. 2017). Peptidoglycan precursor pathways are also found in *Wolbachia* and lipid II is required for proper cell division, although mature peptidoglycan has not been detected (Vollmer et al. 2013).

Finally, several species in the genus *Rickettsia* encode homologs for the full set of peptidoglycan predictor proteins, and therefore are expected to synthesize classical peptidoglycan (Otten et al. 2018). The notion that closely related bacteria have different categories of peptidoglycan may be supported by the fact that these species occupy different intracellular niches. For example, *A. phagocytophilum,* which lacks peptidoglycan, localizes to monocytes and macrophages, while *A. marginale*, which synthesizes an intermediate peptidoglycan, inhabits erythrocytes. The bioinformatics approach has provided predictions for the likelihood of peptidoglycan biosynthesis within the *Rickettsiales* (Otten et al. 2018), which can be tested. Observations of cell growth patterning using FDAAs is a powerful tool to determine how the synthesis of intermediate and classical peptidoglycan is spatially and temporally regulated in the *Rickettsiales*.

7 Conservation of Core Elongation Machinery in the Alphaproteobacteria

The bacterial cell wall is a stress-bearing meshwork of proteins and sugars that is considered a major determinant of bacterial cell shape. Construction of peptidoglycan requires scaffolding, synthesis, and remodeling proteins that are well conserved across the domain Bacteria. The expansion of genes that encode PBPs and LDTs and the evolution of prosthecae-specific proteins are just a few examples of the diversification of the core cell elongation machinery that impacts bacterial morphology within the Alphaproteobacteria. Furthermore, the novel growth modes and phenotypic variation observed within the Alphaproteobacteria suggest that these bacteria must have evolved novel mechanisms to direct and regulate the activities of enzymes belonging to the core elongation machinery. Some regulators of peptidoglycan biosynthesis such as LpoA/B have been identified in *E. coli* (Paradis-Bleau et al. 2010; Typas et al. 2010); however, in the Alphaproteobacteria these regulatory elements remain largely unknown. Here, we provide a brief survey of the predicted peptidoglycan biosynthesis machinery found in genomes of Alphaproteobacteria and consider some possible mechanisms to regulate peptidoglycan biosynthesis.

7.1 Survey of Peptidoglycan Biosynthesis Machinery

The rigid structure of the bacterial cell wall is largely responsible for cell shape, and expansion of new cell wall requires two main types of enzymes: glycosyltransferases to attach glycan chains, and DD-transpeptidases to polymerize peptide crosslinks. In *E. coli*, bifunctional (Class A) and monofunctional (Class B) PBPs along with their cognate SEDS protein, either function during elongation (PBP2/RodA), during division (PBP3/FtsW), or in maintaining cell wall integrity (PBP1a and PBP1b), while the function of PBP1c is not well understood (Vollmer and Bertsche 2008; Cho et al. 2016; Vigouroux et al. 2020). Since the high molecular weight (HMW) PBPs synthesize new peptidoglycan during elongation, division, and prosthecae growth, any gene loss, duplication or evolutionary divergence of HMW PBPs may impact cell growth patterning and morphology. Thus, we carried out a phylogenetic analysis comparing the HMW PBPs from representative species shown in Fig. 1a, which encompasses all six orders of Alphaproteobacteria. Most PBPs group into clades with their respective *E. coli* homologs, suggesting that they may perform similar functions to the well-characterized *E. coli* enzymes (Fig. 3a). A comparative analysis shows that homologs of PBP1a, PBP1b, PBP1c, PBP2, and PBP3 form distinct clades and PBPs largely group together based on bacterial order (Fig. 3a). This analysis confirms impact of genome reduction on the presence of class A PBPs (PBP1a/b) within the *Rickettsiales* (Fig. 3a) (see Sect. 6).

Remarkably, within the *Rhizobiales*, the distribution of PBPs is distinct from other Alphaproteobacterial orders. First, the elongation-specific PBP2 is universally absent (Fig. 3a). This observation coupled with the absence of MreBCD and RodA

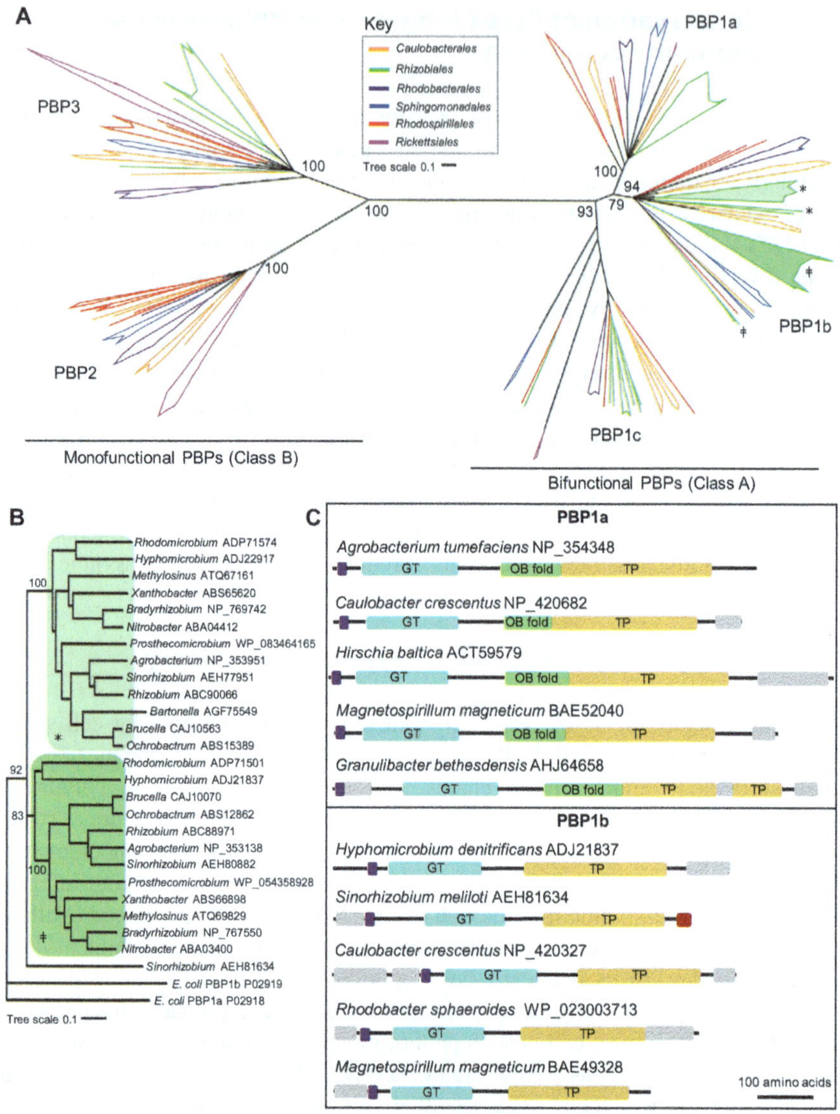

Fig. 3 Distribution and domain comparison of the high molecular weight (HMW) penicillin-binding proteins (PBPs). Phylogenetic trees from A and B were constructed using a ClustalW protein alignment (Larkin et al. 2007) and creating a neighbor joining tree using Geneious Tree Builder with 100 bootstrap datasets. Tree scales represent the branch length with 0.1 substitutions per site. (**a**) Phylogeny of HMW PBPs from 41 representative species of Alphaproteobacteria was formatted using Dendroscope (Huson et al. 2007). PBPs are color coded by order as listed in the key. Leaves contain multiple species of the same order, and the size of a leaf represents the number of species in that leaf. The symbols next to the solid green leaves represent the species used to make the tree in B. (**b**) Phylogeny of PBP1b genes of 12 species of *Rhizobiales*, and *E. coli* PBP1a and PBP1b was formatted using iTOL (Letunic and Bork 2016). The genus and NCBI Accession

is consistent with the hypothesis that a novel pathway for cell elongation has evolved in the *Rhizobiales*. Second, the gene encoding PBP1b has been universally duplicated. Phylogenetic analysis of the PBP1b proteins from *Rhizobiales* species indicates the presence of two distinct clades (Fig. 3b). Each clade contains a PBP1b protein from each bacterial species, suggesting an ancient duplication event. Remarkably, the two PBP1b clades are more similar to one another than either is to the PBP1b clade from *E. coli* (Fig. 3b; Cameron et al. 2014), perhaps suggesting functional divergence. Functional studies of the Alphaproteobacterium-specific PBP1b proteins may provide insights into novel species-specific or environment-specific growth patterns and morphologies. It will be of particular interest to determine if the Alphaproteobacterial PBP1b proteins are division-specific or have an alternative function in peptidoglycan biosynthesis. Finally, two PBP3 proteins are observed within a subset of *Rhizobiales* species (Fig. 3a). Notably, most species with a duplication in the gene encoding PBP3 interact with plants (*Agrobacterium*, *Sinorhizobium*), raising the possibility that there may be a function for PBP3 when the bacterium is host-associated.

In addition to surveying the PBP distribution, it is also necessary to consider the distribution and function of other enzymes which may contribute to peptidoglycan biosynthesis. Currently, the number and distribution of peptidoglycan remodeling enzymes across the Alphaproteobacteria remains poorly characterized. There has been an obvious expansion of LD-transpeptidases (LDTs), particularly within the *Rhizobiales* and *Rhodobacterales* orders (Cameron et al. 2014). In *E. coli*, LDTs are typically considered to be peptidoglycan recycling enzymes with a limited role in peptidoglycan biosynthesis (Uehara and Park 2008). As well as homologs of *E. coli* LDTs, there are *Rhizobiales*- and *Rhodobacterales*-specific clades of LDTs (Cameron et al. 2014). Consistent with this observation, compositional analysis of muropeptides from some species of *Rhizobiales* and *Rhodobacterales* reveals an increase in crosslinks formed by LDTs (Brown et al. 2012; Espaillat et al. 2016). *C. crescentus* stalks are also enriched in crosslinks formed by LDTs, suggesting that LDTs may contribute to specialized forms of peptidoglycan biosynthesis (Billini et al. 2019; Stankeviciute et al. 2019). These observations raise the possibility that LDTs may have a more significant role in peptidoglycan biosynthesis in some Alphaproteobacteria. Furthermore, there is a *Rhizobiales*-specific clade of low molecular weight PBPs (LMW-PBPs; Cameron et al. 2014). LMW-PBPs typically function as peptidoglycan hydrolases (Holtje 1998; Romeis and Holtje 1994; Palomeque-Messia et al. 1991); however, a functionally unique family of LMW-PBPs with transpeptidase activity has been described (Welsh et al. 2017) raising the possibility of functional divergence of LMW-PBPs. Overall,

Fig. 3 (continued) number for each gene are labeled. (**c**) Domain analysis of PBP1a and PBP1b genes. Transmembrane domain is purple, Glycosyltransferase (GT) is blue, Transpeptidase (TP) is orange, regions of intrinsic disorder are gray and were annotated using MobiDB (Piovesan et al. 2018), and BA14K-like Protein domain is red

there is a clear need for more bioinformatic and molecular studies to identify and characterize peptidoglycan remodeling enzymes in the Alphaproteobacteria. Certainly, understanding the landscape of peptidoglycan biosynthesis enzymes will help direct future studies of growth patterning.

7.2 Regulation of Peptidoglycan Biosynthesis Machinery

In addition to considering the function of enzymes involved in peptidoglycan synthesis, it is necessary to identify and characterize the proteins that regulate the activity of these enzymes. The presence of species-specific domains in peptidoglycan synthesis and remodeling enzymes suggests that the activity of some of these enzymes may be regulated directly. The crystal structure of *Acinetobacter baumannii* PBP1a revealed a non-catalytic oligonucleotide–oligosaccharide binding (OB)-fold domain near the TP domain (Han et al. 2011) that is also present in some Alphaproteobacterial PBP1a proteins (Fig. 3c, green). The functional significance of this domain remains to be explored, but it is intriguing to speculate that this domain may be involved in regulation of PBP1a activity. Furthermore, intrinsically disordered (ID) domains exhibit conformational flexibility (Oldfield and Dunker 2014; Schlessinger et al. 2011) and as such are candidate regions to facilitate the regulation of PBP activity. Alphaproteobacterial PBP1a and PBP1b proteins are enriched for regions of intrinsic disorder (Fig. 3c, gray). While ID domains are present mainly at the C-terminus of PBP1a, the location of ID domains in PBP1b genes is more diverse (Fig. 3c). The presence of ID domains before the transmembrane domain suggests that PBP1b may have novel binding partners in the cytoplasm and N-terminal ID domains may bind periplasmic proteins. The majority of ID domains in PBPs are uncharacterized in the Alphaproteobacteria, and a better understanding of these domains has the potential to further our understanding of the regulation of PBP activity.

The localization of PBPs to specific sites of growth by interactions with a widely conserved set of core cytoskeletal proteins is a common theme in peptidoglycan regulation (Shih and Rothfield 2006). During *E. coli* elongation, MreB targets PBP2 to patches along the lateral wall. Similarly, during division, FtsZ filaments provide a scaffold for PBP3 and other cell division machinery. Bactofilins are widely distributed albeit less well-characterized cytoskeletal elements that may play a role in spatial regulation of peptidoglycan biosynthesis. In *Caulobacter*, BacA and BacB localize to the stalked cell pole during the swarmer-to-stalked cell transition and assemble into polymers that recruit an Alphaproteobacterium-specific PBP1b homolog (PBPC) to the pole (Kuhn et al. 2010). Bactofilins are widely distributed across the Alphaproteobacteria with most species having 1 homolog. In contrast, the *Rhizobiales* typically have no bactofilin homologs with the exception of the prosthecate tip budding bacteria *Hyphomicrobium* and *Rhodomicrobium*, which have 4 and 3 bactofilins, respectively. Understanding how the distribution of core cytoskeletal elements contributes to the localization and regulation of peptidoglycan biosynthesis enzymes may provide insight into the microbe-specific patterns of peptidoglycan insertion that generate diversity in cell shape. Several key questions regarding the regulation of peptidoglycan biosynthesis remain open-ended. For example, how are peptidoglycan biosynthesis enzymes targeted to the pole in the *Rhizobiales*, which lack both MreB and bactofilin cytoskeletal elements? The identification of additional growth pole proteins in both *A. tumefaciens* and *S. meliloti* is beginning to hint at the presence of a dedicated growth pole complex which restricts growth to the pole during elongation (Zupan et al. 2019; Krol et al. 2020; Schaper et al. 2018). Do species with multiple bactofilin homologs have additional regulation of peptidoglycan biosynthesis in time and space, or are their functions largely redundant? Exploring the role of bactofilins, intermediate filaments, and polar landmark proteins will shed light on the mechanisms of subcellular targeting of peptidoglycan synthesis and regulation.

The Alphaproteobacteria have evolved novel growth patterns through the diversification of the peptidoglycan biosynthesis machinery. Alphaproteobacteria-specific proteins, novel domains, and novel pathways for regulation are all factors that contribute to diverse cell shapes. Future studies should seek to understand how the core peptidoglycan machinery is regulated in time and space.

8 Concluding Remarks

The molecular basis that determines bacterial shape has been mainly studied in rods and cocci. Expanding the study of cell growth to non-canonical bacterial models will deepen our understanding of morphological complexity. The Alphaproteobacteria have evolved several novel mechanisms of cell growth patterning. Novel growth patterns of budding bacteria include: prosthecate tip budding, prosthecate cell budding, and unipolar elongation. Additional types of novel growth patterns have also been characterized, which include stalk elongation and pre-septal elongation. Each type of growth patterning correlates to a unique morphology that is specifically adapted to its environment. However, the evolutionary pressures and advantages that drive diversity of shape remain poorly understood. Thus, expanding the repertoire of sequenced genomes will allow for comparative genomic analysis to identify novel growth modes. Evidence suggests multiple mechanisms for arriving at similar morphologies have evolved. As such, a multidisciplinary approach, including imaging, genetic, genomic, and molecular investigations will be required to fully understand the evolution of cell growth patterning. As more diverse species of Alphaproteobacteria continue to be characterized, novel growth patterns may be discovered. For bacteria such as *C. crescentus*, *H. neptunium*, and *A. tumefaciens* we have begun to identify some key mechanistic features that regulate growth patterning. In the *Caulobacterales*, the core elongation machinery is conserved in many species and its activity is regulated in time and space by protein–protein interactions to ensure that peptidoglycan is made at the right time and place. In contrast, in the *Rhizobiales* the core elongation machinery remains elusive. Furthermore, detailed studies of growth patterning in many Alphaproteobacteria such as *Sphingomonadales* and acetic acid bacteria remain wholly unexplored, highlighting the need to survey growth patterns in diverse Alphaproteobacteria. Finally, detailed studies of elongation in model Alphaproteobacteria are necessary to shed light on the evolution of growth patterning in bacteria. Ultimately, it is an exciting time to study bacterial cell growth patterning in such a morphologically diverse class of bacteria that occupy a wide range of habitats and include microbes of agricultural, environmental, and medical significance. Williams et al. (2021) demonstrate that a single aPBP (PBP1a) is essential for polar elongation within the Rhizobiales. In contrast, Atwal et al. (2021) find that Rickettsiales build a minimal peptidoglycan-like structure without aPBPs.

Acknowledgements Research in the Brown lab on *A. tumefaciens* cell growth and division is supported by the National Science Foundation (IOS1557806). DTK was supported by the National Institutes of Health under award number R01GM113121.

References

Aaron M, Charbon G, Lam H, Schwarz H, Vollmer W, Jacobs-Wagner C (2007) The tubulin homologue FtsZ contributes to cell elongation by guiding cell wall precursor synthesis in *Caulobacter crescentus*. Mol Microbiol 64(4):938–952. https://doi.org/10.1111/j.1365-2958.2007.05720.x

Addinall SG, Cao C, Lutkenhaus J (1997) FtsN, a late recruit to the septum in *Escherichia coli*. Mol Microbiol 25(2):303–309

Alain K, Tindall BJ, Intertaglia L, Catala P, Lebaron P (2008) *Hellea balneolensis* gen. nov., sp. nov., a prosthecate alphaproteobacterium from the Mediterranean Sea. Int J Syst Evol Microbiol 58 (Pt 11):2511–2519. https://doi.org/10.1099/ijs.0.65424-0

Ashen JB, Goff LJ (2000) Molecular and ecological evidence for species specificity and coevolution in a group of marine algal-bacterial symbioses. Appl Environ Microbiol 66(7):3024–3030

Atwal S, Giengkam S, Chaemchuen S, Dorling J, Kosaisawe N, VanNieuwenhze M, Sampattavanich S, Schumann P, Salje J (2017) Evidence for a peptidoglycan-like structure in *Orientia tsutsugamushi*. Mol Microbiol 105(3):440–452. https://doi.org/10.1111/mmi.13709

Atwal S, Chuenklin S, Bonder EM, Flores J, Gillespie JJ, Driscoll TP, Salje J (2021) Discovery of a diverse set of bacteria that build their cell walls without the canonical peptidoglycan polymerase aPBP. mBio 12:e0134221

Badger JH, Hoover TR, Brun YV, Weiner RM, Laub MT, Alexandre G, Mrazek J, Ren Q, Paulsen IT, Nelson KE, Khouri HM, Radune D, Sosa J, Dodson RJ, Sullivan SA, Rosovitz MJ, Madupu R, Brinkac LM, Durkin AS, Daugherty SC, Kothari SP, Giglio MG, Zhou L, Haft DH, Selengut JD, Davidsen TM, Yang Q, Zafar N, Ward NL (2006) Comparative genomic evidence for a close relationship between the dimorphic prosthecate bacteria *Hyphomonas neptunium* and *Caulobacter crescentus*. J Bacteriol 188(19):6841–6850. https://doi.org/10.1128/JB.00111-06

Banzhaf M, van den Berg van Saparoea B, Terrak M, Fraipont C, Egan A, Philippe J, Zapun A, Breukink E, Nguyen-Disteche M, den Blaauwen T, Vollmer W (2012) Cooperativity of peptidoglycan synthases active in bacterial cell elongation. Mol Microbiol 85(1):179–194. https://doi.org/10.1111/j.1365-2958.2012.08103.x

Baranova N, Radler P, Hernandez-Rocamora VM, Alfonso C, Lopez-Pelegrin M, Rivas G, Vollmer W, Loose M (2020) Diffusion and capture permits dynamic coupling between treadmilling FtsZ filaments and cell division proteins. Nat Microbiol 5(3):407–417. https://doi.org/10.1038/s41564-019-0657-5

Battilani M, De Arcangeli S, Balboni A, Dondi F (2017) Genetic diversity and molecular epidemiology of *Anaplasma*. Infect Genet Evol 49:195–211. https://doi.org/10.1016/j.meegid.2017.01.021

Batut J, Andersson SG, O'Callaghan D (2004) The evolution of chronic infection strategies in the alpha-proteobacteria. Nat Rev Microbiol 2(12):933–945. https://doi.org/10.1038/nrmicro1044

Bauld J, Bigford R, Staley J (1983) *Prosthecomicrobium litoralum*, a new species from marine habitats. Int J Syst Bacteriol 33(3):613–617

Berne C, Ma X, Licata NA, Neves BR, Setayeshgar S, Brun YV, Dragnea B (2013) Physiochemical properties of *Caulobacter crescentus* holdfast: a localized bacterial adhesive. J Phys Chem B 117(36):10492–10503. https://doi.org/10.1021/jp405802e

Bertsche U, Kast T, Wolf B, Fraipont C, Aarsman ME, Kannenberg K, von Rechenberg M, Nguyen-Disteche M, den Blaauwen T, Holtje JV, Vollmer W (2006) Interaction between two murein (peptidoglycan) synthases, PBP3 and PBP1B, in *Escherichia coli*. Mol Microbiol 61(3):675–690. https://doi.org/10.1111/j.1365-2958.2006.05280.x

Billini M, Biboy J, Kuhn J, Vollmer W, Thanbichler M (2019) A specialized MreB-dependent cell wall biosynthetic complex mediates the formation of stalk-specific peptidoglycan in *Caulobacter crescentus*. PLoS Genet 15(2):e1007897. https://doi.org/10.1371/journal.pgen.1007897

Biondi EG, Skerker JM, Arif M, Prasol MS, Perchuk BS, Laub MT (2006) A phosphorelay system controls stalk biogenesis during cell cycle progression in *Caulobacter crescentus*. Mol Microbiol 59(2):386–401. https://doi.org/10.1111/j.1365-2958.2005.04970.x

Bisson-Filho AW, Hsu YP, Squyres GR, Kuru E, Wu F, Jukes C, Sun Y, Dekker C, Holden S, VanNieuwenhze MS, Brun YV, Garner EC (2017) Treadmilling by FtsZ filaments drives peptidoglycan synthesis and bacterial cell division. Science 355(6326):739–743. https://doi.org/10.1126/science.aak9973

Botella H, Yang G, Ouerfelli O, Ehrt S, Nathan CF, Vaubourgeix J (2017) Distinct spatiotemporal dynamics of peptidoglycan synthesis between *Mycobacterium smegmatis* and *Mycobacterium tuberculosis*. mBio 8(5). https://doi.org/10.1128/mBio.01183-17

Botta GA, Park JT (1981) Evidence for involvement of penicillin-binding protein 3 in murein synthesis during septation but not during cell elongation. J Bacteriol 145(1):333–340

Brinkhoff T, Giebel HA, Simon M (2008) Diversity, ecology, and genomics of the *Roseobacter* clade: a short overview. Arch Microbiol 189(6):531–539. https://doi.org/10.1007/s00203-008-0353-y

Brown PJ, de Pedro MA, Kysela DT, Van der Henst C, Kim J, De Bolle X, Fuqua C, Brun YV (2012) Polar growth in the Alphaproteobacterial order Rhizobiales. Proc Natl Acad Sci USA 109(5):1697–1701. https://doi.org/10.1073/pnas.1114476109

Bruhn JB, Nielsen KF, Hjelm M, Hansen M, Bresciani J, Schulz S, Gram L (2005) Ecology, inhibitory activity, and morphogenesis of a marine antagonistic bacterium belonging to the *Roseobacter* clade. Appl Environ Microbiol 71(11):7263–7270. https://doi.org/10.1128/AEM.71.11.7263-7270.2005

Buchan A, Gonzalez JM, Moran MA (2005) Overview of the marine roseobacter lineage. Appl Environ Microbiol 71(10):5665–5677. https://doi.org/10.1128/AEM.71.10.5665-5677.2005

Caccamo PD, Jacq M, VanNieuwenhze MS, Brun YV (2020) A division of labor in the recruitment and topological organization of a bacterial morphogenic complex. Curr Biol. https://doi.org/10.1016/j.cub.2020.07.063

Caldas P, Lopez-Pelegrin M, Pearce DJG, Budanur NB, Brugues J, Loose M (2019) Cooperative ordering of treadmilling filaments in cytoskeletal networks of FtsZ and its crosslinker ZapA. Nat Commun 10(1):5744. https://doi.org/10.1038/s41467-019-13702-4

Cameron TA, Anderson-Furgeson J, Zupan JR, Zik JJ, Zambryski PC (2014) Peptidoglycan synthesis machinery in *Agrobacterium tumefaciens* during unipolar growth and cell division. MBio 5(3):e01219–e01214. https://doi.org/10.1128/mBio.01219-14

Caparros M, Pisabarro AG, de Pedro MA (1992) Effect of D-amino acids on structure and synthesis of peptidoglycan in *Escherichia coli*. J Bacteriol 174(17):5549–5559

Chertkov O, Brown PJ, Kysela DT, de Pedro MA, Lucas S, Copeland A, Lapidus A, Del Rio TG, Tice H, Bruce D, Goodwin L, Pitluck S, Detter JC, Han C, Larimer F, Chang YJ, Jeffries CD, Land M, Hauser L, Kyrpides NC, Ivanova N, Ovchinnikova G, Tindall BJ, Goker M, Klenk HP, Brun YV (2011) Complete genome sequence of *Hirschia baltica* type strain (IFAM 1418(T)). Stand Genomic Sci 5(3):287–297. https://doi.org/10.4056/sigs.2205004

Chiu SW, Roberts MA, Leake MC, Armitage JP (2013) Positioning of chemosensory proteins and FtsZ through the *Rhodobacter sphaeroides* cell cycle. Mol Microbiol 90(2):322–337. https://doi.org/10.1111/mmi.12366

Cho H, Wivagg CN, Kapoor M, Barry Z, Rohs PD, Suh H, Marto JA, Garner EC, Bernhardt TG (2016) Bacterial cell wall biogenesis is mediated by SEDS and PBP polymerase families functioning semi-autonomously. Nat Microbiol 16172. https://doi.org/10.1038/nmicrobiol.2016.172

Connon SA, Giovannoni SJ (2002) High-throughput methods for culturing microorganisms in very-low-nutrient media yield diverse new marine isolates. Appl Environ Microbiol 68(8):3878–3885

Conti SF, Hirsch P (1965) Biology of budding bacteria. 3. Fine structure of *Rhodomicrobium* and *Hyphomicrobium* spp. J Bacteriol 89:503–512

Cox TL, Sly LI (1997) Phylogenetic relationships and uncertain taxonomy of *Pedomicrobium* species. Int J Syst Bacteriol 47(2):377–380. https://doi.org/10.1099/00207713-47-2-377

Crotti E, Rizzi A, Chouaia B, Ricci I, Favia G, Alma A, Sacchi L, Bourtzis K, Mandrioli M, Cherif A, Bandi C, Daffonchio D (2010) Acetic acid bacteria, newly emerging symbionts of insects. Appl Environ Microbiol 76(21):6963–6970. https://doi.org/10.1128/AEM.01336-10

Cserti E, Rosskopf S, Chang YW, Eisheuer S, Selter L, Shi J, Regh C, Koert U, Jensen GJ, Thanbichler M (2017) Dynamics of the peptidoglycan biosynthetic machinery in the stalked budding bacterium *Hyphomonas neptunium*. Mol Microbiol 103(5):875–895. https://doi.org/10.1111/mmi.13593

Curtis PD (2017) Stalk formation of *Brevundimonas* and how it compares to *Caulobacter crescentus*. PLoS One 12(9):e0184063. https://doi.org/10.1371/journal.pone.0184063

Dai K, Lutkenhaus J (1991) *ftsZ* is an essential cell division gene in *Escherichia coli*. J Bacteriol 173(11):3500–3506

Daniel RA, Errington J (2003) Control of cell morphogenesis in bacteria: two distinct ways to make a rod-shaped cell. Cell 113(6):767–776

Darling AE, Jospin G, Lowe E, Matsen FA, Bik HM, Eisen JA (2014) PhyloSift: phylogenetic analysis of genomes and metagenomes. PeerJ 2:e243. https://doi.org/10.7717/peerj.243

de Pedro MA, Quintela JC, Holtje JV, Schwarz H (1997) Murein segregation in *Escherichia coli*. J Bacteriol 179(9):2823–2834

Divakaruni AV, Baida C, White CL, Gober JW (2007) The cell shape proteins MreB and MreC control cell morphogenesis by positioning cell wall synthetic complexes. Mol Microbiol 66(1):174–188. https://doi.org/10.1111/j.1365-2958.2007.05910.x

Dominguez-Escobar J, Chastanet A, Crevenna AH, Fromion V, Wedlich-Soldner R, Carballido-Lopez R (2011) Processive movement of MreB-associated cell wall biosynthetic complexes in bacteria. Science 333(6039):225–228. https://doi.org/10.1126/science.1203466

Du S, Lutkenhaus J (2019) At the heart of bacterial cytokinesis: the Z ring. Trends Microbiol 27(9):781–791. https://doi.org/10.1016/j.tim.2019.04.011

Ehrle HM, Guidry JT, Iacovetto R, Salisbury AK, Sandidge DJ, Bowman GR (2017) Polar organizing protein PopZ is required for chromosome segregation in *Agrobacterium tumefaciens*. J Bacteriol 199(17). https://doi.org/10.1128/JB.00111-17

Eremeeva ME, Madan A, Shaw CD, Tang K, Dasch GA (2005) New perspectives on rickettsial evolution from new genome sequences of rickettsia, particularly *R. canadensis*, and *Orientia tsutsugamushi*. Ann N Y Acad Sci 1063:47–63. https://doi.org/10.1196/annals.1355.006

Espaillat A, Forsmo O, El Biari K, Bjork R, Lemaitre B, Trygg J, Canada FJ, de Pedro MA, Cava F (2016) Chemometric analysis of bacterial peptidoglycan reveals atypical modifications that empower the cell wall against predatory enzymes and fly innate immunity. J Am Chem Soc 138(29):9193–9204. https://doi.org/10.1021/jacs.6b04430

Faivre D, Schuler D (2008) Magnetotactic bacteria and magnetosomes. Chem Rev 108(11):4875–4898. https://doi.org/10.1021/cr078258w

Fritts RK, LaSarre B, Stoner AM, Posto AL, McKinlay JB (2017) A Rhizobiales-specific unipolar polysaccharide adhesin contributes to *Rhodopseudomonas palustris* biofilm formation across diverse photoheterotrophic conditions. Appl Environ Microbiol 83(4). https://doi.org/10.1128/AEM.03035-16

Fujiwara T, Fukui S (1972) Isolation of morphological mutants of *Agrobacterium tumefaciens*. J Bacteriol 110(2):743–746

Fujiwara T, Fukui S (1974) Unidirectional growth and branch formation of a morphological mutant, *Agrobacterium tumefaciens*. J Bacteriol 120(2):583–589

Garner EC, Bernard R, Wang W, Zhuang X, Rudner DZ, Mitchison T (2011) Coupled, circumferential motions of the cell wall synthesis machinery and MreB filaments in *B. subtilis*. Science 333(6039):222–225. https://doi.org/10.1126/science.1203285

Gonin M, Quardokus EM, O'Donnol D, Maddock J, Brun YV (2000) Regulation of stalk elongation by phosphate in *Caulobacter crescentus*. J Bacteriol 182(2):337–347

Gonzalez JM, Mayer F, Moran MA, Hodson RE, Whitman WB (1997) *Sagittula stellata* gen. nov., sp. nov., a lignin-transforming bacterium from a coastal environment. Int J Syst Bacteriol 47(3):773–780. https://doi.org/10.1099/00207713-47-3-773

Gonzalez JM, Covert JS, Whitman WB, Henriksen JR, Mayer F, Scharf B, Schmitt R, Buchan A, Fuhrman JA, Kiene RP, Moran MA (2003) *Silicibacter pomeroyi* sp. nov. and *Roseovarius nubinhibens* sp. nov., dimethylsulfoniopropionate-demethylating bacteria from marine environments. Int J Syst Evol Microbiol 53(Pt 5):1261–1269. https://doi.org/10.1099/ijs.0.02491-0

Grangeon R, Zupan JR, Anderson-Furgeson J, Zambryski PC (2015) PopZ identifies the new pole, and PodJ identifies the old pole during polar growth in *Agrobacterium tumefaciens*. Proc Natl Acad Sci USA 112(37):11666–11671. https://doi.org/10.1073/pnas.1515544112

Grangeon R, Zupan J, Jeon Y, Zambryski PC (2017) Loss of PopZ At activity in *Agrobacterium tumefaciens* by deletion or depletion leads to multiple growth poles, minicells, and growth defects. MBio 8(6). https://doi.org/10.1128/mBio.01881-17

Han S, Caspers N, Zaniewski RP, Lacey BM, Tomaras AP, Feng X, Geoghegan KF, Shanmugasundaram V (2011) Distinctive attributes of beta-lactam target proteins in *Acinetobacter baumannii* relevant to development of new antibiotics. J Am Chem Soc 133(50):20536–20545. https://doi.org/10.1021/ja208835z

Haney SA, Glasfeld E, Hale C, Keeney D, He Z, de Boer P (2001) Genetic analysis of the *Escherichia coli* FtsZ.ZipA interaction in the yeast two-hybrid system. Characterization of FtsZ residues essential for the interactions with ZipA and with FtsA. J Biol Chem 276(15):11980–11987. https://doi.org/10.1074/jbc.M009810200

Hartmann R, van Teeseling MCF, Thanbichler M, Drescher K (2020) BacStalk: a comprehensive and interactive image analysis software tool for bacterial cell biology. Mol Microbiol 114(1):140–150. https://doi.org/10.1111/mmi.14501

Henriques AC, De Marco P (2015) Complete genome sequences of two strains of "*Candidatus Filomicrobium marinum*," a methanesulfonate-degrading species. Genome Announc 3(3). https://doi.org/10.1128/genomeA.00160-15

Henry CS, DeJongh M, Best AA, Frybarger PM, Linsay B, Stevens RL (2010) High-throughput generation, optimization and analysis of genome-scale metabolic models. Nat Biotechnol 28(9):977–982. https://doi.org/10.1038/nbt.1672

Hirsch P (1974) Budding bacteria. Annu Rev Microbiol 28 (0):391–444. doi:https://doi.org/10.1146/annurev.mi.28.100174.002135

Hirsch P, Schlesner H (1981) The genus *Stella*. In: Starr MP, Stolp H, Trüper HG, Balows A, Schlegel HG (eds) The prokaryotes. Springer, Berlin, pp 461–465. https://doi.org/10.1007/978-3-662-13187-9_31

Holm NC, Gliesche CG, Hirsch P (1996) Diversity and structure of *Hyphomicrobium* populations in a sewage treatment plant and its adjacent receiving lake. Appl Environ Microbiol 62(2):522–528

Holtje JV (1998) Growth of the stress-bearing and shape-maintaining murein sacculus of *Escherichia coli*. Microbiol Mol Biol Rev 62(1):181–203

Howell M, Brown PJ (2016) Building the bacterial cell wall at the pole. Curr Opin Microbiol 34:53–59. https://doi.org/10.1016/j.mib.2016.07.021

Howell M, Aliashkevich A, Salisbury AK, Cava F, Bowman GR, Brown PJB (2017) Absence of the polar organizing protein PopZ results in reduced and asymmetric cell division in *Agrobacterium tumefaciens*. J Bacteriol 199(17). https://doi.org/10.1128/JB.00101-17

Hsu YP, Rittichier J, Kuru E, Yablonowski J, Pasciak E, Tekkam S, Hall E, Murphy B, Lee TK, Garner EC, Huang KC, Brun YV, VanNieuwenhze MS (2017) Full color palette of fluorescent d-amino acids for in situ labeling of bacterial cell walls. Chem Sci 8(9):6313–6321. https://doi.org/10.1039/c7sc01800b

Hsu YP, Booher G, Egan A, Vollmer W, VanNieuwenhze MS (2019a) d-Amino acid derivatives as in situ probes for visualizing bacterial peptidoglycan biosynthesis. Acc Chem Res 52(9):2713–2722. https://doi.org/10.1021/acs.accounts.9b00311

Hsu YP, Hall E, Booher G, Murphy B, Radkov AD, Yablonowski J, Mulcahey C, Alvarez L, Cava F, Brun YV, Kuru E, VanNieuwenhze MS (2019b) Fluorogenic D-amino acids enable real-time monitoring of peptidoglycan biosynthesis and high-throughput transpeptidation assays. Nat Chem 11(4):335–341. https://doi.org/10.1038/s41557-019-0217-x

Hughes HV, Huitema E, Pritchard S, Keiler KC, Brun YV, Viollier PH (2010) Protein localization and dynamics within a bacterial organelle. Proc Natl Acad Sci USA 107(12):5599–5604. https://doi.org/10.1073/pnas.0909119107

Hughes HV, Lisher JP, Hardy GG, Kysela DT, Arnold RJ, Giedroc DP, Brun YV (2013) Co-ordinate synthesis and protein localization in a bacterial organelle by the action of a penicillin-binding-protein. Mol Microbiol 90(6):1162–1177. https://doi.org/10.1111/mmi.12422

Huson DH, Richter DC, Rausch C, Dezulian T, Franz M, Rupp R (2007) Dendroscope: an interactive viewer for large phylogenetic trees. BMC Bioinform 8:460. https://doi.org/10.1186/1471-2105-8-460

Ireland MM, Karty JA, Quardokus EM, Reilly JP, Brun YV (2002) Proteomic analysis of the *Caulobacter crescentus* stalk indicates competence for nutrient uptake. Mol Microbiol 45(4):1029–1041

Jung A, Eisheuer S, Cserti E, Leicht O, Strobel W, Moll A, Schlimpert S, Kuhn J, Thanbichler M (2015) Molecular toolbox for genetic manipulation of the stalked budding bacterium *Hyphomonas neptunium*. Appl Environ Microbiol 81(2):736–744. https://doi.org/10.1128/AEM.03104-14

Jung A, Rassbach A, Pulpetta RL, van Teeseling MCF, Heinrich K, Sobetzko P, Serrania J, Becker A, Thanbichler M (2019) Two-step chromosome segregation in the stalked budding bacterium *Hyphomonas neptunium*. Nat Commun 10(1):3290. https://doi.org/10.1038/s41467-019-11242-5

Kaneshiro T, Baker FL, Johnson DE (1983) Pleomorphism and acetylene-reducing activity of free-living rhizobia. J Bacteriol 153(2):1045–1050

Katzmann E, Muller FD, Lang C, Messerer M, Winklhofer M, Plitzko JM, Schuler D (2011) Magnetosome chains are recruited to cellular division sites and split by asymmetric septation. Mol Microbiol 82(6):1316–1329. https://doi.org/10.1111/j.1365-2958.2011.07874.x

Klein EA, Schlimpert S, Hughes V, Brun YV, Thanbichler M, Gitai Z (2013) Physiological role of stalk lengthening in *Caulobacter crescentus*. Commun Integr Biol 6(4):e24561. https://doi.org/10.4161/cib.24561

Krol E, Yau HCL, Lechner M, Schaper S, Bange G, Vollmer W, Becker A (2020) Tol-Pal system and Rgs proteins interact to promote unipolar growth and cell division in *Sinorhizobium meliloti*. mBio 11(3). https://doi.org/10.1128/mBio.00306-20

Kuhn J, Briegel A, Morschel E, Kahnt J, Leser K, Wick S, Jensen GJ, Thanbichler M (2010) Bactofilins, a ubiquitous class of cytoskeletal proteins mediating polar localization of a cell wall synthase in *Caulobacter crescentus*. EMBO J 29(2):327–339. https://doi.org/10.1038/emboj.2009.358

Kuru E, Hughes HV, Brown PJ, Hall E, Tekkam S, Cava F, de Pedro MA, Brun YV, Van-Nieuwenhze MS (2012) In situ probing of newly synthesized peptidoglycan in live bacteria with fluorescent D-amino acids. Angew Chem 51(50):12519–12523. https://doi.org/10.1002/anie.201206749

Kuru E, Tekkam S, Hall E, Brun YV, Van Nieuwenhze MS (2015) Synthesis of fluorescent D-amino acids and their use for probing peptidoglycan synthesis and bacterial growth in situ. Nat Protoc 10(1):33–52. https://doi.org/10.1038/nprot.2014.197

Kuru E, Radkov A, Meng X, Egan A, Alvarez L, Dowson A, Booher G, Breukink E, Roper DI, Cava F, Vollmer W, Brun Y, VanNieuwenhze MS (2019) Mechanisms of incorporation for D-amino acid probes that target peptidoglycan biosynthesis. ACS Chem Biol 14(12):2745–2756. https://doi.org/10.1021/acschembio.9b00664

Kuykendall L (2005) Order VI *Rhizobiales*. In: Brenner DJ, Krieg NR, Stanley JT, Garrity GM (eds) Bergey's manual of systematic bacteriology. Williams & Wilkins, Baltimore, pp 324–286

Labrenz M, Collins MD, Lawson PA, Tindall BJ, Braker G, Hirsch P (1998) *Antarctobacter heliothermus* gen. nov., sp. nov., a budding bacterium from hypersaline and heliothermal Ekho Lake. Int J Syst Bacteriol 48(Pt 4):1363–1372. https://doi.org/10.1099/00207713-48-4-1363

Labrenz M, Collins MD, Lawson PA, Tindall BJ, Schumann P, Hirsch P (1999) *Roseovarius tolerans* gen. nov., sp. nov., a budding bacterium with variable bacteriochlorophyll a production from hypersaline Ekho Lake. Int J Syst Bacteriol 49(Pt 1):137–147. https://doi.org/10.1099/00207713-49-1-137

Larkin MA, Blackshields G, Brown NP, Chenna R, McGettigan PA, McWilliam H, Valentin F, Wallace IM, Wilm A, Lopez R, Thompson JD, Gibson TJ, Higgins DG (2007) Clustal W and Clustal X version 2.0. Bioinformatics 23(21):2947–2948. https://doi.org/10.1093/bioinformatics/btm404

Larson RJ, Pate JL (1976) Glucose transport in isolated prosthecae of *Asticcacaulis biprosthecum*. J Bacteriol 126(1):282–293

LaSarre B, Kysela DT, Stein BD, Ducret A, Brun YV, McKinlay JB (2018) Restricted localization of photosynthetic intracytoplasmic membranes (ICMs) in multiple genera of purple nonsulfur bacteria. mBio 9(4). https://doi.org/10.1128/mBio.00780-18

Latch JN, Margolin W (1997) Generation of buds, swellings, and branches instead of filaments after blocking the cell cycle of *Rhizobium meliloti*. J Bacteriol 179(7):2373–2381

Lefevre CT, Bazylinski DA (2013) Ecology, diversity, and evolution of magnetotactic bacteria. Microbiol Mol Biol Rev 77(3):497–526. https://doi.org/10.1128/MMBR.00021-13

Leifson E (1964) *Hyphomicrobium neptunium* sp. n. Antonie Van Leeuwenhoek 30:249–256

Letunic I, Bork P (2016) Interactive tree of life (iTOL) v3: an online tool for the display and annotation of phylogenetic and other trees. Nucleic Acids Res 44(W1):W242–W245. https://doi.org/10.1093/nar/gkw290

Li G, Brown PJ, Tang JX, Xu J, Quardokus EM, Fuqua C, Brun YV (2012) Surface contact stimulates the just-in-time deployment of bacterial adhesins. Mol Microbiol 83(1):41–51. https://doi.org/10.1111/j.1365-2958.2011.07909.x

Liao L, Schaefer AL, Coutinho BG, Brown PJB, Greenberg EP (2018) An aryl-homoserine lactone quorum-sensing signal produced by a dimorphic prosthecate bacterium. Proc Natl Acad Sci USA 115(29):7587–7592. https://doi.org/10.1073/pnas.1808351115

Liechti GW, Kuru E, Hall E, Kalinda A, Brun YV, VanNieuwenhze M, Maurelli AT (2014) A new metabolic cell-wall labelling method reveals peptidoglycan in *Chlamydia trachomatis*. Nature 506(7489):507–510. https://doi.org/10.1038/nature12892

Lin W, Pan Y (2011) Snapping magnetosome chains by asymmetric cell division in magnetotactic bacteria. Mol Microbiol 82(6):1301–1304. https://doi.org/10.1111/j.1365-2958.2011.07866.x

Lin TY, Gross WS, Auer GK, Weibel DB (2019) Cardiolipin alters *Rhodobacter sphaeroides* cell shape by affecting peptidoglycan precursor biosynthesis. mBio 10(1). https://doi.org/10.1128/mBio.02401-18

Lutkenhaus J, Du S (2017) *E. coli* cell cycle machinery. Subcell Biochem 84:27–65. https://doi.org/10.1007/978-3-319-53047-5_2

Mahone CR, Goley (eds) (2020) Bacterial cell division at a glance. J Cell Sci 133(7). https://doi.org/10.1242/jcs.237057

Margolin W (2009) Sculpting the bacterial cell. Curr Biol 19(17):R812–R822. https://doi.org/10.1016/j.cub.2009.06.033

Meeske AJ, Riley EP, Robins WP, Uehara T, Mekalanos JJ, Kahne D, Walker S, Kruse AC, Bernhardt TG, Rudner DZ (2016) SEDS proteins are a widespread family of bacterial cell wall polymerases. Nature 537(7622):634–638. https://doi.org/10.1038/nature19331

Merhej V, Angelakis E, Socolovschi C, Raoult D (2014) Genotyping, evolution and epidemiological findings of *Rickettsia* species. Infect Genet Evol 25:122–137. https://doi.org/10.1016/j.meegid.2014.03.014

Merker RI, Smit J (1988) Characterization of the adhesive holdfast of marine and freshwater caulobacters. Appl Environ Microbiol 54(8):2078–2085

Moll A, Thanbichler M (2009) FtsN-like proteins are conserved components of the cell division machinery in proteobacteria. Mol Microbiol 72(4):1037–1053. https://doi.org/10.1111/j.1365-2958.2009.06706.x

Monteiro JM, Pereira AR, Reichmann NT, Saraiva BM, Fernandes PB, Veiga H, Tavares AC, Santos M, Ferreira MT, Macario V, VanNieuwenhze MS, Filipe SR, Pinho MG (2018) Peptidoglycan synthesis drives an FtsZ-treadmilling-independent step of cytokinesis. Nature 554(7693):528–532. https://doi.org/10.1038/nature25506

Moore RL (1981) The biology of *Hyphomicrobium* and other prosthecate, budding bacteria. Annu Rev Microbiol 35:567–594. https://doi.org/10.1146/annurev.mi.35.100181.003031

Moore RL, Hirsch P (1973) First generation synchrony of isolated *Hyphomicrobium* swarmer populations. J Bacteriol 116(1):418–423

Morgenstein RM, Bratton BP, Nguyen JP, Ouzounov N, Shaevitz JW, Gitai Z (2015) RodZ links MreB to cell wall synthesis to mediate MreB rotation and robust morphogenesis. Proc Natl Acad Sci USA 112(40):12510–12515. https://doi.org/10.1073/pnas.1509610112

Muller P, Ewers C, Bertsche U, Anstett M, Kallis T, Breukink E, Fraipont C, Terrak M, Nguyen-Disteche M, Vollmer W (2007) The essential cell division protein FtsN interacts with the murein (peptidoglycan) synthase PBP1B in *Escherichia coli*. J Biol Chem 282(50):36394–36402. https://doi.org/10.1074/jbc.M706390200

Oberhardt MA, Zarecki R, Gronow S, Lang E, Klenk HP, Gophna U, Ruppin E (2015) Harnessing the landscape of microbial culture media to predict new organism-media pairings. Nat Commun 6:8493. https://doi.org/10.1038/ncomms9493

Oldfield CJ, Dunker AK (2014) Intrinsically disordered proteins and intrinsically disordered protein regions. Annu Rev Biochem 83:553–584. https://doi.org/10.1146/annurev-biochem-072711-164947

Otten C, Brilli M, Vollmer W, Viollier PH, Salje J (2018) Peptidoglycan in obligate intracellular bacteria. Mol Microbiol 107(2):142–163. https://doi.org/10.1111/mmi.13880

Palomeque-Messia P, Englebert S, Leyh-Bouille M, Nguyen-Disteche M, Duez C, Houba S, Dideberg O, Van Beeumen J, Ghuysen JM (1991) Amino acid sequence of the penicillin-binding protein/DD-peptidase of *Streptomyces* K15. Predicted secondary structures of the low Mr penicillin-binding proteins of class A. Biochem J 279(Pt 1):223–230

Paradis-Bleau C, Markovski M, Uehara T, Lupoli TJ, Walker S, Kahne DE, Bernhardt TG (2010) Lipoprotein cofactors located in the outer membrane activate bacterial cell wall polymerases. Cell 143(7):1110–1120. https://doi.org/10.1016/j.cell.2010.11.037

Pate JL, Porter JS, Jordan TL (1973) *Asticcacaulis biprosthecum* sp.nov. life cycle, morphology and cultural characteristics. Antonie Van Leeuwenhoek 39(4):569–583

Patzelt D, Wang H, Buchholz I, Rohde M, Grobe L, Pradella S, Neumann A, Schulz S, Heyber S, Munch K, Munch R, Jahn D, Wagner-Dobler I, Tomasch J (2013) You are what you talk: quorum sensing induces individual morphologies and cell division modes in *Dinoroseobacter shibae*. ISME J 7(12):2274–2286. https://doi.org/10.1038/ismej.2013.107

Pechter KB, Gallagher L, Pyles H, Manoil CS, Harwood CS (2015) Essential genome of the metabolically versatile alphaproteobacterium *Rhodopseudomonas palustris*. J Bacteriol 198(5):867–876. https://doi.org/10.1128/JB.00771-15

Perez AJ, Cesbron Y, Shaw SL, Bazan Villicana J, Tsui HT, Boersma MJ, Ye ZA, Tovpeko Y, Dekker C, Holden S, Winkler ME (2019) Movement dynamics of divisome proteins and PBP2x:FtsW in cells of *Streptococcus pneumoniae*. Proc Natl Acad Sci USA 116(8):3211–3220. https://doi.org/10.1073/pnas.1816018116

Pfeiffer D, Toro-Nahuelpan M, Bramkamp M, Plitzko JM, Schuler D (2019) The polar organizing protein PopZ is fundamental for proper cell division and segregation of cellular content in *Magnetospirillum gryphiswaldense*. mBio 10(2). https://doi.org/10.1128/mBio.02716-18

Piovesan D, Tabaro F, Paladin L, Necci M, Micetic I, Camilloni C, Davey N, Dosztanyi Z, Meszaros B, Monzon AM, Parisi G, Schad E, Sormanni P, Tompa P, Vendruscolo M, Vranken WF, Tosatto SCE (2018) MobiDB 3.0: more annotations for intrinsic disorder, conformational diversity and interactions in proteins. Nucleic Acids Res 46(D1):D471–D476. https://doi.org/10.1093/nar/gkx1071

Poindexter JS (1964) Biological properties and classification of the *Caulobacter* group. Bacteriol Rev 28:231–295

Poindexter JLS, Cohen-Bazire G (1964) The fine structure of stalked bacteria belonging to the family *Caulobacteraceae*. J Cell Biol 23:587–607

Porter JS, Pate JL (1975) Prosthecae of *Asticcacaulis biprosthecum*: system for the study of membrane transport. J Bacteriol 122(3):976–986

Puskas A, Greenberg EP, Kaplan S, Schaefer AL (1997) A quorum-sensing system in the free-living photosynthetic bacterium *Rhodobacter sphaeroides*. J Bacteriol 179(23):7530–7537

Randich AM, Brun YV (2015) Molecular mechanisms for the evolution of bacterial morphologies and growth modes. Front Microbiol 6:580. https://doi.org/10.3389/fmicb.2015.00580

Rohs PDA, Buss J, Sim SI, Squyres GR, Srisuknimit V, Smith M, Cho H, Sjodt M, Kruse AC, Garner EC, Walker S, Kahne DE, Bernhardt TG (2018) A central role for PBP2 in the activation of peptidoglycan polymerization by the bacterial cell elongation machinery. PLoS Genet 14(10):e1007726. https://doi.org/10.1371/journal.pgen.1007726

Romeis T, Holtje JV (1994) Penicillin-binding protein 7/8 of *Escherichia coli* is a DD-endopeptidase. Eur J Biochem 224(2):597–604

Sakaguchi T, Arakaki A, Matsunaga T (2002) *Desulfovibrio magneticus* sp. nov., a novel sulfate-reducing bacterium that produces intracellular single-domain-sized magnetite particles. Int J Syst Evol Microbiol 52(Pt 1):215–221. https://doi.org/10.1099/00207713-52-1-215

Sallstrom B, Andersson SG (2005) Genome reduction in the alpha-Proteobacteria. Curr Opin Microbiol 8(5):579–585. https://doi.org/10.1016/j.mib.2005.08.002

Schaper S, Yau HCL, Krol E, Skotnicka D, Heimerl T, Gray J, Kaever V, Sogaard-Andersen L, Vollmer W, Becker A (2018) Seven-transmembrane receptor protein RgsP and cell wall-binding protein RgsM promote unipolar growth in Rhizobiales. PLoS Genet 14(8):e1007594. https://doi.org/10.1371/journal.pgen.1007594

Schleifer KH, Schuler D, Spring S, Weizenegger M, Amann R, Ludwig W, Kohler M (1991) The genus *Magnetospirillum* gen. nov. description of *Magnetospirillum gryphiswaldense* sp. nov. and transfer of *Aquaspirillum magnetotacticum* to *Magnetospirillum magnetotacticum* comb. nov. Syst Appl Microbiol 14(4):379–385. https://doi.org/10.1016/S0723-2020(11)80313-9

Schlesner H, Bartels C, Sittig M, Dorsch M, Stackebrandt E (1990) Taxonomic and phylogenetic studies on a new taxon of budding, hyphal proteobacteria, *Hirschia baltica* gen. nov., sp. nov. Int J Syst Bacteriol 40(4):443–451. https://doi.org/10.1099/00207713-40-4-443

Schlessinger A, Schaefer C, Vicedo E, Schmidberger M, Punta M, Rost B (2011) Protein disorder – a breakthrough invention of evolution? Curr Opin Struct Biol 21(3):412–418. https://doi.org/10.1016/j.sbi.2011.03.014

Schlimpert S, Klein EA, Briegel A, Hughes V, Kahnt J, Bolte K, Maier UG, Brun YV, Jensen GJ, Gitai Z, Thanbichler M (2012) General protein diffusion barriers create compartments within bacterial cells. Cell 151(6):1270–1282. https://doi.org/10.1016/j.cell.2012.10.046

Schmidt JM, Stanier RY (1966) The development of cellular stalks in bacteria. J Cell Biol 28(3):423–436

Segev E, Tellez A, Vlamakis H, Kolter R (2015) Morphological heterogeneity and attachment of *Phaeobacter inhibens*. PLoS One 10(11):e0141300. https://doi.org/10.1371/journal.pone.0141300

Shi H, Bratton BP, Gitai Z, Huang KC (2018) How to build a bacterial cell: MreB as the foreman of *E. coli* construction. Cell 172(6):1294–1305. https://doi.org/10.1016/j.cell.2018.02.050

Shibai A, Maeda T, Kawada M, Kotani H, Sakata N, Furusawa C (2019) Complete genome sequences of three star-shaped bacteria, *Stella humosa*, *Stella vacuolata*, and *Stella* species ATCC 35155. Microbiol Resour Announc 8(32). https://doi.org/10.1128/MRA.00719-19

Shih YL, Rothfield L (2006) The bacterial cytoskeleton. Microbiol Mol Biol Rev 70(3):729–754. https://doi.org/10.1128/MMBR.00017-06

Siegrist MS, Whiteside S, Jewett JC, Aditham A, Cava F, Bertozzi CR (2013) (D)-Amino acid chemical reporters reveal peptidoglycan dynamics of an intracellular pathogen. ACS Chem Biol 8(3):500–505. https://doi.org/10.1021/cb3004995

Slightom RN, Buchan A (2009) Surface colonization by marine roseobacters: integrating genotype and phenotype. Appl Environ Microbiol 75(19):6027–6037. https://doi.org/10.1128/AEM.01508-09

Slovak PM, Wadhams GH, Armitage JP (2005) Localization of MreB in *Rhodobacter sphaeroides* under conditions causing changes in cell shape and membrane structure. J Bacteriol 187(1):54–64. https://doi.org/10.1128/JB.187.1.54-64.2005

Slovak PM, Porter SL, Armitage JP (2006) Differential localization of Mre proteins with PBP2 in *Rhodobacter sphaeroides*. J Bacteriol 188(5):1691–1700. https://doi.org/10.1128/JB.188.5.1691-1700.2006

Spratt BG (1975) Distinct penicillin binding proteins involved in the division, elongation, and shape of *Escherichia coli* K12. Proc Natl Acad Sci USA 72(8):2999–3003

Staley JT (1968) *Prosthecomicrobium* and *Ancalomicrobium*: new prosthecate freshwater bacteria. J Bacteriol 95(5):1921–1942

Staley J (1984) *Prosthecomicrobium hirschii*, a new species in a redefined genus. Int J Syst Bacteriol 34(3):304–308

Stamatakis A (2014) RAxML version 8: a tool for phylogenetic analysis and post-analysis of large phylogenies. Bioinformatics 30(9):1312–1313. https://doi.org/10.1093/bioinformatics/btu033

Staniland SS, Moisescu C, Benning LG (2010) Cell division in magnetotactic bacteria splits magnetosome chain in half. J Basic Microbiol 50(4):392–396. https://doi.org/10.1002/jobm.200900408

Stankeviciute G, Miguel AV, Radkov A, Chou S, Huang KC, Klein EA (2019) Differential modes of crosslinking establish spatially distinct regions of peptidoglycan in *Caulobacter crescentus*. Mol Microbiol 111(4):995–1008. https://doi.org/10.1111/mmi.14199

Taguchi A, Welsh MA, Marmont LS, Lee W, Sjodt M, Kruse AC, Kahne D, Bernhardt TG, Walker S (2019) FtsW is a peptidoglycan polymerase that is functional only in complex with its cognate penicillin-binding protein. Nat Microbiol 4(4):587–594. https://doi.org/10.1038/s41564-018-0345-x

Takacs CN, Hocking J, Cabeen MT, Bui NK, Poggio S, Vollmer W, Jacobs-Wagner C (2013) Growth medium-dependent glycine incorporation into the peptidoglycan of *Caulobacter crescentus*. PLoS One 8(2):e57579. https://doi.org/10.1371/journal.pone.0057579

Tam E, Pate JL (1985) Amino acid transport by prosthecae of *Asticcacaulis biprosthecum*: evidence for a broad-range transport system. J Gen Microbiol 131(10):2687–2699. https://doi.org/10.1099/00221287-131-10-2687

Toro-Nahuelpan M, Corrales-Guerrero L, Zwiener T, Osorio-Valeriano M, Muller FD, Plitzko JM, Bramkamp M, Thanbichler M, Schuler D (2019) A gradient-forming MipZ protein mediating the control of cell division in the magnetotactic bacterium Magnetospirillum gryphiswaldense. Mol Microbiol 112(5):1423–1439. https://doi.org/10.1111/mmi.14369

Tucker JD, Siebert CA, Escalante M, Adams PG, Olsen JD, Otto C, Stokes DL, Hunter CN (2010) Membrane invagination in *Rhodobacter sphaeroides* is initiated at curved regions of the cytoplasmic membrane, then forms both budded and fully detached spherical vesicles. Mol Microbiol 76(4):833–847. https://doi.org/10.1111/j.1365-2958.2010.07153.x

Typas A, Banzhaf M, van den Berg van Saparoea B, Verheul J, Biboy J, Nichols RJ, Zietek M, Beilharz K, Kannenberg K, von Rechenberg M, Breukink E, den Blaauwen T, Gross CA, Vollmer W (2010) Regulation of peptidoglycan synthesis by outer-membrane proteins. Cell 143(7):1097–1109. https://doi.org/10.1016/j.cell.2010.11.038

Typas A, Banzhaf M, Gross CA, Vollmer W (2011) From the regulation of peptidoglycan synthesis to bacterial growth and morphology. Nat Rev Microbiol 10(2):123–136. https://doi.org/10.1038/nrmicro2677

Uebe R, Schuler D (2016) Magnetosome biogenesis in magnetotactic bacteria. Nat Rev Microbiol 14(10):621–637. https://doi.org/10.1038/nrmicro.2016.99

Uehara T, Park JT (2008) Growth of *Escherichia coli*: significance of peptidoglycan degradation during elongation and septation. J Bacteriol 190(11):3914–3922. https://doi.org/10.1128/JB.00207-08

Ursell TS, Nguyen J, Monds RD, Colavin A, Billings G, Ouzounov N, Gitai Z, Shaevitz JW, Huang KC (2014) Rod-like bacterial shape is maintained by feedback between cell curvature and cytoskeletal localization. Proc Natl Acad Sci USA 111(11):E1025–E1034. https://doi.org/10.1073/pnas.1317174111

van Teeffelen S, Wang S, Furchtgott L, Huang KC, Wingreen NS, Shaevitz JW, Gitai Z (2011) The bacterial actin MreB rotates, and rotation depends on cell-wall assembly. Proc Natl Acad Sci USA 108(38):15822–15827. https://doi.org/10.1073/pnas.1108999108

Vasilyeva L (1985) *Stella*, a new genus of soil prosthecobacteria, with proposals for *Stella humosa* sp. nov. and *Stella vacuolata* sp. nov. Int J Syst Bacteriol 35(4):518–521

Venter JC, Remington K, Heidelberg JF, Halpern AL, Rusch D, Eisen JA, Wu D, Paulsen I, Nelson KE, Nelson W, Fouts DE, Levy S, Knap AH, Lomas MW, Nealson K, White O, Peterson J, Hoffman J, Parsons R, Baden-Tillson H, Pfannkoch C, Rogers YH, Smith HO (2004) Environmental genome shotgun sequencing of the Sargasso Sea. Science 304(5667):66–74. https://doi.org/10.1126/science.1093857

Vigouroux A, Cordier B, Aristov A, Alvarez L, Ozbaykal G, Chaze T, Oldewurtel ER, Matondo M, Cava F, Bikard D, van Teeffelen S (2020) Class-A penicillin binding proteins do not contribute to cell shape but repair cell-wall defects. elife 9. https://doi.org/10.7554/eLife.51998

Vollmer W, Bertsche U (2008) Murein (peptidoglycan) structure, architecture and biosynthesis in *Escherichia coli*. Biochim Biophys Acta 1778(9):1714–1734. https://doi.org/10.1016/j.bbamem.2007.06.007

Vollmer W, Blanot D, de Pedro MA (2008) Peptidoglycan structure and architecture. FEMS Microbiol Rev 32(2):149–167. https://doi.org/10.1111/j.1574-6976.2007.00094.x

Vollmer J, Schiefer A, Schneider T, Julicher K, Johnston KL, Taylor MJ, Sahl HG, Hoerauf A, Pfarr K (2013) Requirement of lipid II biosynthesis for cell division in cell wall-less *Wolbachia*, endobacteria of arthropods and filarial nematodes. Int J Med Microbiol 303(3):140–149. https://doi.org/10.1016/j.ijmm.2013.01.002

Wagner JK, Galvani CD, Brun YV (2005) *Caulobacter crescentus* requires RodA and MreB for stalk synthesis and prevention of ectopic pole formation. J Bacteriol 187(2):544–553. https://doi.org/10.1128/JB.187.2.544-553.2005

Wagner JK, Setayeshgar S, Sharon LA, Reilly JP, Brun YV (2006) A nutrient uptake role for bacterial cell envelope extensions. Proc Natl Acad Sci USA 103(31):11772–11777. https://doi.org/10.1073/pnas.0602047103

Wali TM, Hudson GR, Danald DA, Weiner RM (1980) Timing of swarmer cell cycle morphogenesis and macromolecular synthesis by *Hyphomicrobium neptunium* in synchronous culture. J Bacteriol 144(1):406–412

Wang H, Ziesche L, Frank O, Michael V, Martin M, Petersen J, Schulz S, Wagner-Dobler I, Tomasch J (2014) The CtrA phosphorelay integrates differentiation and communication in the marine alphaproteobacterium *Dinoroseobacter shibae*. BMC Genomics 15:130. https://doi.org/10.1186/1471-2164-15-130

Weidel W, Pelzer H (1964) Bagshaped macromolecules – a new outlook on bacterial cell walls. Adv Enzymol Relat Subj Biochem 26:193–232

Weiner RM, Blackman MA (1973) Inhibition of deoxyribonucleic acid synthesis and bud formation by nalidixic acid in *Hyphomicrobium neptunium*. J Bacteriol 116(3):1398–1404

Weiner RM, Melick M, O'Neill K, Quintero E (2000) *Hyphomonas adhaerens* sp. nov., *Hyphomonas johnsonii* sp. nov. and *Hyphomonas rosenbergii* sp. nov., marine budding and prosthecate bacteria. Int J Syst Evol Microbiol 50(Pt 2):459–469. https://doi.org/10.1099/00207713-50-2-459

Welsh MA, Taguchi A, Schaefer K, Van Tyne D, Lebreton F, Gilmore MS, Kahne D, Walker S (2017) Identification of a functionally unique family of penicillin-binding proteins. J Am Chem Soc 139(49):17727–17730. https://doi.org/10.1021/jacs.7b10170

Whittenbury R, Dow CS (1977) Morphogenesis and differentiation in *Rhodomicrobium vannielii* and other budding and prosthecate bacteria. Bacteriol Rev 41(3):754–808

Whittenbury R, McLee AG (1967) *Rhodopseudomonas palustris* and *Rh. viridis* – photosynthetic budding bacteria. Arch Mikrobiol 59(1):324–334. https://doi.org/10.1007/BF00406346

Wilkinson DA, Chacko SJ, Venien-Bryan C, Wadhams GH, Armitage JP (2011) Regulation of flagellum number by FliA and FlgM and role in biofilm formation by *Rhodobacter sphaeroides*. J Bacteriol 193(15):4010–4014. https://doi.org/10.1128/JB.00349-11

Williams M, Hoffman MD, Daniel JJ, Madren SM, Dhroso A, Korkin D, Givan SA, Jacobson SC, Brown PJ (2016) Short-stalked *Prosthecomicrobium hirschii* cells have a *Caulobacter*-like cell cycle. J Bacteriol 198(7):1149–1159. https://doi.org/10.1128/JB.00896-15

Williams MA, Aliashkevich A, Krol E, Kuru E, Bouchier JM, Rittichier J, Brun YV, Van-Nieuwenhze MS, Becker A, Cava F, Brown PJB (2021) Unipolar peptidoglycan synthesis in the rhizobiales requires an essential class A penicillin-binding protein. mBio 12:e0234621

Woronowicz K, Niederman RA (2010) Proteomic analysis of the developing intracytoplasmic membrane in *Rhodobacter sphaeroides* during adaptation to low light intensity. Adv Exp Med Biol 675:161–178. https://doi.org/10.1007/978-1-4419-1528-3_10

Xu Q, Carlton D, Miller MD, Elsliger MA, Krishna SS, Abdubek P, Astakhova T, Burra P, Chiu HJ, Clayton T, Deller MC, Duan L, Elias Y, Feuerhelm J, Grant JC, Grzechnik A, Grzechnik SK, Han GW, Jaroszewski L, Jin KK, Klock HE, Knuth MW, Kozbial P, Kumar A, Marciano D, McMullan D, Morse AT, Nigoghossian E, Okach L, Oommachen S, Paulsen J, Reyes R, Rife CL, Sefcovic N, Trame C, Trout CV, van den Bedem H, Weekes D, Hodgson KO, Wooley J, Deacon AM, Godzik A, Lesley SA, Wilson IA (2009) Crystal structure of histidine phosphotransfer protein ShpA, an essential regulator of stalk biogenesis in *Caulobacter crescentus*. J Mol Biol 390(4):686–698. https://doi.org/10.1016/j.jmb.2009.05.023

Yang X, Lyu Z, Miguel A, McQuillen R, Huang KC, Xiao J (2017) GTPase activity-coupled treadmilling of the bacterial tubulin FtsZ organizes septal cell wall synthesis. Science 355(6326):744–747. https://doi.org/10.1126/science.aak9995

Zerfas PM, Kessel M, Quintero EJ, Weiner RM (1997) Fine-structure evidence for cell membrane partitioning of the nucleoid and cytoplasm during bud formation in *Hyphomonas* species. J Bacteriol 179(1):148–156

Zupan JR, Cameron TA, Anderson-Furgeson J, Zambryski PC (2013) Dynamic FtsA and FtsZ localization and outer membrane alterations during polar growth and cell division in *Agrobacterium tumefaciens*. Proc Natl Acad Sci USA 110(22):9060–9065. https://doi.org/10.1073/pnas.1307241110

Zupan JR, Grangeon R, Robalino-Espinosa JS, Garnica N, Zambryski P (2019) GROWTH POLE RING protein forms a 200-nm-diameter ring structure essential for polar growth and rod shape in *Agrobacterium tumefaciens*. Proc Natl Acad Sci USA 116(22):10962–10967. https://doi.org/10.1073/pnas.1905900116

Cell Cycle and Terminal Differentiation in *Sinorhizobium meliloti*

Sara Dendene, Angela Frascella, Quentin Nicoud, Tatiana Timchenko, Peter Mergaert, Benoit Alunni, and Emanuele G. Biondi

Abstract

Sinorhizobium meliloti of the *Alphaproteobacteria* class has a fascinating spectrum of lifestyles, thriving as a free-living soil saprophyte, as an endophyte, and as a nitrogen-fixing legume symbiont. In symbiosis, it undergoes a striking cellular differentiation process, which is controlled by the host plant through the activity of NCR peptides. NCRs interfere with the cell cycle of *S. meliloti* and transform the regular cycle consisting of strict successions of single DNA replication followed by cell division into an endoreduplication cycle of multiple genome duplications without divisions. This cellular differentiation results in giant and polyploid symbiotic bacterial cells that fix atmospheric nitrogen. Here we discuss the regulation of the free-living cell cycle in *S. meliloti* and present the hypothesis that the master regulator CtrA is the ultimate target of the NCR peptides, provoking the cell cycle switch in symbiosis.

S. Dendene · Q. Nicoud · T. Timchenko · P. Mergaert · B. Alunni
Institute for Integrative Biology of the Cell, CEA, CNRS, University Paris-Saclay, Gif-sur-Yvette, France

A. Frascella
Aix Marseille University, CNRS, IMM, LCB, Marseille, France

E. G. Biondi (✉)
Institute for Integrative Biology of the Cell, CEA, CNRS, University Paris-Saclay, Gif-sur-Yvette, France

Aix Marseille University, CNRS, IMM, LCB, Marseille, France
e-mail: emanuele.biondi@i2bc.paris-saclay.fr

E. Biondi (ed.), *Cell Cycle Regulation and Development in Alphaproteobacteria*,
https://doi.org/10.1007/978-3-030-90621-4_8

1 *Sinorhizobium meliloti* Biology

The alphaproteobacterium *S. meliloti* is a free-living soil saprophyte, an endophyte, and a legume endosymbiont (Poole et al. 2018). The latter lifestyle has made its renown and today it is one of the better-studied symbiotic species among all bacteria. This bacterium has the ability to colonize the roots of leguminous plants of the genus *Medicago*. When it is in contact with plant roots, it induces the formation of new underground organs, called nodules (Fig. 1a), in which it finds the ideal conditions to reduce atmospheric nitrogen into ammonia that is incorporated by the plant and used for its nitrogen needs in growth. In exchange, the bacterium receives nutrients and an exclusive niche inside the nodules where it can establish, starting from a single or very few bacterial cells, a very large population of several millions in the very short time span of a few days, the time it takes to form a mature and fully infected nodule. Thus, unlike the soil environment, where the rate of division is low, the nodule allows the bacteria to multiply rapidly, indicating a considerable evolutionary advantage for the symbiotic lifestyle.

The nodule-forming and nitrogen-fixing symbiosis is widespread in legumes, where it constitutes an ancestral trait of the family although some legume species lost their symbiotic ability (ref: Griesmann et al. 2018). The combined nitrogen fixation activity by all legumes on earth is a key process in the biogeochemical nitrogen cycle and has, therefore, a tremendous impact on the ecology of our planet. Moreover, symbiotic nitrogen fixation by legumes has a considerable interest for its applications in agriculture as an alternative to chemical nitrogen fertilizer. Among the rhizobium–legume interactions, the *S. meliloti-Medicago* interaction has emerged as one of the most productive model systems for the study of the nodule-forming and nitrogen-fixing symbiosis.

During the establishment of the symbiotic organ, *S. meliloti* infects symbiotic plant cells and become intracellular nitrogen-fixing organelle-like structures called bacteroids. The bacteroids undergo a drastic differentiation program, resulting in cells that are unable to divide and produce offspring, resulting therefore in a terminally differentiated state (Fig. 1a). From an evolutionary point of view, this terminal differentiation is puzzling at first sight, because the absence of offspring seems to be incompatible with the natural selection of this process. However, since the nodule bacteria form a clonal or nearly clonal population, even if the majority of them are terminally differentiated, still a large fraction of genetically identical undifferentiated bacteria remain in the nodules and those can replenish the soil rhizobial population from senescing nodules. Moreover, it seems to be difficult to explain why a bacterium, in the context of a beneficial symbiosis, sacrifices billions of siblings for the cause of the plant. In this review, we will explain the underlying causes of the differentiation, which are derived from the plant rather than programmed by the bacterium and we will present evolutionary hypotheses about the role of bacteroid differentiation, which is probably beneficial to the plant rather than to the rhizobia. At its most basic level, the terminal differentiation of *S. meliloti* corresponds to a transformation of its regular cell cycle into an endoreduplication

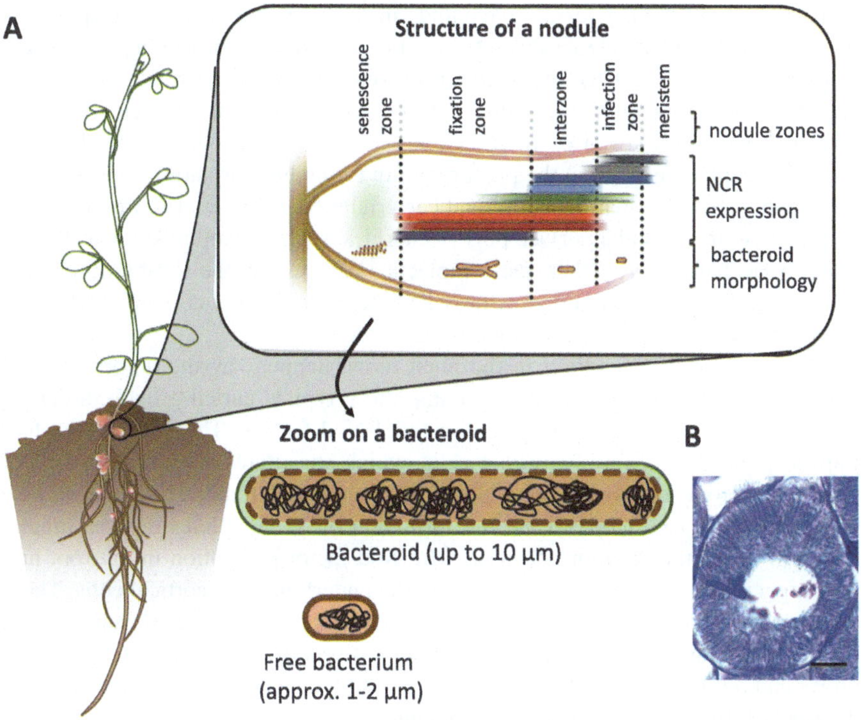

Fig. 1 The symbiosis is a close relationship between two different organisms. (**A**) The symbiosis between the alphaproteobacterium *Sinorhizobium meliloti* and the legume *Medicago sativa* (on the left) produces a new organ called nodule, where an exchange of metabolites occurs: the bacterium reduces atmospheric nitrogen into ammonia that is absorbed by the plant; in return, this latter provides carbon sources and protection to the bacterium. On the top panel, a zoom on the structure of a nodule and, indicated with colored bars, the expression of various NCR (Nodule Cysteine rich) peptides are shown. Once in the infection zone, the bacterium begins a dramatic differentiation process under the influence of the NCR peptides leading to a bacteroid cell. A symbiosome (a zoom is represented in the bottom central part) is composed by a layer of vegetal membrane containing a bacteroid cell, characterized by (a) multiple copies of DNA, (b) a cell enlargement (ten times bigger than the free-living cell also showed here), (c) an inability to divide and (d) a higher permeability, as shown by a dotted envelope. (**B**) Image of a symbiotic plant cell stained with coomassie (vegetal cell wall in blue) full of bacteroids (blue); plant nucleus is white (Peter Mergaert, unpublished). Black bar corresponds to 10 μm

cycle with no offspring generation. In this chapter, we will present our view on how this cell cycle switch can be accomplished mechanistically.

2 Symbiotic Infection and Differentiation

The first bacterial contact with plants consists of an exchange of specific signaling molecules. Chemotaxis, particularly toward the abundant amino acids in *Medicago* exudates, guides the *S. meliloti* bacteria in the soil toward the plant roots (Compton et al. 2020). Plants also secrete flavonoids in the rhizosphere and in response to these plant molecules, rhizobia secrete lipochitooligosaccharidic signals called Nod factors (Poole et al. 2018). Interestingly flavonoids have also growth stimulating activity on the rhizobia, suggesting multiple dose-dependent ecological roles of this plant signal (Nouwen et al. 2019). Nod factors are recognized by the plant through receptors, triggering the plant program for nodule formation and infection. During the nodule formation, additional bacterial molecular patterns are monitored by the plant, in particular surface polysaccharides, such as exopolysaccharides and lipopolysaccharides. Together, these molecular keys direct the symbiotic process and are highly specific, minimizing the risk of infections by non-compatible rhizobia or opportunistic and pathogenic organisms.

The entrance of *S. meliloti* in the plant tissue happens by the deformation of the normally straight growth of root hairs into a typical curled growth direction, which is specifically induced by the bacterial Nod factors. The curling root hair traps a single or very few *S. meliloti* cells, which constitutes the founding cells of what will become the complete nodule population. The entrapped rhizobia are able to penetrate the root hair cell via the formation of an infection thread, a tubular structure containing dividing bacteria. The Nod factor perception in the root hair also triggers cell divisions, at a distance, in the underlying root cortical cells. These dividing plant cells form a nodule primordium that will further develop into a nodule. Simultaneously, the infection thread that was first initiated in the root hair grows and ramifies toward the primordium, thereby conducting the rhizobia toward the newly formed cells of the incipient nodule.

An infection thread that has reached and penetrated a young nodule cell releases rhizobia through an endocytotic process into the plant cell. The endocytotic uptake from an infection thread in a differentiating nodule cell does not release rhizobia freely in the cytosol but inside vesicles, called symbiosomes, which have a plasmalemma-like membrane. Within the symbiosomes, the rhizobia grow and differentiate into their nitrogen-fixing forms called the bacteroids. Repeated infections and the growth of the rhizobia in symbiosomes will ultimately result in a symbiotic nodule cell that is completely packed with intracellular bacteroids (Fig. 1b).

A plant cell that has been infected does not divide anymore but switches into a differentiation path toward a nitrogen-fixing nodule cell. This differentiation includes the activation of an endoreduplication cycle, leading to polyploidy and very strong cell enlargement as well as the activation of a specific transcriptional program that will assure the maintenance and the metabolic integration of the thousands of nitrogen-fixing endosymbionts within each individual mature symbiotic nodule cell (Mergaert et al. 2020). On the other hand, a few distal cells in the incipient nodule that are not penetrated by an infection thread will constitute a nodule meristem and will continue to divide. The formation of this apical meristem in *Medicago* nodules gives the organ an indeterminate state with a continuous growth during the complete lifetime of the nodule.

The nitrogen-fixing *S. meliloti* bacteroids in nodule cells are in a differentiated state, which is in various ways dramatically different from its free-living state in the soil (Fig. 1a). First of all, the bacteroid formation implies a switch in the bacterial physiology that is adapted to the nitrogen fixation process. This switch is made possible by a massive transcriptional activation of a large set of genes encoding nitrogen fixation and associated respiratory functions, which are completely silent in the free-living state (Roux et al. 2014). This transcriptome switch is controlled by a regulatory cascade, composed of the FixLJ two-component regulator that senses the low oxygen concentration prevailing in the nodule cells, and the downstream NifA and FixK transcription factors (Bobik et al. 2006).

The physiological adaptation of the *S. meliloti* bacteroids is in addition accompanied with their above-mentioned remarkable terminal differentiation (Mergaert et al. 2006). This state of *S. meliloti* is characterized by the irreversible loss of capacity to resume growth and to reproduce. The bacteroids also have a partially permeabilized membrane. However, their most striking feature is their metamorphosis into a giant, sometimes branched, bacterial cell of up to ten micrometers long. Moreover, similarly to their host cells, these bacteroids are polyploid.

The terminal bacteroid differentiation is a process that is determined by the host rather than being uniquely encoded in the genetic repertoire of the rhizobia. Indeed, terminal differentiation is not happening in all legumes. It is for example taking place in the Inverted Repeat Lacking Clade (IRLC) and Dalbergioid clade plants to which respectively *Medicago* and *Aeschynomene* species belong but it is not happening in the Robinioid or Millettioid clades containing the well-studied *Lotus* and *Glycine* genera, respectively (Mergaert et al. 2006; Czernic et al. 2015). Broad host range rhizobia or engineered strains that have a switched host range, will form terminally differentiated bacteroids or not according to the host species in which they are found. This suggests that the terminal differentiation is in the first place determined by the plant, although also the bacterial genetic repertoire contributes to the extent of the bacteroid differentiation process (Mergaert et al. 2006; Nicoud et al. 2020).

Based on a phylogenetic analysis of the bacteroid state in the legume family, it was proposed that the ancestral state of bacteroids is the undifferentiated type, the type that is found in the Robinioids or Millettioids (Oono et al. 2011). According to this scenario, terminal bacteroid differentiation has evolved several times in the legumes and appeared independently in for example the IRLC and Dalbergoid legumes.

3 The NCR Peptides, Host Effectors of Bacteroid Differentiation

The identification of the host factors that determine the terminal bacteroid differentiation was based on genomic and transcriptomic comparisons of legumes that display the feature or not (Mergaert et al. 2003; Alunni et al. 2007; Van de Velde et al. 2010). These analyses correlated the formation of terminally differentiated bacteroids with the expression in nodules of a particular family of genes encoding peptides which were called the NCRs for nodule-specific and cysteine-rich peptides. In *M. truncatula*, these peptides are very specifically produced in nodules, in the infected symbiotic cells, and nowhere else in the plant (Guefrachi et al. 2014). Remarkably, many legume species produce a large diversity of them, sometimes over several hundred different ones. *M. truncatula* for example expresses over 600 different *NCR* genes in nodules (Montiel et al. 2017). NCR peptides are small secretory peptides characterized by a pattern of conserved cysteine residues. Importantly, the NCRs are related to antimicrobial peptides, which are innate immunity effectors that are used by eukaryotic hosts, namely plants and animals, to attack and eliminate invading microbes (Mergaert 2018).

All tested species of the IRLC legumes produce NCR peptides in their nodules, which are of the same phylogenetic family and have thus a common ancestor. On the other hand, the *Aeschynomene* legumes of the Dalbergoid clade produce in their nodules NCRs of an unrelated family with distinct sequences and cysteine patterns (Czernic et al. 2015; Gully et al. 2018; Quilbé et al. 2020). The use of different NCR families is in agreement with the independent evolution of bacteroid differentiation in these two clades (Oono et al. 2011).

Several arguments have confirmed the initial phylogenomic correlation between the production of NCRs in the symbiotic nodule cells and the terminal differentiation of bacteroids (Fig. 1a). A recent analysis for example showed that the degree of differentiation of bacteroids in species of the IRLC correlates with the amount of NCRs expressed in nodules and also with the type of peptides they produce. The higher the diversity of peptides and the more cationic the NCR peptides are, the stronger the morphological change of the bacteroids (Montiel et al. 2017).

The majority of the NCR peptides, if not all, are transported to the bacteroids indicating that the endosymbionts are their target. Indeed, the localization of many individual peptides in the bacteroids or symbiosomes have been demonstrated by immunolocalization, by expressing NCR fusion proteins with fluorescent markers as well as by cell fractionations of nodule extracts and purifications of bacteroids followed by western analysis or proteomics (Van de Velde et al. 2010; Haag et al. 2011; Durgo et al. 2015; Czernic et al. 2015; Horváth et al. 2015; Kim et al. 2015; Wang et al. 2017; Yang et al. 2017).

In vitro experiments have shown that pure (synthetic?) NCRs induce features on free-living *S. meliloti* that mimic the terminal bacteroids such as cell elongation and polyploidy (Van de Velde et al. 2010; Haag et al. 2011; Penterman et al. 2014; Montiel et al. 2017). Moreover, transferring *NCR* genes to *Lotus japonicus*, a legume that does not have them and that makes normally reversible bacteroids, leads to new bacteroid features similar to terminal bacteroids (Van de Velde et al. 2010).

Complementary to these "gain-of-function" methods, also loss-of-function experiments are confirming the key role of NCR peptides in bacteroid differentiation. NCR peptides are secretory peptides, which depend on their signal peptide to be taken in charge by the secretory pathway for trafficking to their cellular destination. In the *M. truncatula* mutant of the secretory pathway *dnf1*, NCR transport to the bacteroids is blocked. Thus, NCRs are stuck in the endoplasmic reticulum in the infected nodule cells and this prevents terminal bacteroid differentiation (Van de Velde et al. 2010; Wang et al. 2010). Similarly, downregulation of the orthologous secretory pathway gene *DNF1* in *Aeschynomene* nodules by RNAi blocks bacteroid differentiation (Czernic et al. 2015). More recently, several mutants or allelic variations in particular *NCR* genes were identified in *M. truncatula* that affect the bacteroid differentiation and persistence (Horváth et al. 2015; Kim et al. 2015; Wang et al. 2017, 2018; Yang et al. 2017). These are very surprising findings in light of the high number of *NCR* genes in *M. truncatula*, which would intuitively lead us to suppose a very high level of redundancy, but they provide very strong support for the key role of the peptides in the bacteroid formation.

From the bacterial side, several factors were identified that are involved in the response of bacteroids to this assault of NCR peptides (Mergaert 2018). They can be divided into three broad categories of functions. First, the NCR peptides interfere with several metabolic processes including protein synthesis, energy household, and nitrogen fixation itself (Farkas et al. 2014). Second, as mentioned above NCRs are similar to antimicrobial peptides and they indeed have antimicrobial activity. *S. meliloti* bacteroids use several mechanisms to protect themselves against this harmful activity of the NCRs (Haag et al. 2011; Montiel et al. 2017; Arnold et al. 2018; Nicoud et al. 2020). The most notable among them is mediated by the peptide transporter BacA. Finally, the polyploid state of the bacteroids implies that the terminal bacteroid differentiation is driven by a switch in the bacterial cell cycle whereby the regular cycle composed of sequential steps of a single genome replication followed by cell division is transformed into a process of repeated genome replications without cell divisions (Mergaert et al. 2006). In the next sections, we first discuss the state of the art of our knowledge of the regular cell cycle control in *S. meliloti,* and then we will analyze the available data that highlight how the NCR peptides can interfere with the cell cycle to promote the bacteroid differentiation.

4 The Cell Cycle in *Sinorhizobium meliloti*

S. meliloti division is asymmetrical and always produces two different cell types, a "small" type and a "large" type (Fig. 2). The large cell is able to replicate its genome and to produce new small and large cells. The small cell, on the contrary, does not have the capacity to replicate its DNA and to divide immediately. It must first differentiate into a large cell before initiating a new cell cycle (De Nisco et al. 2014). This morphological asymmetry imposes thus a continuous asynchrony between the subsequent cell cycles of the daughter cells after division. To our knowledge, there is no exception in *S. meliloti* to the rule of one single round of genome replication per cell division as the origin of replication is strictly controlled by multiple regulatory mechanisms that ensure this perfect coordination between DNA replication and cell division (De Nisco et al. 2014; Pini et al. 2015).

The expression of almost 500 genes varies as a function of progression in cell cycle in *S. meliloti*, and these genes show peak expression corresponding to the timing of their cellular function (De Nisco et al. 2014). This time-regulated expression of genes, which are required for specific functions, was analyzed by developing a new method of synchronization of *S. meliloti* bacterial cultures. The method is based on the induction of the stringent response (induced by carbon and nitrogen starvation) able to block cells in the G1 phase by Rel-dependent ppGpp accumulation (De Nisco et al. 2014). G1-blocked cells are then able to proceed

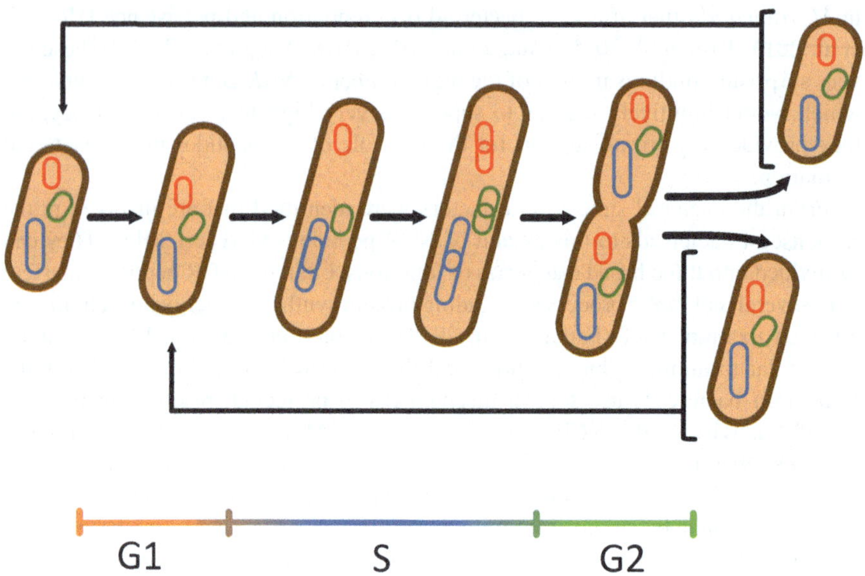

Fig. 2 The cell cycle in *Sinorhizobium meliloti*. *S. meliloti* is a rod-shaped bacterium belonging to the *Alphaproteobacteria* class. The bacterium contains three replicons (circles of different colors) that duplicate only once per cell cycle. DNA replication is followed by an asymmetrical cell division producing a large and a small cell. From the left to the right: during the G1 phase (one copy of each replicon) a small cell differentiates into a large cell, this latter begins the S phase (DNA replication), then in the G2 phase, the pre-divisional cell divides asymmetrically producing a new small cell, unable to replicate, and a large cell that is able to immediately initiate a new round of DNA replication

through a complete and synchronized cell cycle with only one DNA replication round, ultimately leading to an asymmetrical cell division. DNA replication was analyzed further by tracking the origin of replication of the different replicons of *S. meliloti* (Frage et al. 2016). This bacterium possesses three large replicons: a 4 mega-bases circular chromosome with a single DnaA-dependent origin of replication, a second smaller replicon (1.9 mega-bases), named pSymB, that contains several essential genes and many genes involved in the adaptation to environmental niches, and finally a dispensable smaller megaplasmid (I,5 mega-bases), named pSymA, carrying genes mostly associated to symbiosis (Galibert et al. 2001; Capela et al. 2001; Finan et al. 2001). Surprisingly, the initiation of replication of the three origins of replication is temporally and spatially separated in the cell, with the chromosome being the first to be replicated with the newly-replicated origins located very close to the polar regions of the cell. The megaplasmid pSymB follows the chromosome replication with its origin located in the proximity of the pole but shifted toward the center of the cell. Finally, pSymA replication starts after pSymB and its origin at the beginning of its replication is localized almost at mid-cell (Frage et al. 2016). This remarkable spatial and temporal organization suggests that DNA replication in *S. meliloti* is highly organized by mechanisms that are still unknown.

5 Cell Cycle Regulation in *Sinorhizobium meliloti*

5.1 The Conserved Architecture of the Master Cell Cycle Regulatory Circuit

As revealed by the bioinformatic analysis of alphaproteobacterial genomes, almost all factors that regulate the cell cycle in the model system *Caulobacter crescentus,* are also present in *S. meliloti* (Brilli et al. 2010). The conservation obviously suggests the evolution of the cell cycle program from a common ancestor of the two organisms. However, as we will specifically discuss here for *S. meliloti*, every alphaproteobacterial species appears different from the others by displaying variations on the common theme, suggesting that the cell cycle machinery has subsequently diverged in order to adapt to different lifestyles and physiologies. The adaptation to intracellular life or life in host tissues in the case of alphaproteobacterial species interacting with eukaryotes (rhizobia, *Brucella*, *Agrobacterium,* etc.) involves the formation of specific infecting cell types that may have required the evolution of particular cell cycle regulators.

Regulation of cell cycle in *S. meliloti* and other alphaproteobacterial species is based on a small number of conserved master regulators of the cell cycle. These master regulators coordinate most of the genes controlling essential steps in cell cycle progression and together constitute the master regulatory circuit of the cell cycle. Although our knowledge is still preliminary in many bacterial models, it is reasonable to assume that the master regulators DnaA, GcrA, CtrA, and CcrM are well-conserved cell cycle factors in most of the species of the class *Alphaproteobacteria* (Wright et al. 1997; Barnett et al. 2001; Brilli et al. 2010). The four master regulators, at least in *C. crescentus,* where they have been studied since the early 90s, are synthesized in succession to drive sequential steps of the cell cycle but also to directly activate the downstream master regulator (Fig. 3). DnaA activates *gcrA*, GcrA activates *ctrA*, CtrA activates *ccrM*, and finally, the DNA methylase CcrM resets the cycle by completely methylating the chromosome. DnaA is a protein that activates the initiation of DNA replication in bacteria by opening the double helix at the origin of replication and facilitating the action of the helicase DnaB (Sibley et al. 2006; Skarstad and Katayama 2013). However, DnaA has a dual role and is also involved in the transcriptional regulation of the next master regulator gene *gcrA* although the mechanism by which DnaA is able to activate gene expression remains still elusive. GcrA promotes on its turn *ctrA* transcription, most likely indirectly since it is probably not a transcription factor. CtrA (Cell cycle Transcriptional Regulator A) is a DNA-binding response regulator, a member of the two-component signal transduction family. CtrA is the most interconnected regulator of the four master regulators. CtrA controls transcription of the downstream master regulator *ccrM* but it also inhibits *gcrA* transcription, promotes its own transcription, and inhibits DnaA-mediated DNA replication through binding to the replication origin. Finally, the methylase CcrM exerts a negative epigenetic regulation on its own expression and expression of *ctrA* and a positive epigenetic regulation on *dnaA* expression. In addition, it methylates

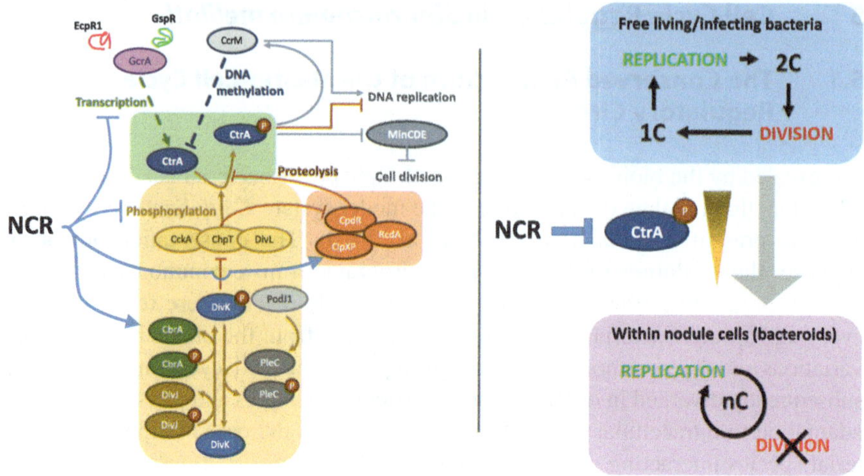

Fig. 3 Schematics of the complex regulation network of the cell cycle in *S. meliloti* and the involvement of NCR peptides during the symbiosis process. (**a**) CtrA is the master regulator of the cell cycle; its active form is phosphorylated and it directly inhibits the DNA replication and indirectly activates the cell division by inhibiting the expression of Min system, an inhibitor of the cell division. CtrA is strictly regulated to ensure a normal cell cycle progression. This regulation occurs at different levels, including a post-translation level, by phosphorylation (yellow box) and proteolysis (orange box), as well as the transcription level through potentially GcrA and CcrM. The potential targets cell cycle of NCRs peptides are indicated with blue arrows. (**b**) The free-living *S. meliloti* replicates its genome only once per cell cycle leading each time to two cell daughters (blue box). However, during the symbiotic process (pink box) the bacterium is targeted by NCRs which lead to a drop in CtrA levels; as a result of this differentiation process, the bacterium becomes a bacteroid characterized by multiple copies of DNA (nC) and a cell division stop

the origin of replication possibly making it competent for a new round of replication, although its precise role with respect to the initiation of DNA replication is not yet uncovered. A second regulatory circuit is integrated in this loop of master regulators at the level of CtrA and is discussed in detail in the following paragraph.

5.2 CtrA: Evolution from Control of Motility to a Cell Cycle Regulator

CtrA can be considered as the most important cell cycle regulator in *S. meliloti*. The crucial role in the regulation of the cell cycle by the CtrA response regulator was demonstrated for the first time in the model species *C. crescentus* (Quon et al. 1996). Response regulators belong to the family of Two-Component systems (TCS) and are generally proteins composed by a receiver domain (REC) with a conserved aspartic residue and an output domain, which usually binds DNA. Phosphorylation of the REC domain leads to dimerization (Gao and Stock 2009), creating an active dimer of the response regulator that is able to bind its consensus sequence (a palindromic sequence composed of two half-sites) located in the promoter region of target genes and activate their expression. CtrA presumably belongs to this class of response regulators suggesting that a dimeric form of phosphorylated CtrA interacts with its palindromic consensus sequence that we can approximate to the sequence AATT(N_7)AATT. This consensus sequence is conserved across alphaproteobacterial genera, spanning from *Rickettsia* to *Caulobacter*, *Sinorhizobium*, *Magnetospirillum*, or *Rhodobacter* (Brassinga et al. 2002; Brilli et al. 2010; Mercer et al. 2010; Greene et al. 2012). Based on the presence of this consensus in the promoter region of genes of alphaproteobacterial genomes, the conservation of functions in alphaproteobacterial species was analyzed in silico, revealing that regulation by CtrA is usually linked to motility, which is probably the ancestral function controlled by CtrA (Brilli et al. 2010; Greene et al. 2012; Mercer et al. 2012). In species belonging to the *Caulobacterales* (including *C. crescentus*) and *Rhizobiales* (*S. meliloti*, *B. abortus*, and *A. tumerfaciens*, for example), CtrA controls in addition to motility also cell cycle-related functions (Brilli et al. 2010). This recruitment of CtrA to the essential function of cell division is obviously associated with the essentiality of the gene *ctrA* in these species (Quon et al. 1996; Barnett et al. 2001; Pini et al. 2015). Conversely, in species in which CtrA controls only motility, CtrA is not essential for bacterial viability and its disruption only affects the flagellum biogenesis and possibly some other non-essential functions (Greene et al. 2012; Mercer et al. 2012).

In cell cycle regulation, CtrA controls DNA replication and cell division. Genome replication is affected in a negative way. In *C. crescentus*, this inhibition is directed by binding of phosphorylated CtrA (CtrA~P) to several CtrA boxes present in the origin of replication, preventing DnaA to initiate replication (Quon et al. 1998). In contrast, in *S. meliloti* this control cannot be direct as there are no CtrA boxes in its origin of replication (Sibley et al. 2006; Pini et al. 2015). However, upon depletion of CtrA, the cell fails to block the reinitiation of DNA replication, resulting in cells with multiple DNA copies. This observation suggests that also in *S. meliloti* some CtrA-dependent mechanism for inhibition of DNA replication exists. Thus even if the molecular mechanisms are different between *C. crecentus* and *S. meliloti*, in both strains the function of CtrA to inhibit replication is conserved. Cell division, on the contrary, is positively regulated by CtrA (Quon et al. 1996; Pini et al. 2015). Although the gene sets regulated by CtrA are strikingly different in the *Alphaproteobacteria*, among them are motility and chemotaxis functions, DNA methylation, and cell division (Laub et al. 2002; De Nisco et al. 2014; Pini et al. 2015). For example, in *S. meliloti*, CtrA represses the Min system, which inhibits septum formation and division by preventing FtsZ polymerization and Z-ring formation, while in *C. crescentus*, which lacks the Min system, *ftsZ* transcription is positively regulated by CtrA.

The dual and opposite activity on replication and division places CtrA at the center of the strict cell cycle control in *Alphaproteobacteria*. It further suggests that CtrA levels must change during the cell cycle: at the onset of DNA replication, CtrA must be inactive in order to activate DNA replication, while in the pre-divisional step CtrA must be present in order to activate crucial division functions. This observation implies that CtrA activity must be highly regulated. In the next section, we will review these CtrA regulatory mechanisms (Fig. 3).

5.3 The CtrA Regulatory Circuit: Multiple Levels of Regulation of CtrA

CtrA has to be phosphorylated to be active and this is mediated by the DivL, CckA, and ChpT phosphorelay cascade in *C. crescentus* (Fig. 3) (Xue and Biondi 2019). Although the orthologous genes of this cascade are present in *S. meliloti* (Brilli et al. 2010), their characterization has not yet been carried out.

On the contrary, the module of the CtrA-inhibitor DivK, a single receiver domain of the two-component system protein family, similar to CheY, has been intensively investigated in *S. meliloti*, together with its complex kinase/phosphatase module, composed of the kinases DivJ and CbrA and the phosphatase PleC (Fig. 3) (Lam et al. 2003; Gibson et al. 2006, 2007; Sadowski et al. 2013; Pini et al. 2013; Schallies et al. 2015). In *C. crescentus* phosphorylated DivK blocks, by protein–protein interactions, the DivL, CckA, and ChpT and thereby prevents CtrA phosphorylation and activation. DivK in *C. crescentus* is an essential factor for cell cycle progression as loss of function mutants of *divK* are arrested at the G1 phase (Hecht et al. 1995). DivK is also essential in *S. meliloti* acting as the main negative regulator of CtrA (Pini et al. 2013, 2015). The absence of DivK or its inability to be phosphorylated results in a stable and constitutively active CtrA that in its turn blocks the origin of replication.

In *S. meliloti*, the active form of DivK, responsible for CtrA inhibition, is phosphorylated by two kinases, DivJ and CbrA (Pini et al. 2013), which both contribute to the pool of DivK~P. Deletion of either of the two kinases leads to a severe cell cycle defect showing elongated and branched cells with a slow growth rate. The double deletion of *divJ* and *cbrA* is lethal, unambiguously demonstrating that phosphorylation of DivK is absolutely necessary for a proper cell cycle progression (Pini et al. 2013). Conversely, the ability to remove the phosphate group from DivK~P at specific stages of the cell cycle is also essential as the only known DivK phosphatase, PleC, is equally indispensable in *S. meliloti* (Fields et al. 2012; Pini et al. 2013). Surprisingly, in *C. crescentus* the deletion of *divJ*, a gene encoding the only known DivK kinase, the deletion of the phosphatase-encoding *pleC* are possible, including the double deletion, while the mutation of the phosphorylation site in DivK is not tolerated by *C. crescentus* cells. This observation suggests an unknown redundant function that may compensate for the absence of DivK phosphorylation or an alternative pathway that phosphorylates DivK. The CbrA alternative kinase of DivK in *S. meliloti* does not exist in *C. crescentus*, illustrating the evolution of unique architectural features of the cell cycle network in different species.

CtrA protein levels in *S. meliloti* are modulated during cell cycle progression with a minimum at the G1-S transition (initiation of the chromosome replication) (Pini et al. 2015). Presumably, this decrease of CtrA levels depends on a mechanism of active degradation of the protein by the protease ClpXP, assisted by several other proteins that are present and well-characterized in *C. crescentus* (Fig. 3). Specifically, the single receiver domain protein CpdR, active in the non-phosphorylated form, is required for CtrA degradation (Kobayashi et al. 2009; Pini et al. 2015; Schallies et al. 2015). CpdR is itself phosphorylated and inactivated by ChpT of the above-mentioned DivL, CckA, and ChpT phosphorelay cascade (Fig. 3) (Schallies et al. 2015). Moreover, the protein RcdA is essential in *S. meliloti* and it is required for CtrA degradation, as a conditional mutant of *rcdA* showed high levels of CtrA and a lethal block of the cell cycle (Pini et al. 2015).

Besides the post-translational control by phosphorylation and protein degradation of the CtrA protein activity, the *ctrA* gene also has a complex transcriptional regulation (Fig. 3). Its transcription is driven by a compound promoter region with at least two different promoters, named P1 and P2 (Barnett et al. 2001). The *ctrA* promoter in *C. crescentus* has two equivalent P1 and P2 promoters. The upstream master regulator GcrA activates the P1 promoter but indirectly. The subsequently produced phosphorylated CtrA blocks then the P1 promoter but activates transcription from the P2 promoter. In *S. meliloti* however, CtrA has only a mild positive regulation of its own promoter at P1 (Pini et al. 2015). Thus other factors than those operating in *C. crescentus* are probably involved in the transcriptional regulation of *S. meliloti ctrA*. The role of *S. meliloti* GcrA in the activation of *ctrA* expression has not been studied yet but the phenotype of the genetic depletion of GcrA is compatible with such an activity (Robledo et al. 2015). Furthermore, the *ctrA* gene in *C. crescentus* is also epigenetically regulated by CcrM methylation. In *S. meliloti*, this mechanism needs to be examined. In addition, two small non-coding RNAs (sRNA) were recently identified that may post-transcriptionally repress the expression of the *ctrA* gene. The sRNA EcpR1 was predicted to target multiple cell cycle genes in *S. meliloti*, including *ctrA*. However, further experimental validation with a GFP reporter assay, involving wild-type and mutant sRNA and mRNA pairs, did confirm regulation by EcpR1 only for *dnaA* and *gcrA* but not for *ctrA* (Robledo et al. 2015). Another sRNA, named GspR, was confirmed with the GFP reporter assay to post-transcriptionally downregulated *ctrA* expression (Robledo et al. 2018). Thus, sRNAs may directly or indirectly through *gcrA* fine-tune *ctrA* expression and modulate the cell cycle regulation, potentially in response to external factors.

6 Symbiosis and the *S. meliloti* Cell Cycle

The polyploidy of the *S. meliloti* bacteroids is a deviation of the single round of genome replication per cell division rule, which is governed as discussed above by CtrA and other master regulators (Fig. 3). Thus, the cell cycle switch underlying the bacteroid differentiation should perturb this cascade. Most tellingly, the cell cycle switch is clearly observable by analyzing inside nodule tissues the expression of an extended set of *S. meliloti* cell cycle regulators (Fig. 4). The meristem of *M. truncatula* nodules continuously generates new cells, which become infected and

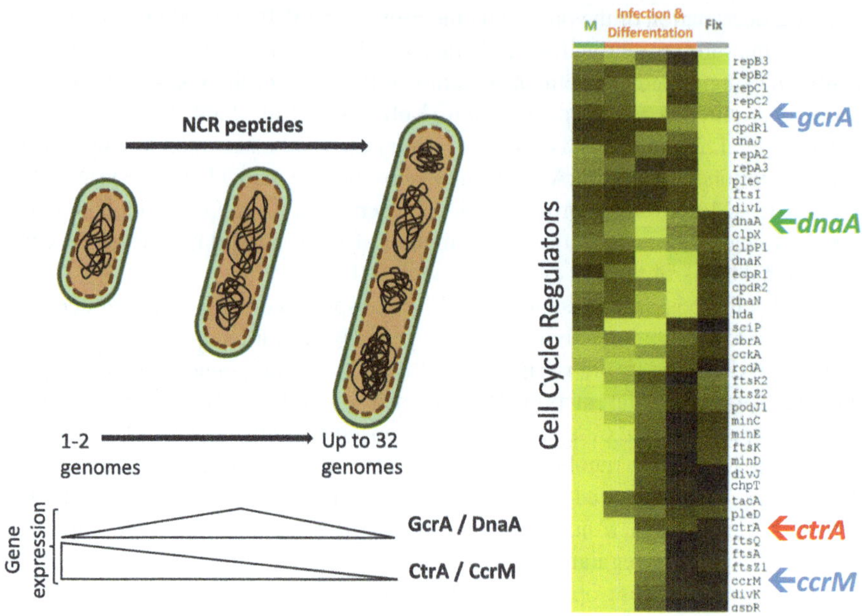

Fig. 4 The gene expression of cell cycle regulators during the differentiation. Under the impact of NCRs peptides, the bacterium differentiates into a bacteroid (upper left schematics). This differentiation consists of an endoreduplication and an absence of division resulting in a larger cell with, up to 16–32 DNA copies. Transcriptomic data of the cell cycle regulators genes in different nodule zones (on the right), shows a variation of the expression of cell cycle genes compared to the free-living growth condition. In particular, DnaA/GcrA and CtrA/CcrM patterns are indicated in the lower-left part of the figure

in which bacteria differentiate. This differentiation takes place gradually along the longitudinal axis of the nodule. The expression of plant and bacterial genes was analyzed by combining laser-capture microdissection of nodule tissues along this longitudinal axis with RNA-seq (Roux et al. 2014). A uniform expression of the cell cycle genes in all tissues would be expected in case the cell cycle stays unaffected during bacteroid differentiation. However, this is not what is observed (Fig. 4). Rather, the relative expression of genes greatly varies in the different tissues and thus according to the stage of bacterial differentiation. This modulation of expression is gene- and cell cycle-dependent. For example, the expression of *ctrA* as well as of many CtrA-regulated genes or genes encoding CtrA phosphorylation regulators are very rapidly downregulated when differentiation starts. Accordingly, Western blot analysis confirmed the absence of the CtrA and FtsZ proteins from bacteroids extracted from *Medicago* nodules (Pini et al. 2013; Farkas et al. 2014). The DNA replication-associated genes *dnaA*, *dnaN*, and *hdaA* on the other hand show the strongest relative expression in the nodule tissues where differentiation and genome amplification is taking place.

The cell cycle regulatory cascade is a robust machinery that allows the strict respect of the cycle: haploid state—replication—diploid state—division, characterizing wild type *S. meliloti* growth. However, genetic or pharmacological interference with this cascade can disrupt the regular cell cycle and induce bacteroid-like cells, which are strongly enlarged and branched and have a multiplied genome while at the same time further growth and cell divisions are blocked. For example, the depletion of *ctrA* has such an effect (Pini et al. 2015). Also, depletion of the *ctrA*-transcriptional regulators *gcrA* or overexpression of *ccrM* or the sRNA EcpR1 provoke the same phenomenon (Wright et al. 1997; Robledo et al. 2015). Furthermore, the overexpression of *divJ*, depletion of *pleC* or mutation of *podJ1* lead to the accumulation of phosphorylated DivK which blocks CtrA phosphorylation. This prevents CtrA activation and results in bacteroid-like cells (Fields et al. 2012; Pini et al. 2015). The expression of a constitutively active form of CpdR1 that stimulates CtrA degradation has a similar effect (Kobayashi et al. 2009). Moreover, mutation or overexpression of the septum-formation controlling genes *ftsZ* and *minCDE* or pharmacological inhibition of septum formation again provoke the same cell elongation and branching effect (Latch and Margolin 1997; Cheng et al. 2007). Finally, overexpression of the *dnaA* and *hdaA* genes, encoding the replication machinery, have also such a cellular effect, although in that case the balance between the three replicons, the chromosome, pSymA and pSymB, is not maintained in the amplified genomes (Sibley et al. 2006; Frage et al. 2016). Taken together, perturbing the CtrA pathway leads thus systematically to cellular changes that mimic partially or strongly the bacteroid state.

Thus, CtrA, because of its key position in the cell cycle regulation, is a likely target for the NCR peptides in bacteroid differentiation: its elimination would be compatible with the inhibition of cell division and the continued DNA replication (Fig. 3). As suggested by the above-cited genetic studies, CtrA could be directly targeted or it could be inactivated and eliminated via its transcriptional or post-translational regulators. In agreement with the key position of CtrA for bacteroid differentiation, *S. meliloti* mutants in the genes encoding negative regulators of CtrA, such as *cbrA*, *divJ*, and *cpdR* genes, make non-functional nodules without bacteroid differentiation (Gibson et al. 2006; Kobayashi et al. 2009; Pini et al. 2013).

Another strong argument in favor of CtrA being the ultimate target of the NCR peptides to trigger the bacteroid differentiation is the demonstration that in NCR247-treated synchronized *S. meliloti* cells the expression of the CtrA-controlled genes are not properly activated during the progression of the cell cycle (De Nisco et al. 2014).

7 Benefits of Terminal Bacteroid Differentiation

S. meliloti lives in the soil as a free-living saprophyte even without the presence of legumes (Carelli et al. 2000). This suggests that the capability to establish a symbiosis is not an essential function of the species, as it is further revealed by the discovery of *S. meliloti* strains unable to induce and infect nodules. A recent study has highlighted that *S. meliloti* colonizes the plant as an endophyte, being recovered from leaves and other tissues from *Medicago* plants (Pini et al. 2012). This observation suggests an interesting scenario about the evolutionary origin of bacteroid formation. The plant may have evolved a way to prevent uncontrolled infection of the endophyte, by blocking bacterial duplication and inducing a "terminal differentiation." Possibly, the NCR peptides that are now only active in nodules are derived from immune peptides that were originally employed to control endophytic bacteria including *S. meliloti* strains.

The multiple independent origins of terminal bacteroid differentiation in the legume family is a strong suggestion that the process provides benefits to the host plant (Oono et al. 2011). The benefits should be indeed on the plant side because it is the host that imposes the process by the production of NCR peptides and the process limits very strongly the bacterial reproduction. Several studies have provided experimental confirmations that the terminal bacteroid differentiation improves the efficiency of the symbiosis by increasing the plant biomass production per investment in the symbiosis. These analyses were comparative studies in which either a particular rhizobium strain was compared on two host plants, or alternatively, one host plant nodulated with different rhizobium strains. When different hosts were compared, one host induced terminal bacteroid differentiation and the other host undifferentiated bacteroids (Sen and Weaver 1981, 1984; Oono et al. 2011), or the two hosts induced both terminal bacteroid differentiation but to different levels (Lamouche et al. 2019a, b). In the opposite type of comparison, *Medicago* hosts were nodulated with a panel of strains displaying contrasted bacteroid differentiation levels (different levels of genome amplification and cell enlargement in the bacteroids), correlating well with the efficiency of the interaction (Kazmierczak et al. 2017). However, these comparisons although consistent with what we predicted, can be criticized as comparing apples with oranges because it is not possible to determine how much of the differences is due to bacteroid differentiation, and how much is due to other differences between the compared plant species or bacterial species. Moreover, they only show correlations, which do not mean causality. To go beyond these correlations, an experimental system would be required that uses one particular host in interaction with one rhizobium strain and in which the bacteroid differentiation can be manipulated. Possibly, the *in planta* modification of the expression of cell cycle regulators, as discussed above in in vitro studies, can offer such opportunities.

Such an approach could provide a firm proof for the improvement of the symbiotic functioning of the bacteroids when terminally differentiated. However, this would still not explain why this type of bacteroids is performing better. This is at present an unresolved question but we can speculate about some of the consequences of the bacteroid differentiation that could impact the functioning of the bacteroids. The first possibility is that the advantage is linked to the cell size and that the cell enlargement makes bacteroids better nitrogen-fixing machines. Could larger bacteroids be more energy-efficient than small non-differentiated ones? This is not self-evident. Larger bacterial cells means a higher volume-to-surface ratio but since respiratory energy production is a membrane process, an increased volume-to-surface ratio is energetically not advantageous. On the other hand, if energy production would not be rate-limiting for nitrogen fixation in bacteroids, a larger volume could favor protein synthesis (for example for the massive production of the nitrogenase complex and its metal cofactors) by reducing its cost. Cell enlargement could also be viewed as a form of cell compartmentalization that physically separates the oxygen-requiring respiratory complexes from the oxygen-sensitive nitrogenase machinery. In a large cell, the nitrogenase could be located in the center of the cell, far away (on an atomic scale) from respiration on the cell membranes.

Additionally, large cells could dampen functional heterogeneity between bacteroids as recently demonstrated. Indeed, many bacterial genes show cell-to-cell fluctuations due to noise in gene expression, leading to phenotypic diversity between cells (Ackermann 2015). Heterogeneity in a cell population can be advantageous to bacteria in certain circumstances, by providing adaptability to unpredictably changing environments. In the nodule, however, it could be detrimental for the symbiosis and be associated with suboptimal performance of a subpopulation of bacteroids. In large cells such as terminally differentiated bacteroids, gene expression noise could be reduced by effectively averaging cell contents, as has been shown in polyploid division-blocked *Bacillus subtilis* mutants, resulting in a decreased functional heterogeneity between cells (Süel et al. 2007).

Another possibility is that the polyploidy state may provide the improvement of bacteroid functioning. The respiration of bacteroids and the nitrogen fixation process itself by the nitrogenase are inevitable sources of reactive oxygen species (Matamoros et al. 2003). Reactive oxygen species may induce deleterious mutations, which in the long term may affect the functioning of the bacteroids. Polyploid bacteroids could be less sensitive to DNA damage because they have multiple gene copies. Thus, the polyploid state of bacteroids could increase their longevity, which would imply a delayed senescence. The polyploid chromosomes in bacteroids could bring along also a benefit at another level. More condensed than the chromosomes in free-growing rhizobia (Mergaert et al. 2006), the polyploid chromosomes in bacteroids could function differently. Their compaction could have an epigenetic impact on for example gene expression.

On the other hand, the cell cycle switch with the ensuing polyploidy and cell enlargement could be only side effects of another important function of the NCRs on the bacteroids. The NCR peptides, as many other antimicrobial peptides, disturb the membrane integrity of bacteria (Van de Velde et al. 2010; Mikuláss et al. 2016) and this correlates with the known enhanced membrane permeability in terminally differentiated bacteroids (Mergaert et al. 2006). The membrane permeabilization of the bacteroids could enhance the metabolic exchanges between the symbionts thereby favoring optimally the nitrogen fixation metabolism of the bacteroids with the metabolism of the host cell (Mergaert et al. 2017). Moreover, metabolite exchange can also be favored in the terminally differentiated bacteroids because they are individually enclosed in a symbiosome and have a much closer contact with the symbiosome membrane than undifferentiated bacteroids. In the last case, a single symbiosome harbors multiple bacteria, and thus the direct contact of the bacteria with the symbiosome membrane is reduced. Furthermore, NCR peptides were reported to interact directly with several metabolic enzymes, including the ribosomes, chaperones, enzymes of the energy metabolism, and the nitrogenase (Farkas et al. 2014). Thus, the primary effect of the NCRs could be the manipulation of the metabolism of the endosymbiont in order to mold the bacterial metabolism for optimal nitrogen fixation (Kereszt et al. 2011; Farkas et al. 2014). For example, to maintain a redox balance during nitrogen fixation, bacteroids channel part of their carbon sources in lipid and polyhydroxybutyrate electron sinks (Terpolilli et al. 2016). From the plant perspective, this accumulation of carbon by the bacteroids is a net loss of resources. It is striking that undifferentiated bacteroids accumulate much larger amounts of these storage compounds than terminally differentiated bacteroids (Lodwig et al. 2005).

A final hypothesis is related to the terminally differentiated state of the bacteroids. The terminal differentiation could limit the release of rhizobia from senescing nodules thereby moderating the impact of the symbiosis on the rhizosphere and endophyte microbiota. Moreover, the plant recovers during senescence the bacterial biomass from terminally differentiated bacteroids, which are entirely digested during nodule senescence (Van de Velde et al. 2006) whereas undifferentiated bacteroids largely survive nodule senescence (Müller et al. 2001).

8 Conclusions

The regulation of the cell cycle in *S. meliloti* can be largely modeled on *C. crescentus,* which has been extensively studied in recent decades. At the same time, the study of the *S. meliloti* cell cycle has revealed that despite the large conservation, many *Sinorhizobium* specificities do exist and those might be very important because they are likely specific adaptations to the particularities of the lifestyle of this bacterium. One of the most striking of these particularities is the cell cycle (de)regulation that happens during the terminal bacteroid differentiation.

As argued here, the CtrA master cell cycle regulator is the preferred suspect to be targeted by the NCR peptides and to direct this major cell cycle event. CtrA has a complex regulation on its own implementing transcriptional control, small RNA post-transcriptional regulation, epigenetic mechanisms by DNA methylation, and posttranslational regulation by phosphorylation and by targeted proteolysis. In principle, NCRs could interfere with any of these regulations (Fig. 3). Future studies will have to decipher at which regulatory stage this cascade is affected. A complication in this challenge is the large number and diversity of NCR peptides that are produced by the plant. It is very well possible that not a single peptide does the full job but that several peptides act synergistically, sequentially, or redundantly. Moreover, the NCRs could be interacting directly with these intracellular regulators but they could also act indirectly at the level of the bacterial membrane and interfere with the cell cycle via a signal transduction process.

We can furthermore hope that deepening our understanding of this particular cell cycle operating in bacteroid differentiation will uncover novel aspects of the regular cell cycle in *S. meliloti* and by extension in other members of the class *Alphaproteobacteria*.

Acknowledgments/Funding Collaborative research on the *S. meliloti* cell cycle in our laboratories is funded by the Agence National de la Recherche (ANR) grant ANR-17-CE20-0011. SD and QN were supported by a Ph.D. fellowship from the Université Paris-Saclay.

References

Ackermann M (2015) A functional perspective on phenotypic heterogeneity in microorganisms. Nat Rev Microbiol 13:497–508. https://doi.org/10.1038/nrmicro3491

Alunni B, Kevei Z, Redondo-Nieto M, Kondorosi A, Mergaert P, Kondorosi E (2007) Genomic organization and evolutionary insights on GRP and NCR genes, two large nodule-specific gene families in *Medicago truncatula*. Mol Plant-Microbe Interact 20:1138–1148. https://doi.org/10.1094/MPMI-20-9-1138

Arnold MFF, Penterman J, Shabab M, Chen EJ, Walker GC (2018) Important late-stage symbiotic role of the *Sinorhizobium meliloti* exopolysaccharide Succinoglycan. J Bacteriol 200. https://doi.org/10.1128/JB.00665-17

Barnett MJ, Hung DY, Reisenauer A, Shapiro L, Long SR (2001) A homolog of the CtrA cell cycle regulator is present and essential in *Sinorhizobium meliloti*. J Bacteriol 183:3204–3210. https://doi.org/10.1128/JB.183.10.3204-3210.2001

Bobik C, Meilhoc E, Batut J (2006) FixJ: a major regulator of the oxygen limitation response and late symbiotic functions of *Sinorhizobium meliloti*. J Bacteriol 188:4890–4902. https://doi.org/10.1128/JB.00251-06

Brassinga AKC, Siam R, McSween W, Winkler H, Wood D, Marczynski GT (2002) Conserved response regulator CtrA and IHF binding sites in the alpha-proteobacteria *Caulobacter crescentus* and *Rickettsia prowazekii* chromosomal replication origins. J Bacteriol 184:5789–5799

Brilli M, Fondi M, Fani R, Mengoni A, Ferri L, Bazzicalupo M, Biondi EG (2010) The diversity and evolution of cell cycle regulation in alpha-proteobacteria: a comparative genomic analysis. BMC Syst Biol 4:52. https://doi.org/10.1186/1752-0509-4-52

Capela D, Barloy-Hubler F, Gouzy J, Bothe G, Ampe F, Batut J, Boistard P, Becker A, Boutry M, Cadieu E, Dréano S, Gloux S, Godrie T, Goffeau A, Kahn D, Kiss E, Lelaure V, Masuy D, Pohl T, Portetelle D, Pühler A, Purnelle B, Ramsperger U, Renard C, Thébault P, Vandenbol M, Weidner S, Galibert F (2001) Analysis of the chromosome sequence of the legume symbiont *Sinorhizobium meliloti* strain 1021. Proc Natl Acad Sci USA 98:9877–9882. https://doi.org/10.1073/pnas.161294398

Carelli M, Gnocchi S, Fancelli S, Mengoni A, Paffetti D, Scotti C, Bazzicalupo M (2000) Genetic diversity and dynamics of *Sinorhizobium meliloti* populations nodulating different alfalfa cultivars in Italian soils. Appl Environ Microbiol 66:4785–4789

Cheng J, Sibley CD, Zaheer R, Finan TM (2007) A *Sinorhizobium meliloti* minE mutant has an altered morphology and exhibits defects in legume symbiosis. Microbiol Read Engl 153:375–387. https://doi.org/10.1099/mic.0.2006/001362-0

Compton KK, Hildreth SB, Helm RF, Scharf BE (2020) An updated perspective on *Sinorhizobium meliloti* chemotaxis to alfalfa flavonoids. Front Microbiol 11:581482. https://doi.org/10.3389/fmicb.2020.581482

Czernic P, Gully D, Cartieaux F, Moulin L, Guefrachi I, Patrel D, Pierre O, Fardoux J, Chaintreuil C, Nguyen P, Gressent F, Da Silva C, Poulain J, Wincker P, Rofidal V, Hem S, Barrière Q, Arrighi J-F, Mergaert P, Giraud E (2015) Convergent evolution of endosymbiont differentiation in Dalbergioid and inverted repeat-lacking clade legumes mediated by nodule-specific cysteine-rich peptides. Plant Physiol 169:1254–1265. https://doi.org/10.1104/pp.15.00584

De Nisco NJ, Abo RP, Wu CM, Penterman J, Walker GC (2014) Global analysis of cell cycle gene expression of the legume symbiont *Sinorhizobium meliloti*. Proc Natl Acad Sci USA. https://doi.org/10.1073/pnas.1400421111

Durgo H, Klement E, Hunyadi-Gulyas E, Szucs A, Kereszt A, Medzihradszky KF, Kondorosi E (2015) Identification of nodule-specific cysteine-rich plant peptides in endosymbiotic bacteria. Proteomics. https://doi.org/10.1002/pmic.201400385

Farkas A, Maróti G, Durgő H, Györgypál Z, Lima RM, Medzihradszky KF, Kereszt A, Mergaert P, Kondorosi E (2014) Medicago truncatula symbiotic peptide NCR247 contributes to bacteroid differentiation through multiple mechanisms. Proc Natl Acad Sci USA 111:5183–5188. https://doi.org/10.1073/pnas.1404169111

Fields AT, Navarrete CS, Zare AZ, Huang Z, Mostafavi M, Lewis JC, Rezaeihaghighi Y, Brezler BJ, Ray S, Rizzacasa AL, Barnett MJ, Long SR, Chen EJ, Chen JC (2012) The conserved polarity factor podJ1 impacts multiple cell envelope-associated functions in *Sinorhizobium meliloti*. Mol Microbiol 84:892–920. https://doi.org/10.1111/j.1365-2958.2012.08064.x

Finan TM, Weidner S, Wong K, Buhrmester J, Chain P, Vorhölter FJ, Hernandez-Lucas I, Becker A, Cowie A, Gouzy J, Golding B, Pühler A (2001) The complete sequence of the 1,683-kb pSymB megaplasmid from the N2-fixing endosymbiont *Sinorhizobium meliloti*. Proc Natl Acad Sci USA 98:9889–9894. https://doi.org/10.1073/pnas.161294698

Frage B, Döhlemann J, Robledo M, Lucena D, Sobetzko P, Graumann PL, Becker A (2016) Spatiotemporal choreography of chromosome and megaplasmids in the *Sinorhizobium meliloti* cell cycle. Mol Microbiol 100:808–823. https://doi.org/10.1111/mmi.13351

Galibert F, Finan TM, Long SR, Puhler A, Abola P, Ampe F, Barloy-Hubler F, Barnett MJ, Becker A, Boistard P, Bothe G, Boutry M, Bowser L, Buhrmester J, Cadieu E, Capela D, Chain P, Cowie A, Davis RW, Dreano S, Federspiel NA, Fisher RF, Gloux S, Godrie T, Goffeau A, Golding B, Gouzy J, Gurjal M, Hernandez-Lucas I, Hong A, Huizar L, Hyman RW, Jones T, Kahn D, Kahn ML, Kalman S, Keating DH, Kiss E, Komp C, Lelaure V, Masuy D, Palm C, Peck MC, Pohl TM, Portetelle D, Purnelle B, Ramsperger U, Surzycki R, Thebault P, Vandenbol M, Vorholter FJ, Weidner S, Wells DH, Wong K, Yeh KC, Batut J (2001) The composite genome of the legume symbiont *Sinorhizobium meliloti*. Science 293:668–672. https://doi.org/10.1126/science.1060966

Gao R, Stock AM (2009) Biological insights from structures of two-component proteins. Annu Rev Microbiol 63:133–154. https://doi.org/10.1146/annurev.micro.091208.073214

Gibson KE, Barnett MJ, Toman CJ, Long SR, Walker GC (2007) The symbiosis regulator CbrA modulates a complex regulatory network affecting the flagellar apparatus and cell envelope proteins. J Bacteriol 189:3591–3602. https://doi.org/10.1128/JB.01834-06

Gibson KE, Campbell GR, Lloret J, Walker GC (2006) CbrA is a stationary-phase regulator of cell surface physiology and legume symbiosis in *Sinorhizobium meliloti*. J Bacteriol 188:4508–4521. https://doi.org/10.1128/JB.01923-05

Greene SE, Brilli M, Biondi EG, Komeili A (2012) Analysis of the CtrA pathway in Magnetospirillum reveals an ancestral role in motility in alphaproteobacteria. J Bacteriol 194:2973–2986. https://doi.org/10.1128/JB.00170-12

Griesmann M, Chang Y, Liu X, Song Y, Haberer G, Crook MB, Billault-Penneteau B, Lauressergues D, Keller J, Imanishi L, Roswanjaya YP, Kohlen W, Pujic P, Battenberg K, Alloisio N, Liang Y, Hilhorst H, Salgado MG, Hocher V, Gherbi H, Svistoonoff S, Doyle JJ, He S, Xu Y, Xu S, Qu J, Gao Q, Fang X, Fu Y, Normand P, Berry AM, Wall LG, Ané J-M, Pawlowski K, Xu X, Yang H, Spannagl M, Mayer KFX, Wong GK-S, Parniske M, Delaux P-M, Cheng S (2018) Phylogenomics reveals multiple losses of nitrogen-fixing root nodule symbiosis. Science (New York, NY). https://doi.org/10.1126/science.aat1743

Guefrachi I, Nagymihaly M, Pislariu CI, Van de Velde W, Ratet P, Mars M, Udvardi MK, Kondorosi E, Mergaert P, Alunni B (2014) Extreme specificity of NCR gene expression in *Medicago truncatula*. BMC Genomics 15:712. https://doi.org/10.1186/1471-2164-15-712

Gully D, Czernic P, Cruveiller S, Mahé F, Longin C, Vallenet D, François P, Nidelet S, Rialle S, Giraud E, Arrighi J-F, DasGupta M, Cartieaux F (2018) Transcriptome profiles of nod factor-independent symbiosis in the tropical legume *Aeschynomene evenia*. Sci Rep 8:10934. https://doi.org/10.1038/s41598-018-29301-0

Haag AF, Baloban M, Sani M, Kerscher B, Pierre O, Farkas A, Longhi R, Boncompagni E, Hérouart D, Dall'angelo S, Kondorosi E, Zanda M, Mergaert P, Ferguson GP (2011) Protection of Sinorhizobium against host cysteine-rich antimicrobial peptides is critical for symbiosis. PLoS Biol 9:e1001169. https://doi.org/10.1371/journal.pbio.1001169

Hecht GB, Lane T, Ohta N, Sommer JM, Newton A (1995) An essential single domain response regulator required for normal cell division and differentiation in *Caulobacter crescentus*. EMBO J 14:3915–3924

Horváth B, Domonkos Á, Kereszt A, Szűcs A, Ábrahám E, Ayaydin F, Bóka K, Chen Y, Chen R, Murray JD, Udvardi MK, Kondorosi É, Kaló P (2015) Loss of the nodule-specific cysteine rich peptide, NCR169, abolishes symbiotic nitrogen fixation in the *Medicago truncatula* dnf7 mutant. Proc Natl Acad Sci USA 112:15232–15237. https://doi.org/10.1073/pnas.1500777112

Kazmierczak T, Nagymihály M, Lamouche F, Barrière Q, Guefrachi I, Alunni B, Ouadghiri M, Ibijbijen J, Kondorosi É, Mergaert P, Gruber V (2017) Specific host-responsive associations between *Medicago truncatula* accessions and Sinorhizobium strains. Mol Plant-Microbe Interact 30:399–409. https://doi.org/10.1094/MPMI-01-17-0009-R

Kereszt A, Mergaert P, Kondorosi E (2011) Bacteroid development in legume nodules: evolution of mutual benefit or of sacrificial victims? Mol Plant-Microbe Interact 24:1300–1309. https://doi.org/10.1094/MPMI-06-11-0152

Kim M, Chen Y, Xi J, Waters C, Chen R, Wang D (2015) An antimicrobial peptide essential for bacterial survival in the nitrogen-fixing symbiosis. Proc Natl Acad Sci USA 112:15238–15243. https://doi.org/10.1073/pnas.1500123112

Kobayashi H, De Nisco NJ, Chien P, Simmons LA, Walker GC (2009) *Sinorhizobium meliloti* CpdR1 is critical for coordinating cell cycle progression and the symbiotic chronic infection. Mol Microbiol 73:586–600. https://doi.org/10.1111/j.1365-2958.2009.06794.x

Lam H, Matroule J-Y, Jacobs-Wagner C (2003) The asymmetric spatial distribution of bacterial signal transduction proteins coordinates cell cycle events. Dev Cell 5:149–159

Lamouche F, Bonadé-Bottino N, Mergaert P, Alunni B (2019a) Symbiotic efficiency of spherical and elongated bacteroids in the Aeschynomene-Bradyrhizobium symbiosis. Front Plant Sci 10:377. https://doi.org/10.3389/fpls.2019.00377

Lamouche F, Chaumeret A, Guefrachi I, Barrière Q, Pierre O, Guérard F, Gilard F, Giraud E, Dessaux Y, Gakière B, Timchenko T, Kereszt A, Mergaert P, Alunni B (2019b) From intracellular bacteria to differentiated bacteroids: transcriptome and metabolome analysis in Aeschynomene nodules using the Bradyrhizobium sp. strain ORS285 bclA mutant. J Bacteriol 201. https://doi.org/10.1128/JB.00191-19

Latch JN, Margolin W (1997) Generation of buds, swellings, and branches instead of filaments after blocking the cell cycle of *Rhizobium meliloti*. J Bacteriol 179:2373–2381. https://doi.org/10.1128/jb.179.7.2373-2381.1997

Laub MT, Chen SL, Shapiro L, McAdams HH (2002) Genes directly controlled by CtrA, a master regulator of the Caulobacter cell cycle. Proc Natl Acad Sci USA 99:4632–4637. https://doi.org/10.1073/pnas.062065699

Lodwig EM, Leonard M, Marroqui S, Wheeler TR, Findlay K, Downie JA, Poole PS (2005) Role of polyhydroxybutyrate and glycogen as carbon storage compounds in pea and bean bacteroids. Mol Plant-Microbe Interact 18:67–74. https://doi.org/10.1094/MPMI-18-0067

Matamoros MA, Dalton DA, Ramos J, Clemente MR, Rubio MC, Becana M (2003) Biochemistry and molecular biology of antioxidants in the rhizobia-legume symbiosis. Plant Physiol 133:499–509. https://doi.org/10.1104/pp.103.025619

Mercer RG, Callister SJ, Lipton MS, Pasa-Tolic L, Strnad H, Paces V, Beatty JT, Lang AS (2010) Loss of the response regulator CtrA causes pleiotropic effects on gene expression but does not affect growth phase regulation in *Rhodobacter capsulatus*. J Bacteriol 192:2701–2710. https://doi.org/10.1128/JB.00160-10

Mercer RG, Quinlan M, Rose AR, Noll S, Beatty JT, Lang AS (2012) Regulatory systems controlling motility and gene transfer agent production and release in *Rhodobacter capsulatus*. FEMS Microbiol Lett 331:53–62. https://doi.org/10.1111/j.1574-6968.2012.02553.x

Mergaert P (2018) Role of antimicrobial peptides in controlling symbiotic bacterial populations. Nat Prod Rep 35:336–356. https://doi.org/10.1039/c7np00056a

Mergaert P, Kereszt A, Kondorosi E (2020) Gene expression in nitrogen-fixing symbiotic nodule cells in *Medicago truncatula* and other nodulating plants. Plant Cell 32:42–68. https://doi.org/10.1105/tpc.19.00494

Mergaert P, Kikuchi Y, Shigenobu S, Nowack ECM (2017) Metabolic integration of bacterial endosymbionts through antimicrobial peptides. Trends Microbiol 25:703–712. https://doi.org/10.1016/j.tim.2017.04.007

Mergaert P, Nikovics K, Kelemen Z, Maunoury N, Vaubert D, Kondorosi A, Kondorosi E (2003) A novel family in *Medicago truncatula* consisting of more than 300 nodule-specific genes coding for small, secreted polypeptides with conserved cysteine motifs. Plant Physiol 132:161–173. https://doi.org/10.1104/pp.102.018192

Mergaert P, Uchiumi T, Alunni B, Evanno G, Cheron A, Catrice O, Mausset A-E, Barloy-Hubler F, Galibert F, Kondorosi A, Kondorosi E (2006) Eukaryotic control on bacterial cell cycle and differentiation in the Rhizobium-legume symbiosis. Proc Natl Acad Sci USA 103:5230–5235. https://doi.org/10.1073/pnas.0600912103

Mikuláss KR, Nagy K, Bogos B, Szegletes Z, Kovács E, Farkas A, Váró G, Kondorosi É, Kereszt A (2016) Antimicrobial nodule-specific cysteine-rich peptides disturb the integrity of bacterial outer and inner membranes and cause loss of membrane potential. Ann Clin Microbiol Antimicrob 15:43. https://doi.org/10.1186/s12941-016-0159-8

Montiel J, Downie JA, Farkas A, Bihari P, Herczeg R, Bálint B, Mergaert P, Kereszt A, Kondorosi É (2017) Morphotype of bacteroids in different legumes correlates with the number and type of symbiotic NCR peptides. Proc Natl Acad Sci USA 114:5041–5046. https://doi.org/10.1073/pnas.1704217114

Müller J, Wiemken A, Boller T (2001) Redifferentiation of bacteria isolated from Lotus japonicus root nodules colonized by Rhizobium sp. NGR234. J Exp Bot 52:2181–2186. https://doi.org/10.1093/jexbot/52.364.2181

Nicoud Q, Barrière Q, Busset N, Bourge M, Bars RL, Boulogne C, Lecroël M, Jenei S, Kereszt A, Kondorosi E, Timtchenko T, Alunni B, Mergaert P (2020) Sinorhizobium meliloti functions required for resistance to antimicrobial NCR peptides and bacteroid differentiation bioRxiv 2020.12.04.412775. https://doi.org/10.1101/2020.12.04.412775

Nouwen N, Gargani D, Giraud E (2019) The modification of the flavonoid naringenin by Bradyrhizobium sp. strain ORS285 change the nod genes inducer function to a growth stimulator. Mol Plant-Microbe Interact. https://doi.org/10.1094/MPMI-05-19-0133-R

Oono R, Anderson CG, Denison RF (2011) Failure to fix nitrogen by non-reproductive symbiotic rhizobia triggers host sanctions that reduce fitness of their reproductive clonemates. Proc Biol Sci 278:2698–2703. https://doi.org/10.1098/rspb.2010.2193

Penterman J, Abo RP, De Nisco NJ, Arnold MFF, Longhi R, Zanda M, Walker GC (2014) Host plant peptides elicit a transcriptional response to control the *Sinorhizobium meliloti* cell cycle during symbiosis. Proc Natl Acad Sci USA 111:3561–3566. https://doi.org/10.1073/pnas.1400450111

Pini F, De Nisco NJ, Ferri L, Penterman J, Fioravanti A, Brilli M, Mengoni A, Bazzicalupo M, Viollier PH, Walker GC, Biondi EG (2015) Cell cycle control by the master regulator CtrA in *Sinorhizobium meliloti*. PLoS Genet 11:e1005232. https://doi.org/10.1371/journal.pgen.1005232

Pini F, Frage B, Ferri L, De Nisco NJ, Mohapatra SS, Taddei L, Fioravanti A, Dewitte F, Galardini M, Brilli M, Villeret V, Bazzicalupo M, Mengoni A, Walker GC, Becker A, Biondi EG (2013) The DivJ, CbrA and PleC system controls DivK phosphorylation and symbiosis in *Sinorhizobium meliloti*. Mol Microbiol 90:54–71. https://doi.org/10.1111/mmi.12347

Pini F, Frascella A, Santopolo L, Bazzicalupo M, Biondi EG, Scotti C, Mengoni A (2012) Exploring the plant-associated bacterial communities in *Medicago sativa* L. BMC Microbiol 12:78. https://doi.org/10.1186/1471-2180-12-78

Poole P, Ramachandran V, Terpolilli J (2018) Rhizobia: from saprophytes to endosymbionts. Nat Rev Microbiol 16:291–303. https://doi.org/10.1038/nrmicro.2017.171

Quilbé J, Lamy L, Brottier L, Leleux P, Fardoux J, Rivallan R, Benichou T, Guyonnet R, Becana M, Villar I, Garsmeur O, Hufnagel B, Delteil A, Gully D, Chaintreuil C, Pervent M, Cartieaux F, Bourge M, Valentin N, Martin G, Fontaine L, Droc G, Dereeper A, Farmer A, Libourel C, Nouwen N, Gressent F, Mournet P, D'Hont A, Giraud E, Klopp C, Arrighi J-F (2020) Genetics of nodulation in Aeschynomene evenia uncovers new mechanisms of the rhizobium-legume symbiosis bioRxiv 2020.11.26.399428. https://doi.org/10.1101/2020.11.26.399428

Quon KC, Marczynski GT, Shapiro L (1996) Cell cycle control by an essential bacterial two-component signal transduction protein. Cell 84:83–93

Quon KC, Yang B, Domian IJ, Shapiro L, Marczynski GT (1998) Negative control of bacterial DNA replication by a cell cycle regulatory protein that binds at the chromosome origin. Proc Natl Acad Sci USA 95:120–125

Robledo M, Frage B, Wright PR, Becker A (2015) A stress-induced small RNA modulates alpha-rhizobial cell cycle progression. PLoS Genet 11:e1005153. https://doi.org/10.1371/journal.pgen.1005153

Robledo M, Schlüter J-P, Loehr LO, Linne U, Albaum SP, Jiménez-Zurdo JI, Becker A (2018) An sRNA and cold shock protein homolog-based feedforward loop post-transcriptionally controls cell cycle master regulator CtrA. Front Microbiol 9:763. https://doi.org/10.3389/fmicb.2018.00763

Roux B, Rodde N, Jardinaud M-F, Timmers T, Sauviac L, Cottret L, Carrère S, Sallet E, Courcelle E, Moreau S, Debellé F, Capela D, de Carvalho-Niebel F, Gouzy J, Bruand C, Gamas P (2014) An integrated analysis of plant and bacterial gene expression in symbiotic root nodules using laser-capture microdissection coupled to RNA sequencing. Plant J Cell Mol Biol 77:817–837. https://doi.org/10.1111/tpj.12442

Sadowski C, Wilson D, Schallies K, Walker G, Gibson KE (2013) The *Sinorhizobium meliloti* sensor histidine kinase CbrA contributes to free-living cell cycle regulation. Microbiol Read Engl. https://doi.org/10.1099/mic.0.067504-0

Schallies KB, Sadowski C, Meng J, Chien P, Gibson KE (2015) *Sinorhizobium meliloti* CtrA stability is regulated in a CbrA-dependent manner that is influenced by CpdR1. J Bacteriol 197:2139–2149. https://doi.org/10.1128/JB.02593-14

Sen D, Weaver RW (1981) A comparison of nitrogen-fixing ability of peanut, cowpea and siratro plants nodulated by different strains of rhizobium. Plant Soil 60:317–319. https://doi.org/10.1007/BF02374116

Sen D, Weaver RW (1984) A basis for different rates of N2-fixation by the same strains of rhizobium in peanut and cowpea root nodules. Plant Sci Lett 34:239–246. https://doi.org/10.1016/S0304-4211(84)80002-4

Sibley CD, MacLellan SR, Finan T (2006) The *Sinorhizobium meliloti* chromosomal origin of replication. Microbiol Read Engl 152:443–455. https://doi.org/10.1099/mic.0.28455-0

Skarstad K, Katayama T (2013) Regulating DNA replication in bacteria. Cold Spring Harb Perspect Biol 5:a012922. https://doi.org/10.1101/cshperspect.a012922

Süel GM, Kulkarni RP, Dworkin J, Garcia-Ojalvo J, Elowitz MB (2007) Tunability and noise dependence in differentiation dynamics. Science 315:1716–1719. https://doi.org/10.1126/science.1137455

Terpolilli JJ, Masakapalli SK, Karunakaran R, Webb IUC, Green R, Watmough NJ, Kruger NJ, Ratcliffe RG, Poole PS (2016) Lipogenesis and redox balance in nitrogen-fixing pea bacteroids. J Bacteriol 198:2864–2875. https://doi.org/10.1128/JB.00451-16

Van de Velde W, Guerra JCP, De Keyser A, De Rycke R, Rombauts S, Maunoury N, Mergaert P, Kondorosi E, Holsters M, Goormachtig S (2006) Aging in legume symbiosis. A molecular view on nodule senescence in *Medicago truncatula*. Plant Physiol 141:711–720. https://doi.org/10.1104/pp.106.078691

Van de Velde W, Zehirov G, Szatmari A, Debreczeny M, Ishihara H, Kevei Z, Farkas A, Mikulass K, Nagy A, Tiricz H, Satiat-Jeunemaître B, Alunni B, Bourge M, Kucho K, Abe M, Kereszt A, Maroti G, Uchiumi T, Kondorosi E, Mergaert P (2010) Plant peptides govern terminal differentiation of bacteria in symbiosis. Science 327:1122–1126. https://doi.org/10.1126/science.1184057

Wang D, Griffitts J, Starker C, Fedorova E, Limpens E, Ivanov S, Bisseling T, Long S (2010) A nodule-specific protein secretory pathway required for nitrogen-fixing symbiosis. Science 327:1126–1129. https://doi.org/10.1126/science.1184096

Wang Q, Liu J, Li H, Yang S, Körmöczi P, Kereszt A, Zhu H (2018) Nodule-specific cysteine-rich peptides negatively regulate nitrogen-fixing symbiosis in a strain-specific manner in *Medicago truncatula*. Mol Plant-Microbe Interact MPMI 31:240–248. https://doi.org/10.1094/MPMI-08-17-0207-R

Wang Q, Yang S, Liu J, Terecskei K, Ábrahám E, Gombár A, Domonkos Á, Szűcs A, Körmöczi P, Wang T, Fodor L, Mao L, Fei Z, Kondorosi É, Kaló P, Kereszt A, Zhu H (2017) Host-secreted antimicrobial peptide enforces symbiotic selectivity in *Medicago truncatula*. Proc Natl Acad Sci USA 114:6854–6859. https://doi.org/10.1073/pnas.1700715114

Wright R, Stephens C, Shapiro L (1997) The CcrM DNA methyltransferase is widespread in the alpha subdivision of proteobacteria, and its essential functions are conserved in *Rhizobium meliloti* and *Caulobacter crescentus*. J Bacteriol 179:5869–5877

Xue S, Biondi EG (2019) Coordination of symbiosis and cell cycle functions in *Sinorhizobium meliloti*. Biochim Biophys Acta Gene Regul Mech 1862:691–696. https://doi.org/10.1016/j.bbagrm.2018.05.003

Yang S, Wang Q, Fedorova E, Liu J, Qin Q, Zheng Q, Price PA, Pan H, Wang D, Griffitts JS, Bisseling T, Zhu H (2017) Microsymbiont discrimination mediated by a host-secreted peptide in *Medicago truncatula*. Proc Natl Acad Sci USA 114:6848–6853. https://doi.org/10.1073/pnas.1700460114

Integration of the Cell Cycle and Development in *Agrobacterium tumefaciens*

Kevin C. Failor, Priya Aryal, Michelle C. Brock, Cavanagh Gohlich, and Jason E. Heindl

Abstract

A bacterium's ability to properly sense and respond to its internal and external environments is crucial for the survival of the bacterium. As such, alphaproteobacteria like *Agrobacterium tumefaciens* have evolved key two-component systems and phosphorelays to quickly adapt to these changes. Several of these key phosphotransfer systems regulate cell cycle progression and affect key developmental phenotypes including attachment and biofilm formation, motility, cell morphology, and virulence. In this chapter, we describe the primary *A. tumefaciens* cell cycle regulatory pathway: the DivJ/PleC/PdhS-DivK/PleD and CckA-ChpT-CpdR/CtrA phosphorelays and their associated proteins. We aim to focus on each protein involved in these phosphorelays, describing the current state of *Agrobacterium* research while providing a fundamental background of the history of these proteins in other alphaproteobacteria. This chapter covers the gene regulation, protein biosynthesis and degradation, and downstream effects caused by the overexpression or deletion of these cell cycle-associated proteins. Finally, we address current shortcoming in *Agrobacterium* cell cycle research while highlighting emerging data and trends.

1 Introduction

An organism's ability to reproduce efficiently and at the appropriate time is essential for the survival of the organism and, in broader context, the species. In bacteria, like the plant pathogen *Agrobacterium tumefaciens*, the cell cycle is defined at its

K. C. Failor · P. Aryal · M. C. Brock · C. Gohlich · J. E. Heindl (✉)
Department of Biological Sciences, University of the Sciences in Philadelphia, Philadelphia, PA, USA
e-mail: j.heindl@usciences.edu

E. Biondi (ed.), *Cell Cycle Regulation and Development in Alphaproteobacteria*,
https://doi.org/10.1007/978-3-030-90621-4_9

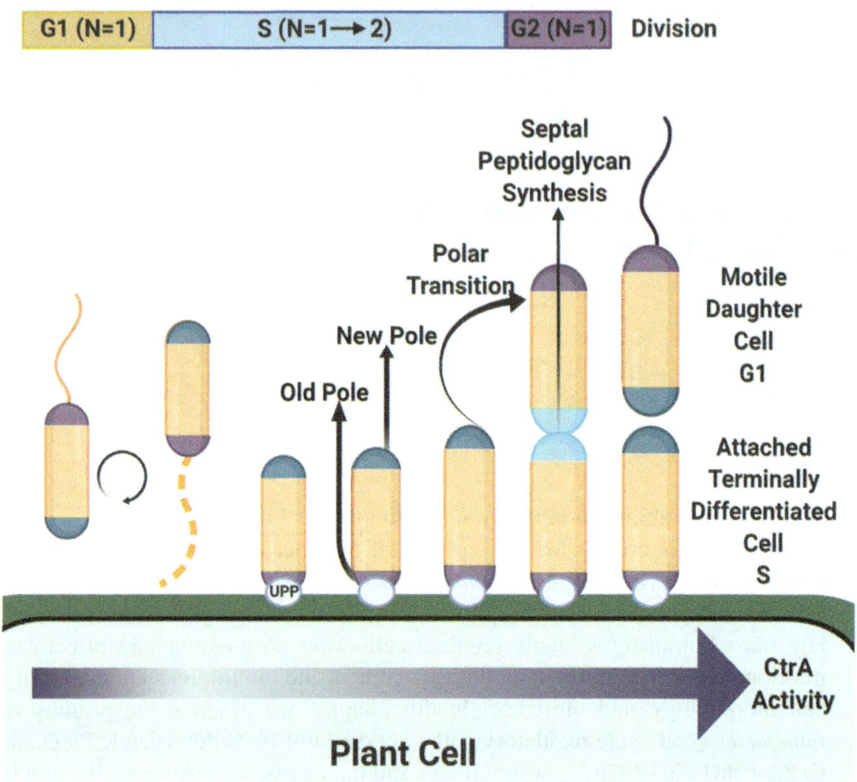

Fig. 1 Schematic of *A. tumefaciens* cell cycle progression. The image follows the bacterium as it progresses from its initial growth phase (G1) as a motile cell on the left of the image; through differentiation of this motile cell to a non-motile cell entering synthesis phase (S); and proceeding through genome duplication (indicated by N) to a second growth phase (G2) prior to cell division on the right of the image. At the completion of septal peptidoglycan biosynthesis and cell division two daughter cells are produced, one of which is motile and in G1, and the other of which is in S phase and is in this case still attached to the host plant cell substrate. Also indicated is relative pole age, with old poles colored purple and relatively younger, new poles colored blue. At the bottom of the image the purple shaded arrow indicates predicted CtrA activity during cell cycle progression. Created with BioRender.com

most basic level as a series of three fundamental stages: the duplication of genetic material, the separation of said genetic material to distinct regions of the cell, and the division of the cell into two, fully functional, daughter cells, with concomitant growth of each daughter cell (Fig. 1).

A. tumefaciens, also sometimes called *A. fabrum* (though this name has not been given standing nomenclature), is a heterotypic synonym of *A. radiobacter*, with the two type strains varying minutely on overall G+C content and genome size (Hordt et al. 2020; Lassalle et al. 2011). It is a gram-negative, phytopathogenic alphaproteobacterium capable of inducing tumor formation that was first isolated at the turn of the twentieth century (Young et al. 2005). Like some other alphaproteobacteria, *A. tumefaciens* grows and divides via unipolar growth and asymmetrical division. The resultant daughter cells, though genetically identical, are morphologically distinct and have separate transcriptional programs governing their form and function. While experimental data on the *A. tumefaciens* cell cycle is relatively sparse, there are several well-studied processes, such as motility and surface attachment that are ultimately dependent on an efficient, regulated, dynamic cell cycle.

Much of the *A. tumefaciens* cell cycle regulatory process is inferred based on observations in the model alphaproteobacterium *Caulobacter crescentus*. While *C. crescentus* undergoes a differentiation between vegetative state "swarmer" cells and a "stalked" replicating cell morphology that *A. tumefaciens* does not exhibit, both species do present a pattern of G1-S-G2 phases of development, controlled by conserved elements, suggesting an early evolutionary origin of these cell cycle control processes (Brilli et al. 2010; Greene et al. 2012). Much of the cell cycle is controlled by several ubiquitous proteins (GcrA, DnaA, and CtrA, among others) that have been rigorously studied in *C. crescentus* and provide a cyclical genetic circuit for successful growth, replication, and division (Tan et al. 2010). These genetic components are further controlled by stringent regulatory pathways and localization patterns that allow for dynamic responses to environmental stimuli.

The central cell cycle process is mediated, in part, by the PdhS-DivK-CtrA regulatory pathway that ultimately influences gene expression controlling chromosomal duplication, polar protein localization, and cell division. In addition to cell intrinsic regulatory control of complex phenotypes such as biofilm formation and pathogenesis, such as that mediated by the PdhS-DivK-CtrA regulatory pathway, bacteria are responsive to cell extrinsic factors such as substrate availability, temperature, and nutrient availability, to name a few. Several recent studies have detailed how certain cell extrinsic stimuli affect biofilm formation, swimming motility, and gene expression by *A. tumefaciens*. Together these studies provide a more complete picture of signal inputs that must be integrated with the bacterial cell cycle, possibly through the PdhS-DivK-CtrA pathway. The activity of many of the pathway-associated proteins can be observed through their effects on biofilm formation and exopolysaccharide production, motility, and virulence. Below we first outline the known and inferred architecture and function of the PdhS-DivK-CtrA pathway of *A. tumefaciens*, using the more well-studied *C. crescentus* as a guide. Following this we discuss additional key cellular and environmental inputs that are known to affect the cell cycle and cell cycle-dependent phenotypes.

2 Global Regulation of Cell Cycle Progression

2.1 Intro to DNA Replication/Modification/Gene Regulation

In *C. crescentus*, chromosomal replication and polar morphogenesis are largely controlled by a suite of global regulators (CcrM, CtrA, DnaA, GcrA, MucR, and SciP) and associated interacting partners. These global regulators work in a modular fashion to entrain the cyclical progression of the cell cycle. Pre-replicative cells exist in an initial primary growth phase, G1, under transcriptional control of $CtrA_{Cc}$ and MucR. Cells transition to the synthesis, S, phase during which genome replication occurs, primarily under transcriptional control of GcrA and CcrM (with input, in some cases, from $CtrA_{Cc}$ and DnaA DNA-binding activities). Following genome duplication and segregation, cells enter a relatively brief secondary growth phase, G2, with gene expression regulated by SciP and $CtrA_{Cc}$ (Fumeaux et al. 2014; Panis et al. 2015).

2.1.1 Chromosomal Replication Initiator Protein, AAA$^+$ Protein DnaA (Atu0324)

In *C. crescentus* DNA replication is largely controlled by the ATPase DnaA (Holtzendorff et al. 2006). *A*TPases *a*ssociated with diverse cellular *a*ctivities (AAA+ proteins) are responsible for a variety of roles within the cell, including protein folding and recruitment, biosynthesis, and DNA replication as is the case in *C. crescentus* (Snider et al. 2008). This central protein is responsible for forming an oligomeric structure at the origin of replication, unwinding the DNA, and recruiting additional, necessary proteins for replication. DnaA itself is relatively unstable, requiring constant degradation and replenishment to control intracellular concentrations; start of the G1-S transition, DnaA levels rise and DNA replication is initiated (Gorbatyuk and Marczynski 2005; Holtzendorff et al. 2006). DnaA degradation is performed by Lon protease, which also degrades CcrM, SciP, and other related proteins, when both are bound to the cell's DNA (Wright et al. 1996; Zhou et al. 2019). Under stress conditions, increased degradation rates of DnaA by the ATP-dependent serine peptidase Lon protease can arrest cellular division, allowing the cell to divert energy into alternative pathways (Jonas et al. 2013).

While *A. tumefaciens* does possess a DnaA homolog (Atu0324), little research has been done on the interactions between this replication initiator and its associated proteins. Due to its homology with the *C. crescentus* homolog (Tables 1), it can be inferred that many of the interactions and controls act similarly in this bacterium.

2.1.2 Cell Cycle-*R*egulated *M*ethyltransferase CcrM (Atu0794)

CcrM is an orphan methyltransferase similar to the well-studied Dam methyltransferase identified in gammaproteobacteria that functions as an essential methyltransferase in *C. crescentus*. The *C. crescentus* CcrM ($CcrM_{Cc}$) is necessary to efficiently transcribe S-phase promoters and transfers the methyl group from the S-adenosylmethionine to the N-6 position of a target sequence adenine (Gora et al. 2010; Kahng and Shapiro 2001; Stephens et al. 1996). CcrM's transcription is controlled by $CtrA_{Cc}$ but the protein is, however, rapidly degraded by the Lon protease mediated pathway that requires both $CcrM_{Cc}$ and Lon to be bound to DNA for the degradation to occur (Stephens et al. 1996; Wright et al. 1996; Zhou et al. 2019). This rapid degradation of $CcrM_{Cc}$ leads *C. crescentus* to exhibit a semi-methylated genome until late in its cell cycle when $CcrM_{Cc}$ synthesis is amplified, leading to full DNA methylation (Kahng and Shapiro 2001). This methylation does affect the *ctrA* promoter, P1, preventing transcription of *ctrA* during the late stages

Table 1 Homology of cell cycle regulation-associated proteins between *Agrobacterium tumefaciens* and related alphaproteobacteria

A. tumefaciens C58[a]				Homolog						
Protein	Locus	Length (AA)	Essential	Organism	Locus	Length (AA)	Identical (%)	Identical + similar (%)	Gaps (%)	Essential
CckA	Atu1362	861	Yes	*B. abortus* 2308	BAB1_1059	945	52.03	66.94	11.68	–
				C. crescentus NA1000[b]	CCNA_01132	691	31.99	44.00	25.81	Yes
				S. meliloti 1021	SMc00471	869	74.29	86.17	2.29	–
CcrM	Atu0794	381	Yes	*B. abortus* 2308	BAB1_0516	377	75.45	84.50	4.13	–
				C. crescentus NA1000	CCNA_00382	358	59.58	71.92	6.04	Yes
				S. meliloti 1021	SMc00021	376	82.68	91.34	1.31	–
ChpT	Atu2438	215	Yes	*B. abortus* 2308	BAB1_1613	209	58.80	74.07	3.70	–
				C. crescentus NA1000	CCNA_03584	225	33.62	49.14	10.34	Yes
				S. meliloti 1021	SMc00652	221	72.85	82.35	2.71	–
ClpA	Atu1364	836	No	*B. abortus* 2308	BAB1_1191	824	77.66	87.23	3.78	–
				C. crescentus NA1000	CCNA_02553	776	66.35	79.00	7.64	No
				S. meliloti 1021	SMc02109	838	82.71	90.24	3.06	–
ClpP1	Atu1258	210	No	*B. abortus* 2308	BAB1_1132	209	75.23	85.98	4.21	–
				C. crescentus NA1000	CCNA_02041	213	75.70	84.58	2.34	Yes
				S. meliloti 1021	SMc01903	208	87.62	93.81	0.95	–
ClpP2	Atu2270	193	No	*B. abortus* 2308	–	–	–	–	–	–
				C. crescentus NA1000	–	–	–	–	–	–
				S. meliloti 1021	SMc02720	195	80.00	91.28	1.03	–
ClpP3	Atu1627	202	No	*B. abortus* 2308	–	–	–	–	–	–
				C. crescentus NA1000	–	–	–	–	–	–
				S. meliloti 1021	SMc03841	205	88.29	92.20	1.46	–
ClpS1	Atu1363	117	No	*B. abortus* 2308	BAB1_1192	116	73.50	84.62	0.85	–
				C. crescentus NA1000	CCNA_02552	109	63.87	77.31	10.08	No
				S. meliloti 1021	SMc02110	117	83.76	90.60	0.00	–

(continued)

Table 1 (continued)

A. tumefaciens C58[a]				Homolog						
Protein	Locus	Length (AA)	Essential	Organism	Locus	Length (AA)	Identical (%)	Identical + similar (%)	Gaps (%)	Essential
ClpS2	Atu2232	103	No	*B. abortus* 2308	–	–	–	–	–	–
				C. crescentus NA1000	–	–	–	–	–	–
				S. meliloti 1021	SMc02694	101	74.76	86.41	1.94	–
ClpX	Atu1259	425	Yes	*B. abortus* 2308	BAB1_1131	424	89.41	95.76	0.24	–
				C. crescentus NA1000	CCNA_02039	420	79.53	89.88	1.18	–
				S. meliloti 1021	SMc01904	425	95.06	98.12	0.00	–
CpdR1	Atu3883	120	Yes	*B. abortus* 2308	BAB2_0042	120	83.47	93.39	1.65	–
				C. crescentus NA1000	CCNA_00781	118	59.17	75.00	1.67	No
				S. meliloti 1021	SMc04044	120	93.33	95.83	0.00	–
CpdR2	Atu3603	119	No	*B. abortus* 2308	–	–	–	–	–	–
				C. crescentus NA1000	–	–	–	–	–	–
				S. meliloti 1021	SMc00720	119	79.83	88.24	0.00	–
CtrA	Atu2434	234	Yes	*B. abortus* 2308	BAB1_1614	232	91.88	95.30	0.85	Yes
				C. crescentus NA1000	CCNA_03130	225	79.75	89.03	3.80	Yes
				S. meliloti 1021	SMc00654	233	97.86	99.57	0.43	Yes
DgcA	Atu1257	415	No	*B. abortus* 2308	–	–	–	–	–	–
				C. crescentus NA1000	CCNA_03394	237	16.70	27.23	50.80	No
				S. meliloti 1021	SMc01464	426	28.10	42.70	16.78	–
DgcB	Atu1691	356	No	*B. abortus* 2308	–	–	–	–	–	–
				C. crescentus NA1000	CCNA_01926	353	30.05	45.74	11.44	No
				S. meliloti 1021	–	–	–	–	–	–

Table 1 (continued)

A. tumefaciens C58[a]				Homolog						
Protein	Locus	Length (AA)	Essential	Organism	Locus	Length (AA)	Identical (%)	Identical + similar (%)	Gaps (%)	Essential
DgcC	Atu2179	379	No	*B. abortus* 2308	–	–	–	–	–	–
				C. crescentus NA1000	–	–	–	–	–	–
				S. meliloti 1021	–	–	–	–	–	–
DivJ	Atu0921	510	Yes	*B. abortus* 2308	BAB1_0601	609	29.46	42.33	26.51	–
				C. crescentus NA1000	CCNA_01116	585	26.82	39.39	34.09	No
				S. meliloti 1021	SMc00059	525	52.04	65.19	8.33	–
DivK	Atu1296	123	No	*B. abortus* 2308	BAB2_0628	123	86.99	94.31	0.00	–
				C. crescentus NA1000	CCNA_02547	125	72.80	84.80	1.60	Yes
				S. meliloti 1021	SMc01371	123	97.56	98.37	0.00	Yes
DivL	Atu0027	881	Yes	*B. abortus* 2308	BAB1_2101	859	46.30	61.66	7.73	–
				C. crescentus NA1000	CCNA_03598	769	27.07	42.64	22.77	Yes
				S. meliloti 1021	SMc02756	826	63.83	76.87	6.46	–
DnaA	Atu0324	520	Yes	*B. abortus* 2308	BAB1_0001	496	64.46	75.80	7.94	–
				C. crescentus NA1000	CCNA_00008	490	35.78	50.09	9.79	Yes
				S. meliloti 1021	SMc01167	507	74.76	82.92	5.12	–
GcrA	Atu0426[c]	177	–	*B. abortus* 2308	BAB1_0329	180	58.15	69.02	5.98	–
				C. crescentus NA1000	CCNA_02328	173	42.63	53.16	15.79	No
				S. meliloti 1021	SMc02139	173	67.42	80.34	3.37	–
Lon	Atu1261	805	Yes	*B. abortus* 2308	BAB1_1130	812	80.10	89.31	1.35	–
				C. crescentus NA1000	CCNA_02037	799	67.25	80.77	0.99	–
				S. meliloti 1021	SMc01905	806	87.97	93.55	0.12	–

(continued)

Table 1 (continued)

A. tumefaciens C58[a]				Homolog						
Protein	Locus	Length (AA)	Essential	Organism	Locus	Length (AA)	Identical (%)	Identical + similar (%)	Gaps (%)	Essential
PdhS1	Atu0614	1275	No	*B. abortus* 2308	BAB1_1621	1035	39.15	51.85	22.31	Yes
				C. crescentus NA1000	–	–	–	–	–	–
				S. meliloti 1021	SMc00776	1134	51.46	51.46	15.26	–
PdhS2	Atu1888	511	No	*B. abortus* 2308	–	–	–	–	–	–
				C. crescentus NA1000	–	–	–	–	–	–
				S. meliloti 1021	SMc04212	511	63.80	80.04	0.00	–
PleC	Atu0982	778	No	*B. abortus* 2308	BAB1_0640	783	44.21	61.21	7.76	–
				C. crescentus NA1000	CCNA_02567	842	31.03	45.80	25.43	No
				S. meliloti 1021	SMc02369	772	63.72	75.77	1.28	–
CelR/PleD	Atu1297	456	No	*B. abortus* 2308	BAB2_0630	456	49.46	70.28	2.17	–
				C. crescentus NA1000	CCNA_02546	454	51.30	68.40	3.03	No
				S. meliloti 1021	SMc01370	455	71.93	84.87	0.22	–
PodJ	Atu0499	1248	No	*B. abortus* 2308	BAB2_0413	913	27.29	39.80	36.16	–
				C. crescentus NA1000	CCNA_02125	974	22.94	34.78	33.43	No
				S. meliloti 1021	SMc02230	960	34.90	46.84	29.89	–
PopZ	Atu1720	333	No	*B. abortus* 2308	BAB1_0964	243	31.39	38.89	40.00	–
				C. crescentus NA1000	CCNA_01380	177	22.85	32.94	48.66	Yes
				S. meliloti 1021	SMc02081	251	36.44	46.65	29.74	–
RcdA	Atu3742	171	Yes	*B. abortus* 2308	BAB1_1717	168	53.80	67.25	1.75	–
				C. crescentus NA1000	CCNA_03404	169	42.22	55.00	11.11	No
				S. meliloti 1021	SMc03989	196	73.47	79.59	12.76	–
SciP	Atu2430	91	No	*B. abortus* 2308	BAB1_1609	91	92.31	98.90	0.00	–
				C. crescentus NA1000	CCNA_00948	93	74.47	81.91	4.26	No
				S. meliloti 1021	SMc00657	91	97.80	100.00	0.00	–

[a] *A. tumefaciens* C58 is cataloged in the National Center for Biotechnology Information (NCBI) protein database as *A. fabrum* C58

[b] *C. crescentus* NA1000 is cataloged in the NCBI protein database as *C. vibrioides* NA1000

[c] An unannotated protein at Atu0426 with high homology to GcrA is overlapping and antisense to the annotated Atu0426

All sequences were analyzed utilizing a Needleman–Wunsch global alignment with a BLOSUM62 similarity matrix and default parameters. Proteins with absent homology data had no identified homolog present in the genome

of the cell cycle (Reisenauer and Shapiro 2002). The fully methylated P1 promoter is present at the initiation of replication and it can be transcribed after the replication fork passes resulting in two hemi-methylated copies of the gene (Collier et al. 2006).

The *A. tumefaciens* homolog ($CcrM_{At}$) shares a high degree of sequence homology to $CcrM_{Cc}$ (59.6% identical and 71.9% identical plus positives, Tables 1), suggesting that it functions much the same way (Kahng and Shapiro 2001; Wright et al. 1997). Like its *C. crescentus* homolog, $CcrM_{At}$ is essential for the survival of the bacterium and its promoter region has an *A. tumefaciens* CtrA ($CtrA_{At}$) binding motif, indicating transcriptional control by the global regulator. If $CcrM_{At}$ is overexpressed in *A. tumefaciens*, there is an excess of DNA methylation resulting in morphological and flow-cytometric abnormalities such as branching, elongated cells, and increased DNA content. Furthermore, as in *C. crescentus*, $CcrM_{At}$ exhibits an increase in activity late in the S-phase of growth, just before cellular division (Kahng and Shapiro 2001).

2.1.3 Small CtrA *I*nhibitory *P*rotein SciP (Atu2430)

In *Caulobacter*, the helix-turn-helix transcription factor, SciP ($SciP_{Cc}$), is a component in the cyclical control circuit with the global regulators that drive the cell cycle (Tan et al. 2010). SciP is cell cycle controlled and is co-conserved with CtrA in alphaproteobacteria. $CtrA_{Cc}$ activates *sciP* late in the cell cycle with $SciP_{Cc}$ accumulating in the daughter swarmer cells (Gora et al. 2010; Tan et al. 2010). These $SciP_{Cc}$ levels are tightly regulated and restricted to the G1 swarmer cells and quickly disappear as the cell transitions from the G1-S phase (Gora et al. 2010). Accumulation of this protein during the G1 phase represses *ctrA* and the subsequent $CtrA_{Cc}$ target genes by preventing $CtrA_{Cc}$ from recruiting the RNA polymerase (Gora et al. 2010; Tan et al. 2010). Overall, $SciP_{Cc}$ is an additional layer of control that represses late pre-divisional cell transcription and the swarmer genes that are activated by CtrA. The flagella and chemotaxis genes activated by $CtrA_{Cc}$ are subsequently repressed by $SciP_{Cc}$ as it preferentially accumulates in the swarmer cell. The impacted flagella and chemotaxis gene promoters have the $SciP_{Cc}$ binding motif upstream of the $CtrA_{Cc}$ binding motif, the primary function of $SciP_{Cc}$ is to enhance the overall robustness of the core cell cycle control circuit (Tan et al. 2010).

The *A. tumefaciens* SciP ($SciP_{At}$) homolog shares 74.5% identity (81.9% identical plus positives) with $SciP_{Cc}$ (Tables 1) and maintains the helix-turn-helix motif, suggesting that both proteins function in a similar manner (Mohari et al. 2018). However, $SciP_{At}$ was determined to not be essential in *A. tumefaciens* leading to uncertainty regarding its role in Ctr regulation (Curtis and Brun 2014). SciP has been shown to suppress swim motility by interfering with flagellar gene expression while suppression of *sciP* results in increased flagellar protein production, possibly through increased *ctrA* expression (Mohari et al. 2018).

2.1.4 Cell Cycle Regulator GcrA

Identified alongside CtrA, GcrA in *C. crescentus* ($GcrA_{Cc}$) is a 174 amino acid cell-cycle transcription factor (Holtzendorff et al. 2006; Quon et al. 1996). $GcrA_{Cc}$ is found predominantly during the swarmer-to-stalked cell transition as it is activated by $DnaA_{Cc}$ after the start of DNA replication. It controls approximately 125 genes, though the affected genes may be directly or indirectly regulated (Haakonsen et al. 2015; Holtzendorff et al. 2006). While no specific binding site has been identified for $GcrA_{Cc}$, it does preferentially bind a GANTC site with a methylated N6-adenine in vitro. $GcrA_{Cc}$ appears to function as a molecular effector for the regulation of gene expression during the cell cycle utilizing a methylation-dependent signal (Fioravanti et al. 2013).

While there is no annotated GcrA homolog present in *A. tumefaciens* (Curtis and Brun 2014), a homologous locus was recently identified in *A. tumefaciens* C58 overlapping and antisense to the hypothetical protein Atu0426 (Wang et al. 2012). The *A. tumefaciens* GcrA ($GcrA_{At}$) homolog is a 177 amino acid protein that shares 42.6% sequence similarity (53.2% identical plus positives) with $GcrA_{Cc}$ (Tables 1). Furthermore, this *A. tumefaciens* homolog is missing a 13 amino acid C-terminal tail containing a predicted N-myristoylation site that is present in *C. crescentus*. Myristoylation of proteins has previously been implicated in protein–membrane and protein–protein interactions, suggesting that the deletion of this region may affect *A. tumefaciens* $GcrA_{At}$ interactions (Martin et al. 2011). As methylation- dependent DNA binding was observed in *C. crescentus* utilizing *Sinorhizobium meliloti* and *Brucella abortus* orthologs, it can be inferred that *A. tumefaciens* $GcrA_{At}$ would be functionally similar (Fioravanti et al. 2013).

2.2 Introduction to Phosphorylation, Two-Component Systems, and Phosphorelays

An organism's ability to sense its environment and quickly adapt to changing inputs is critical for the survival of both the individual organism and the larger community. *Agrobacterium*, like many other alphaproteobacteria, and prokaryotes as a whole, utilizes complex phosphorylation pathways to relay signals throughout the cell. These signals can communicate broader commands, including initiation of cell division, attachment to a substrate, or production of defense molecules (Esser et al. 2016; Kobir et al. 2011; Kyriakis 2014). Many phosphorylation cascades rely on the phosphorylation and dephosphorylation of key threonine, tyrosine, or serine residues which ultimately causes a downstream effect (Deutscher and Saier 2005). However, most bacterial species also possess two-component systems; these pathways are a key mechanism for sensing and responding to both internal and external variation (Ryan 2006). These two-component systems are activated by the recognition of a signal by the sensory kinase. Efficient phosphatase activity of many sensor kinases, including histidine kinases, requires a conserved threonine residue downstream of the phosphor-accepting histidine residue. This conserved threonine is located approximately one α-helical turn away from the histidine (Deutscher and Saier 2005; Gao and Stock 2017; Huynh and Stewart 2011). Following signal recognition, the kinase activates its own autokinase activity before transferring the phosphate to its response regulator partner. This transfer ultimately leads to a cellular response typically through the binding of the response regulator to DNA and activation of transcription (Kobir et al. 2011).

Typically, histidine kinases are encoded in the genome nearby the partner response regulator, however, some two-component genes, such as the *Streptomyces* AmfR and *E. coli* FimZ, are orphans and have no functional partner coded in proximity (Gao et al. 2006; Ryan 2006). In some cases, these two-component systems have evolved a more complex mechanism to relay the phosphate between the sensor kinase and the response regulator. These phosphorelays rely on the kinase and phosphatase activity of intermediate proteins to carry out the phosphate transfer (Ryan 2006).

2.2.1 The PdhS-DivK-CtrA Pathway in *Agrobacterium tumefaciens*

One such multicomponent phosphorelay system is the highly conserved PdhS-DivK-CtrA pathway present in *Agrobacterium* spp. and other alphaproteobacterial genera. In *A. tumefaciens*, the PdhS-DivK-CtrA pathway has been thus far shown to affect cellular motility, biofilm formation, and the regulation of the cell cycle, but also demonstrates effects on other developmental phenotypes, such as cell morphology and virulence (Barnhart et al. 2014; Heindl et al. 2014; Howell et al. 2017; Kim et al. 2013; Su et al. 2006). Many of the key regulatory elements were first identified in *Caulobacter crescentus* (Wheeler and Shapiro 1999) and are highly conserved among other alphaproteobacteria, including *A. tumefaciens*, *Brucella abortus*, and *Sinorhizobium meliloti* (Brilli et al. 2010). The main elements of the pathway can be organized into two key phosphorelays: the DivJ/PleC/PdhS-DivK/PleD phosphorelay and the CckA-ChpT-CpdR/CtrA phosphorelay. CckA, DivJ, PdhS1 and 2, and PleC comprise the key histidine kinases in this pathway, while CpdR1 and 2, CtrA, DivK, and PleD serve as the response regulators, with CtrA acting as a global response regulator. Several other proteins, including DivL, PodJ, and PopZ, all contribute to the localization or regulation of these phosphorelay proteins, resulting in increased control over the regulation of the *A. tumefaciens* cell cycle.

Phosphorelay 1: DivJ/PleC/PdhS-DivK/PleD and Associated Proteins

Sensor Histidine Kinases *DivJ* (Atu0921) and *PleC* (Atu0982)

Two proteins, DivJ and PleC have been identified as key sensor histidine kinases present in *Caulobacter*, *Agrobacterium*, and other alphaproteobacteria (Curtis and Brun 2014; Kim et al. 2013; Ohta et al. 1992; Pini et al. 2013; Wang et al. 1993). The phosphorylation states of both PleD and DivK are regulated collectively by the sensor histidine kinases DivJ and PleC in *C. crescentus* which localizes to either pole, resulting in a cell cycle-dependent localization of the response regulators (Matroule et al. 2004; Wheeler and Shapiro 1999).

DivJ localizes to the stalked cell pole of *C. crescentus* where it phosphorylates its response regulator, DivK, ultimately leading to localized decreased levels of phosphorylated $CtrA_{Cc}$ (Viollier et al. 2002; Wheeler and Shapiro 1999). DivJ localization to the stalked pole relies on several factors, including polar localizing protein PopZ migration to the same pole prior to cellular division (Bowman et al.

2008; Ebersbach et al. 2008). This interaction proceeds through an intermediate protein, SpmX. Both SpmX and DivJ are produced at the start of the swarmer to stalked cell transition at which point SpmX binds PopZ and then binds DivJ, anchoring it to the pole (Radhakrishnan et al. 2008). Disruptions in either SpmX or DivJ result in ectopic pole formation and disruption of DivK localization (Jacobs et al. 2001; Perez et al. 2017). After the swarmer-to-stalked transition, DivJ replaces PleC at the stalked pole, resulting in a surge of DivK phosphorylation and degradation of $CtrA_{Cc}$ (Lasker et al. 2016; Tsokos et al. 2011).

Similarly, PleC plays a key role in polar development during cell division where it localizes to the new/younger pole (distal to the stalk), opposite of DivJ; this localization depends on the polar organelle development protein PodJ. PleC acts as a bifunctional histidine-modifying enzyme with both kinase and phosphatase activity, though it is primarily responsible for the dephosphorylation of phosphorylated DivK. This dephosphorylation leads to the release of DivK from the younger, flagellated pole and ultimately an increase in phosphorylated $CtrA_{Cc}$ levels and activity (Heindl et al. 2019; Laub et al. 2000, 2002; Matroule et al. 2004; Quon et al. 1998; Subramanian et al. 2015). Interestingly, PleC does appear briefly alone with dispersed, dephosphorylated DivK in post-divisional cells when DivJ migrates under the control of PopZ and SpmX (Matroule et al. 2004).

In *A. tumefaciens*, DivJ is an essential gene responsible for numerous phenotypic characteristics resulting from its influence on the CckA-ChpT-CtrA pathway (Curtis and Brun 2014). Like its *C. crescentus* homolog, it localizes to the old pole where it interacts with its direct targets, DivK and PleD (Ehrle et al. 2017). However, *A. tumefaciens* lacks an SpmX homolog, therefore it is unclear how DivJ localizes to the old pole and interacts with PopZ. One intriguing possibility is a putative peptidoglycan binding protein whose coding sequence overlaps with that of *divJ*. This locus, Atu8087, is not present in *C. crescentus*, but is conserved in several Rhizobiales. Related to this point, locus Atu0923, just downstream of *divJ* (Atu0921), encodes a hypothetical protein with a DUF2336 domain. In *C. crescentus*, two DUF2336-containing proteins have been shown to participate in cell cycle progression, SpbR and SpmY (Janakiraman et al. 2016; Wang and Bowman 2019). Although *A. tumefaciens* Atu1742 is a more direct homolog to both SpbR and SpmY, it is possible that Atu0923 likewise participates in cell cycle regulation. As with *C. crescentus* interaction with its targets ultimately results in the deactivation of CtrA, likely through CckA (Heindl et al. 2019). Depletion of the gene results in an increase in overall biofilm formation, production of branched cells, and a reduction in swim motility (unpublished results) and these phenotypes are also seen in Δ*divJ C. crescentus* cells (Pierce et al. 2006). Recent data suggests that DivJ acts antagonistically with another sensor kinase, PdhS2, to control the development of *A. tumefaciens* through the generation of phosphorylated CtrA gradient within the cell. This spatiotemporal gradient is linked to the localization patterns of both proteins (Heindl et al. 2019).

In *A. tumefaciens*, PleC is hypothesized to localize to the same (younger) pole as *C. crescentus* under the control of PodJ, however, it has not yet been demonstrated (Anderson-Furgeson et al. 2016; Ehrle et al. 2017; Heindl et al. 2019; Kim et al. 2013). However, Kim et al. have shown that flagellar localization is affected by the deletion of *pleC*, providing circumstantial evidence for this localization pattern. Strains lacking PleC have demonstrated phenotypic abnormalities including elongated and branching rods, ectopic localization of flagella, and slight alterations in unipolar polysaccharide placement and, by extension, weaker biofilm formation (Kim et al. 2013).

PleC/DivJ Homolog Sensor Kinases PdhS1 (Atu0614) and PdhS2 (Atu1888)
Identified in *A. tumefaciens* by Hallez et al. in 2004, after previously being found in *B. abortus*, *pdhS* genes exhibit a high degree of similarity to the *C. crescentus* genes *pleC* and *divJ*. It is suggested that, since the *pdhS* genes were found in several host-associated bacteria (*A. tumefaciens*, *B. abortus*, *Mesorhizobium loti*, and *S. meliloti*), these genes may be responsible for the sensing and response of diversified signals (Hallez et al. 2004). Characterization of these genes has demonstrated their role in regulation of cell growth, division, and development of the poles. Furthermore, deletion of the genes crucial to the coordination of division and development regulatory pathways, specifically *pleC*, *pdhS1*, and *divJ*, leads to aberrant branching due to localization errors of associated division proteins such as FtsZ (Brown et al. 2012; Kim et al. 2013).

The *Agrobacterium pdhS1*, unlike its *B. abortus* homolog, is a non-essential gene coding for a large (100 kDa) cytoplasmic protein with no transmembrane domains (Hallez et al. 2007; Kim et al. 2013). While the other PdhS homologs (PleC, DivJ, and PdhS2) contain a transmembrane domain, it is currently unclear what interacts with PdhS1 to hold it at the old pole. PdhS1, along with DivJ, localizes to the old pole independent of PopZ where it phosphorylates its target proteins such as DivK and PleD. These target proteins ultimately control the inactivation of CtrA (Ehrle et al. 2017; Heindl et al. 2019). Deletion of *pdhS1* in *A. tumefaciens* results in similar phenotypic abnormalities also seen in Δ*divK* and Δ*pleC* mutant strains; specifically, diminished biofilm formation, diminished swim motility, and atypical cellular morphology (Kim et al. 2013). PdhS, the *B. abortus* homolog, has been shown to localize to the old pole, as it does in *Agrobacterium*. While there, it recruits the fumarase FumC to the pole. This activity, however, is not shared by either *C. crescentus* or *S. meliloti*, and therefore may not occur in *A. tumefaciens* or other alphaproteobacteria (Mignolet et al. 2010).

Characterization of PdhS2 suggests that it is a non-essential gene involved in regulating the transition between the free-floating, planktonic state, and the non-motile state of *A. tumefaciens* primarily through its phosphatase activity and regulation of CtrA (Heindl et al. 2019; Kim et al. 2013). PdhS2 localizes primarily to the new pole where it dephosphorylates its target proteins leading to the stability and activation of CtrA (Ehrle et al. 2017; Heindl et al. 2019). Strains missing PdhS2 do not exhibit any abnormal growth or division phenotypes similar to those seen with Δ*popZ* mutations, but do exhibit increased biofilm formation, attachment and, subsequently, significantly decreased motility (Ehrle et al. 2017; Heindl et al. 2019; Kim et al. 2013). This increased attachment and decreased motility is also noted in ci-d-GMP regulation, suggesting that PdhS2 and ci-d-GMP may coordinate the regulation of biofilm formation and motility. The deletion of the diguanylate cyclase DgcA, DgcB, or PleD has been shown to reduce the overall biofilm formation present in a Δ*pdhS2* mutant. This suggests that the ci-d-GMP generated by these cyclases is necessary for the attachment phenotype observed in this mutant (Heindl et al. 2019).

Single-Domain Response Regulator, Cell *Div*ision Protein DivK (Atu1296) and Diguanylate Cyclase PleD/*Ce*llulose *R*egulator CelR (Atu1297)

DivK is an essential, single-domain response regulator involved in cellular division and polar differentiation in *C. crescentus* (Hecht et al. 1995; Jacobs et al. 2001; Sommer and Newton 1991). Transcription of *divK* occurs primarily during the late, pre-divisional cell, but does exist transiently during all remaining cell stages where its localization is controlled by both DivJ and PleC (Jacobs et al. 2001). After the transition from swarmer to stalked cell, DivK is rapidly phosphorylated by DivJ (Lasker et al. 2016; Tsokos et al. 2011). Phosphorylated DivK in *C. crescentus* has been demonstrated to directly interact with DivL, leading to the inactivation of CckA, and by extension, $CtrA_{Cc}$ and CpdR (Biondi et al. 2006; Iniesta et al. 2006; Tsokos et al. 2011). Likewise, *B. abortus* DivK has been demonstrated to directly interact with four key histidine kinases: DivJ, DivL, PdhS, and PleC (Hallez et al. 2007).

Like DivK, PleD interacts with both DivJ and PleC via the activity of both histidine kinases, leading to the transition to a stalked cell in *C. crescentus* (Aldridge et al. 2003; Paul et al. 2004). Deletion of *pleD* results in hypermotility, with the cells becoming unable to eject the flagellum or properly form the stalk during transition into the stalked-cell state (Aldridge and Jenal 1999; Hecht and Newton 1995). Conversely, overexpression of PleD results in diminished motility and elongated cell morphologies (Aldridge et al. 2003). These phenotypes are likely due to PleD's primary role in the coordination of polar morphogenesis. Paul et al. demonstrated that PleD dynamically localizes to the stalked pole when phosphorylated, but does not localize otherwise (Paul et al. 2004). In addition to its role in polar morphogenesis, PleD is required for the synthesis of c-di-GMP through two response regulator receiver domains (REC) coupled to a GGDEF c-di-GMP biosynthesis domain (Paul et al. 2004; Römling et al. 2013). The activity of this domain is largely controlled by DivK, and does require an active PleC for synthesis of ci-d-GMP (Paul et al. 2008).

In *A. tumefaciens*, DivK is a small, CheY-like receiver protein that appears to be non-essential, a key difference from its *C. crescentus* homolog (Barnhart et al. 2013; Kim et al. 2013). DivK functions to control several processes important for the division of *A. tumefaciens*, including polar localization of several cell division proteins and flagellar motility. Deletion of DivK leads to altered flagellar placement, branching of the cell, affected biofilm formation, and incorrect localization of FtsZ during division (Barnhart et al. 2014; Kim et al. 2013).

PleD, also called CelR for its role in cellulose production through the production of c-di-GMP (Barnhart et al. 2013, 2014), is located directly downstream of *divK*, as it is in *C. crescentus*, and appears to be non-essential in *A. tumefaciens*. *Agrobac-*

terium PleD exhibits several key functional domains including an N-terminal two-component response regulator domain and, downstream of this, a diguanylate cyclase (DGC) domain with a GGDEF motif (Thompson et al. 2018; Xu et al. 2013). PleD has been shown to be involved with unipolar polysaccharide (UPP) and cellulose production, with the overexpression of *pleD* leading to increased biofilm formation and polysaccharide formation, as well as branching and elongated rod cell morphologies. Interestingly, the overexpression of *pleD* results in attenuated virulence against *Kalanchoe daigremontiana* and tomato (Barnhart et al. 2013; Xu et al. 2013). While *divK* and *pleD* are part of the same operon and both affect polysaccharide production and cell cycle regulation, PleD forms an antiparallel homodimer with its GGDEF domains rather than demonstrating any interaction with DivK (Barnhart et al. 2014; Römling et al. 2013). Furthermore, a DGC homolog, Atu1060, has similar domain structures as seen with PleD and leads to increased polysaccharide production when overexpressed, but mutation of the gene does not affect the regulation of cellulose production as is seen with mutated PleD (Barnhart et al. 2013). These data suggest that elevated levels of both PleD and Atu1060 affect overall polysaccharide production through an increase in c-di-GMP levels throughout the cell, but only PleD affects the overall regulation of this biosynthesis (Thompson et al. 2018).

Phosphorelay 2: CckA-ChpT-CpdR/CtrA

Cell Cycle *K*inase CckA (Atu1362)

In *C. crescentus*, CckA functions as either a kinase or phosphatase of $CtrA_{Cc}$ via the intermediate ChpT (Biondi et al. 2006; Chen et al. 2009; Jacobs et al. 1999). In pre-division cells, CckA exists transiently throughout the cell in both the kinase and phosphatase states, with the swarmer pole exhibiting a greater concentration of CckA in the kinase state and the stalked pole containing greater phosphatase-state CckA concentrations (Chen et al. 2011; Jacobs et al. 1999). This variability in activity ultimately leads to a concentration gradient of phosphorylated $CtrA_{Cc}$ (Chen et al. 2006; Tsokos et al. 2011). Furthermore, it has been demonstrated that CckA is recruited by and co-localizes with DivL to the swarmer pole prior to division and is regulated by both DivK and DivL (Iniesta et al. 2010; Jacobs et al. 1999; Tsokos et al. 2011). In this system, DivL binds to phosphorylated DivK which, in turn, biases CckA toward its phosphatase mode. Unphosphorylated DivK, therefore, does not ultimately inhibit CckA activity.

In *A. tumefaciens*, *cckA* cannot be deleted; as such, it, along with *chpT* and *ctrA*, is considered essential (Kim et al. 2013). Since *A. tumefaciens* possesses a complete suite of CckA-ChpT-CtrA phosphorelay genes, the assumption can be made that these genes behave in a similar manner, including regulation by DivK. While working to characterize the coordination of division and development pathway, which comprises both the PleC/DivJ-DivK/PleD and the CckA-ChpT-CtrA/CpdR phosphorelays, a single CckA mutant ($CckA^{Y674A}$) was found to have increased autophosphorylation levels than the wild type, suggesting that the mutant has a lower sensitivity to DivK activity. This ultimately results in the phosphorylation of the downstream proteins CpdR and CtrA. When examined in Δ*pdhS1* and Δ*pdhs2* mutants, expression of CckA resulted in only modest effect on swim motility. This is contrasted to the results seen in the Δ*pleC* mutant where the impaired swim motility exhibited in the mutant is restored by the plasmid-borne $CckA^{Y674A}$ (Kim et al. 2013). These data are supported by a similar study done in a *C. crescentus* Δ*divJ* mutant that demonstrated an increase in activity and phosphorylation of CtrA (Pierce et al. 2006). Interestingly, the *A. tumefaciens* CckA did not consistently complement a *cckA* mutant of *Ruegeria* sp. KLH11, a sponge symbiont, suggesting that additional signals may control the activity of the *A. tumefaciens* protein (Zan et al. 2013).

Pseudokinase DivL (Atu0027)

While not much work has been done on DivL in *Agrobacterium*, the *C. crescentus* homolog is an essential protein involved with the activation of the CckA-ChpT-CtrA pathway (Reisinger et al. 2007; Wu et al. 1999). While DivL demonstrates a histidine kinase folding pattern, it does not function by phosphorylating or dephosphorylating DivK as it lacks both kinase and phosphatase activity. Instead, DivL preferentially binds to the phosphorylated DivK, stabilizing the activation state and inactivating the CckA-ChpT-CtrA signaling cascade (Childers and Shapiro 2014; Childers et al. 2014; Tsokos et al. 2011). The lack of kinase and phosphatase activity is due to DivL utilizing a tyrosine in place of a histidine at its site of autophosphorylation (Tyr-550) (Wu et al. 1999). In swarmer cells, DivL is largely diffuse throughout the cell, but ultimately localizes to one or both poles prior to division (Kowallis et al. 2020). DivL is also involved in the recruitment and autophosphorylation of CckA to the swarmer pole (Iniesta et al. 2010; Mann and Shapiro 2018; Tsokos et al. 2011). Furthermore, DivL does not appear to affect the localization of DivK (Reisinger et al. 2007).

Based on TnSeq data gathered by Curtis and Brun, DivL is essential in *A. tumefaciens* (Curtis and Brun 2014). It can also be assumed, based on sequence homology (Tables 1), and conserved functions within the alphaproteobacteria, that DivL functions in a similar manner in *A. tumefaciens* to bind phosphorylated DivK and regulate CckA activity (Childers and Shapiro 2014; Mann and Shapiro 2018). However, as the DivJ and DivK essentialities are reversed, with DivK being non-essential, there may be other proteins that control the interactions between DivL and CckA (Curtis and Brun 2014; Kim et al. 2013).

Cell Cycle *H*istidine *P*hosphotransferase ChpT (Atu2438)

ChpT is an essential pathway component and histidine phosphotranserase associated with the CckA-ChpT-CtrA and CckA-ChpT-CpdR phosphorelays first identified in *C. crescentus*. It was determined that the CckA receiver domain is highly preferred by ChpT, suggesting that CckA acts as the protein's only input (Biondi et al. 2006). After the autophosphorylation of CckA, a phosphate group is passed to ChpT; this group ultimately is passed to either $CtrA_{Cc}$ or CpdR (Biondi et al. 2006; Reisinger et al. 2007). Under certain conditions, namely when CckA is dephosphorylating, ChpT can act as a phosphate sink, dephosphoylating $CtrA_{Cc}$ and CpdR, passing the phosphoryl group back to CckA. As such, ChpT is a direct partner to CckA's kinase and phosphatase activity (Chen et al. 2009). As ChpT is reliant on CckA for its phosphorylation, any gene that regulates the activity or expression of CckA will ultimately lead to a change in expression of ChpT, most notably DivK and DivL.

ChpT has been identified as an essential gene in *A. tumefaciens* (Kim et al. 2013). However, while it does share >30% homology with the *C. crescentus* ChpT (Tables 1) and likely behaves in a similar manner, experimental confirmation of its role in the CckA-ChpT-CtrA/CpdR1 pathway has not yet been performed.

Cell Cycle *T*ranscriptional *R*egulator CtrA (Atu2434)

First identified in *C. crescentus*, $CtrA_{Cc}$ belongs to the response regulator superfamily of proteins, which typically act as transcription factors within two-component systems. During its initial characterization, the *Caulobacter* $CtrA_{Cc}$ was found to play a role in cell cycle regulation and was determined to be essential to the survival of the bacterium (Quon et al. 1996). $CtrA_{Cc}$ controls DNA replication and its degradation by ClpXP protease is involved in the transition from G1 to S phase of growth (Jenal and Fuchs 1998). Phosphorylation of $CtrA_{Cc}$ occurs on a conserved aspartate residue and is largely controlled by CckA, but CckA ultimately does not have any influence over the localization of $CtrA_{Cc}$ (Angelastro et al. 2010; Domian et al. 1997; Quon et al. 1996) When phosphorylated, $CtrA_{Cc}$ interacts directly with a CtrA box (TTAA-N7-TTAA) located in the promoter region of numerous cell-cycle proteins, such as CcrM (Holtzendorff et al. 2006; Marczynski and Shapiro 1992; Stephens et al. 1996). In *C. crescentus*, phosphorylated $CtrA_{Cc}$ regulates the transcription of 95 genes across 55 operons, including DNA methylation, chemotaxis, and cell division genes (Laub et al. 2002).

In an extensive study on the essential genes of several alphaproteobacterial genera, CtrA was found to be essential in *A. tumefaciens*, as it is in several other genera including *C. crescentus* and *Brevundimonas subvibrioides*, confirming what had been reported previously in *Agrobacterium* (Curtis and Brun 2014; Kim et al. 2013). *A. tumefaciens* CtrA ($CtrA_{At}$) is predicted to control the expression of more than 200 genes, including *chpT*, *cpdR1*, *ctrA*, *pleC*, *rcdA*, and the putative *gcrA*. Many of the genes enriched by CtrA regulation include those involved with cell cycle control and cellular division, cell wall synthesis, motility, and signal transduction. This enrichment pattern can also be seen in CtrA-controlled genes in *C. crescentus* (Brilli et al. 2010). As in *C. crescentus*, $CtrA_{At}$ regulation is accomplished through integrated phosphorelays that ultimately result in phosphorylation of either $CtrA_{At}$ or CpdR1 (Kim et al. 2013). Recent data shows that both DivK and PdhS2 regulate the downstream $CtrA_{At}$ resulting in effects on both attachment and motility (Heindl et al. 2019). Additionally, *ctrA* transcription in *A. tumefaciens* is autoregulated due to the presence of two CtrA binding sites in the promoter, similar to *C. crescentus* (Heindl et al. 2019). Finally, $CtrA_{At}$ protein levels decrease following translational arrest with chloramphenicol, suggesting the protein is not stable and is a target of proteolysis in this organism (Peter Chien, personal communication). Due to the essentiality of $CtrA_{At}$, the effect of $CtrA_{At}$ on cellular viability and morphology was performed utilizing a depleted $CtrA_{At}$ cell line. The data obtained from these experiments demonstrated that the depletion of $CtrA_{At}$ leads to the failure of the cells to divide, leading to the elongation and swelling of non-viable cells (Figueroa-Cuilan et al. 2016).

Cell Cycle Response Regulators CpdR1 (Atu3883) and CpdR2 (Atu3603)

In *C. crescentus*, CpdR ($CpdR_{Cc}$) is a non-essential, single-domain, response regulator protein active in the CckA-ChpT-CpdR phosphorelay. This phosphorelay ultimately causes the phosphorylation of $CpdR_{Cc}$ by a phosphorylated $ChpT_{Cc}$ (Biondi et al. 2006; Curtis and Brun 2014). Phosphorylation of $CpdR_{Cc}$ results in the stability of $CtrA_{Cc}$ due to a suppression of $CpdR_{Cc}$ triggering $CtrA_{Cc}$ proteolysis (Biondi et al. 2006; Iniesta et al. 2006). $CpdR_{Cc}$ remain inactive in the pre-divisional cell until transition between swarmer and stalked cell is activated. At this point, the active (dephosphorylated) $CpdR_{Cc}$ triggers ClpXP to proteolyze both $CtrA_{Cc}$ and $CpdR_{Cc}$ (Iniesta et al. 2006; Iniesta and Shapiro 2008).

Agrobacterium species possess a CpdR homolog, CpdR1 as well as a second CpdR protein, CpdR2. The *Agrobacterium* CpdR1 homolog shares 59.2% amino acid identity (75.0% identical plus similar) with the *Caulobacter* CpdR (Tables 1); as such, it is reasonable to hypothesize the same effect in this genus and it has been demonstrated to share the same non-essentiality as the *C. crescentus* homolog discussed by Iniesta et al. (Curtis and Brun 2014; Iniesta et al. 2006). Furthermore, CpdR1 appears to be important for cell cycle progression, but specific data on this interaction has not yet been generated. Interestingly, screening for Δ*cpdR1* mutants has proven ineffective, suggesting that either Cpdr1 is essential to *A. tumefaciens* but did not flag in the Tn-mutagenesis data due to the large number of unresolved genes (Curtis and Brun 2014) or that its essentiality is tied to the growth and selection conditions incurred by growth on minimal media (unpublished results).

Preliminary data on CpdR2 suggests that this locus is not essential for growth but is wired into the PdhS-DivK-CtrA pathway. Screening of a saturated mutant library generated by transposon mutagenesis of an *A. tumefaciens* strain C58 Δ*pdhS2* Δ*pleD* background identified the *cpdR2* locus as suppressing this strain's swimming motility phenotype. Moreover, deletion of *cpdR2* in a background lacking *pdhS2* also suppresses both the biofilm formation and motility phenotypes of the Δ*pdhS2* mutant. Although these data are preliminary and more work is needed for confirmation, it is tempting to speculate that CpdR2 is the hypothetical "RR-X" response regulator proposed earlier by our group (unpublished results; Heindl et al. 2019).

2.3 Proteolysis

Most bacteria utilize AAA+ proteases, such as ClpAP, ClpXP, or Lon for ATP-dependent protein degradation (Sauer and Baker 2011; Striebel et al. 2009). In the Clp family proteases, ClpA and ClpX both utilize a hexameric AAA+ ring containing six identical multi-domain subunits. Similarly, Lon proteases contain hexamers comprised of six identical subunits, each with an AAA+ domain, a protease domain, and several additional domains. The ClpP protein, however, forms a double ring structure comprised of 14 subunits. Among the proteolytic enzymes, ClpAP (and ClpCP in some bacteria) contain two AAA+ domains while ClpXP and Lon contain a single AAA+ domain. The AAA+ domain in Lon differs from the Clp proteins in that it is directly linked to the proteolytic domains as opposed to being distinct oligomeric complexes (Sauer and Baker 2011; Striebel et al. 2009).

In ClpAP, the AAA+ ClpA unfoldase pairs with the ClpP protease, creating a double ring structure with a long axial pore (Olivares et al. 2016; Sauer and Baker 2011; Striebel et al. 2009). This structure allows for efficient degradation of ssrA-tagged proteins, but can be inhibited by ClpS interaction (Sauer and Baker 2011). Like ClpAP, ClpXP is comprised of the ClpP protease and an unfoldase, in this case ClpX (Olivares et al. 2016). Orientation of the large and small ClpX subunits generates two conformations. Only the "loadable" L conformation presents a nucleotide binding pocket between the two domains (Stinson et al. 2013). Like the Clp proteins, Lon forms a multi-subunit enzyme which typically forms a hexamer. Due to the hexamer arrangement, Lon can form a dodecamer, resulting in varied protein degradation. The Lon substrates also demonstrate greater control over the AAA+ unfoldase ring than those of other AAA+ unfoldase substrates (Gur and Sauer 2009).

2.3.1 Caseinolytic Proteases ClpA (Atu1364), ClpP1 (Atu1258), ClpP2 (Atu2270), ClpP3 (Atu1627), ClpS1 (Atu1363), ClpS2 (Atu2232), and ClpX (Atu1259)

The caseinolytic proteases ClpP and ClpX have both been shown to be essential for the survival of *C. crescentus* (Jenal and Fuchs 1998). While initially thought to be involved with the degradation of $SciP_{Cc}$ during G1 to S phase transition, the ATP-dependent serine peptidase Lon Protease was instead the primary proteolytic enzyme responsible for $SciP_{Cc}$ degradation. However, the *Caulobacter* ClpP ($ClpP_{Cc}$) and ClpX ($ClpX_{Cc}$) both play a role in $CtrA_{Cc}$ proteolysis. In order for successful proteolysis, the two-component ClpXP protease ($ClpXP_{Cc}$) must be localized via $CpdR_{Cc}$, which itself is degraded by $ClpXP_{Cc}$, and degradation of $CtrA_{Cc}$ can be prevented by the formation of a CtrA–SciP–DNA complex (Gora et al. 2013; Iniesta et al. 2006; Iniesta and Shapiro 2008). $SciP_{Cc}$ may be similarly protected against Lon degradation (Gora et al. 2013). $ClpXP_{Cc}$ also degrades $DnaA_{Cc}$ in a manner that suggests that $DnaA_{Cc}$ proteolysis is performed more efficiently than that of $CtrA_{Cc}$ (Gorbatyuk and Marczynski 2005). The degradation of $DnaA_{Cc}$ via $ClpP_{Cc}$ requires the addition of the chaperone proteins $ClpA_{Cc}$ (the *C. crescentus* ClpA) and $ClpX_{Cc}$ while $CtrA_{Cc}$ degradation requires only $ClpX_{Cc}$ and $ClpP_{Cc}$ (Gorbatyuk and Marczynski 2005; Jenal and Fuchs 1998). During the transition from G1 to S phase, proteolytic activity of $ClpXP_{Cc}$ results in $DgcB_{Cc}$ activity. This activity, paired with $PleD_{Cc}$, initiates stalk formation (Abel et al. 2011). While ClpXP has been shown to degrade FtsZ in *E. coli*, it does not regulate FtsZ proteolysis in *C. crescentus* (Camberg et al. 2009; Kelly et al. 1998).

Agrobacterium Clp-mediated proteolysis functions much the same as its *C. crescentus* homologs, acting to maintain specific intracellular protein levels (Kaewnum et al. 2012). *A. tumefaciens* expresses one ClpA ($ClpA_{At}$) and ClpX ($ClpX_{At}$), two ClpS proteins and three ClpP proteins (Costa et al. 2012; Stein et al. 2016). In the grape (*Vitis vinifera*) pathogen *A. vitis*, mutation in the *clpA* gene resulted in diminished necrotic lesions on the host, suggesting that the mutations prevented $ClpA_{At}$ from binding and delivering the marked proteins to ClpP for degradation (Kaewnum et al. 2012). The two *A. tumefaciens* ClpS proteins ($ClpS1_{At}$ and $ClpS2_{At}$), like many bacterial ClpS proteins, are involved with binding the N-end-rule degradation signals and recruiting the ClpAP complex (Erbse et al. 2006; Rivera-Rivera et al. 2014; Roman-Hernandez et al. 2011; Stein et al. 2016; Tobias et al. 1991). Canonically, after the successful recruitment of ClpAP, ClpS is modified by ClpA and the bound substrate is transferred to ClpAP (Rivera-Rivera et al. 2014; Roman-Hernandez et al. 2011; Stein et al. 2016). In *Agrobacterium*, and many other alphaproteobacteria, the $ClpS1_{At}$ protein resembles the homologous ClpS found in many other bacterial taxa while $ClpS2_{At}$ is more divergent; this suggests that the alphaproteobacteria employ various ClpS proteins with differing binding specificities under certain growth conditions. Furthermore, $ClpS1_{At}$ can be found at robust levels throughout exponential and stationary phase while $ClpS2_{At}$ is largely restricted to the stationary phase (Stein et al. 2016).

2.3.2 *R*egulator of *C*trA *D*egradation RcdA (Atu3742)

C. crescentus RcdA ($RcdA_{Cc}$) plays a key role in the polar localization and degradation of global regulator $CtrA_{Cc}$. For successful proteolysis to occur, $CtrA_{Cc}$ must localize to the cell pole and interact with ClpXP. For this to occur, $RcdA_{Cc}$ aids in the mediation of $CtrA_{Cc}$ and $ClpX_{Cc}$ interaction. $RcdA_{Cc}$ appears to bind specifically to $CtrA_{Cc}$, as no interaction between $RcdA_{Cc}$ and the *C. crescentus* McpA chemoreceptor was detected. However, the localization of $RcdA_{Cc}$ with ClpXP at the division plane suggests that $RcdA_{Cc}$ may degrade other proteins during cellular division (McGrath et al. 2006). Just prior to the swarmer-to-stalked cell transition, $RcdA_{Cc}$ levels increase dramatically. This surge of protein is regulated by its primary target, $CtrA_{Cc}$, and is necessary for proper cellular development (Laub et al. 2002). While $CtrA_{Cc}$ degradation by ClpXP can occur without RcdA and another associated protein PopA in vitro, inclusion of both chaperone proteins drastically increases the rate of $CtrA_{Cc}$ degradation. This interaction is not observed in vivo, where RcdA and PopA are required for $CtrA_{Cc}$ proteolysis (Chien et al. 2007; Mahmoud and Chien 2018; Smith et al. 2014).

A. tumefaciens does possess an essential RcdA homolog (Atu3742), however, this protein has not been studied in detail. Homology data suggests that the putative RcdA homolog in *A. tumefaciens* could function in a similar manner to $RcdA_{Cc}$ and other RcdA homologs due to similar homology values shared by other cell cycle-associated genes (Tables 1). A separate protein, Atu5090, has been named RcdA in *A. tumefaciens* but appears to be involved in the synthesis of an unidentified exopolysaccharide homopolymer, as it shares homology with cellulose synthase CelA and curdlan synthase CrdS (Curtis and Brun 2014; Matthysse 2018).

2.3.3 ATP-Dependent Serine Peptidase Lon Protease (Atu1261)

In *C. crescentus*, Lon protease interacts with CcrM, DnaA, and SciP all of which play a role in the cell cycle, while Lon in other prokaryotes affects biofilm formation, motility, stress tolerance, and/or virulence (Breidenstein et al. 2012; Gora et al. 2013; Jonas et al. 2013; Matsui et al. 2003; Rogers et al. 2016; Wright et al. 1996). In strains absent of the *lon* gene, CcrM degradation is considerably slower than in wild-type when complemented with a DNA binding mutant (LonQM). This depression of degradation rate, however, may not lead to any cell cycle defects due to additional regulation of CcrM by other proteins (Zhou et al. 2019).

In *A. tumefaciens*, mutations in the Lon protease lead to several morphologic and phenotypic mutations, including swollen and branching cells, irregular timing of DNA replication, and loss of phytopathogenicity. While Lon mutants in other organisms produce different cell shapes than those seen in *Agrobacterium* (e.g., filamentous cells in *Escherichia coli* (Howard-Flanders et al. 1964)), it can be determined that the failure of the Lon protease to properly degrade its target has an observable effect on the cell. As noted previously, CcrM is actively degraded by the Lon protease in all stages of the cell cycle except for late S phase, where CcrM is upregulated. Accumulation of this methyltransferase leads to abnormal timing of DNA replication and, as a result, altered cell morphology (Kahng and Shapiro 2001; Su et al. 2006). It is likely that both DnaA and SciP are equally degraded by Lon protease based on homology to *C. crescentus.*

2.4 Asymmetrical Cellular Division and Polar Differentiation in *Agrobacterium*

Many Rhizobiales, including *Agrobacterium*, and *Brucella* species, exhibit asymmetrical division and unipolar growth as part of their complex life cycle, unlike *Bacillus subtilis* or *E. coli* (Ardissone and Viollier 2012; Brown et al. 2011, 2012). This life cycle is comprised of (1) an increase in length at the growth pole via addition of peptidoglycan, (2) elongation and transition of the growth pole to a non-growing pole, (3) assembly of the divisome and initiation of peptidoglycan synthesis at the mid-cell, (4) constriction of the divisome resulting in two daughter cells, and (5) generation of new growth poles at the division site (Brown et al. 2012; Cameron et al. 2014; Grangeon et al. 2015, 2017). The presence of the peptidoglycan is essential not only for the division of the cells, but also in the role of some cytoskeletal proteins, like the bacterial actin, MreB, present in *E. coli* and some, but not all, alphaproteobacteria (Ebersbach et al. 2008; van Teeffelen et al. 2011). While *A. tumefaciens* lacks MreB, several proteins, including PodJ and PopZ, play a role in the organization of polar proteins necessary for the asymmetrical division (Grangeon et al. 2015).

As in *C. crescentus*, genome replication in *A. tumefaciens* occurs once per cell cycle, maintaining a relative copy number of one for the large circular chromosome (chromosome I) and the linear chromosome (chromosome II), and relatively low copy numbers of both the pTi virulence megaplasmid, and likely the pAt accessory megaplasmid (Barton et al. 2018; Goodner et al. 2001; Kahng and Shapiro 2001; Kahng and Shapiro 2003; Lee et al. 2006; Pappas 2008; Suzuki et al. 2001; Wood et al. 2001). This is despite the circular chromosome replicating using the DnaA-dependent *Cori* mechanism and the other replicons using plasmid-type *repABC*-dependent mechanism. Not only is genome replication coordinated across all four replicons, but genome segregation is likewise coordinated (Ehrle et al. 2017).

Polar growth via the incorporation of new peptidoglycan is achieved by co-opting the usual role of certain divisome components, including FtsA, FtsW, and FtsZ, as well as by maintaining a relatively large number of novel peptidoglycan-synthesizing enzymes, notably L,D-transpeptidases (Brown et al. 2012; Cameron et al. 2014; Howell et al. 2019; John 2013). This novel variation on the canonical cell division known from *C. crescentus* is further necessitated by the lack of standard elongasome machinery, including MreB and associated proteins. Details on this process have been covered in several excellent recent reviews (Figueroa-Cuilan and Brown 2018).

2.4.1 *Polar Organelle Development* Protein PodJ (Atu0499)

The PodJ homolog found in *C. crescentus* ($PodJ_{Cc}$) is synthesized during the swarmer-to-stalked phase of growth and localizes to the new, flagellated pole. $CtrA_{Cc}$ plays a critical role in the modulation and expression of $PodJ_{Cc}$, and by extension, the recruitment of polar proteins during division (Crymes et al. 1999). $PodJ_{Cc}$ is necessary for the recruitment of PleC and for pili formation. $PodJ_{Cc}$ undergoes further processing via a periplasmic protease (PerP) after which it becomes necessary for chemotaxis and holdfast formation (Ardissone and Viollier 2012; Lawler et al. 2006; Viollier et al. 2002). *Caulobacter* cells lacking $PodJ_{Cc}$ have shown a reduction in expression of many $CtrA_{Cc}$-regulated genes, possibly due to its recruitment of PleC (Ardissone and Viollier 2012; Chen et al. 2006; Viollier and Shapiro 2003).

A. tumefaciens's PodJ ($PodJ_{At}$) is ~300 amino acids longer and shares 22.9% identity (34.8% identical plus positives) with its *Caulobacter* homolog (Tables 1). Both proteins share similar coiled-coil regions and contain both a cytoplasmic and a C-terminal periplasmic domain (Grangeon et al. 2015; Lawler et al. 2006). The *Agrobacterium* protein localizes primarily to the old pole during much of the cell cycle but does migrate to the growth pole late in the cell cycle. This migration suggests that $PodJ_{At}$ functions in the transition of the growth pole to an old pole as the cells age (Anderson-Furgeson et al. 2016). Interestingly, however, $PodJ_{At}$, as well as the pole organizing protein $PopZ_{At}$, is not strictly required for polar growth (Anderson-Furgeson et al. 2016; Ehrle et al. 2017; Grangeon et al. 2015, 2017; Howell et al. 2017). Deletion of $podJ_{At}$ results in formation of ectopic poles, improper localization of cell division proteins (e.g., $PopZ_{At}$), and asymmetric cell division (Anderson-Furgeson et al. 2016; Howell et al. 2017). These growth defects suggest that, in cells lacking $PodJ_{At}$, termination of polar growth is not performed efficiently (Anderson-Furgeson et al. 2016). Furthermore, in these cells, both FtsA and FtsZ form their respective rings at failed cell division sites in the

Δ*podJ*$_{At}$ strains. It is unclear why this phenomenon occurs, but it is suggested that peptidoglycan biosynthesis may play a role in the coordination of the transition from new-to-old pole (Figueroa-Cuilan and Brown 2018).

2.4.2 Pole Organizing Protein PopZ (Atu1720)

Another protein involved in the organization of the poles in *C. crescentus* is PopZ (PopZ$_{Cc}$), a proline-rich, polymeric anchoring protein. PopZ$_{Cc}$ is present at both poles throughout much of the cell cycle, but does migrate to the flagellar pole prior to division (Bowman et al. 2010; Ebersbach et al. 2008; Laloux and Jacobs-Wagner 2013; Ptacin et al. 2010). It is responsible for binding and anchoring the origin of replication (*ori*) region of chromosome partitioning protein ParB at the poles and works in tandem with a membrane protein, TipN, that is only found at the new pole (Ardissone and Viollier 2012; Ptacin et al. 2010; Schofield et al. 2010). PopZ$_{Cc}$ is necessary for the localization of seven different stalked pole proteins (CckA, ClpX, CpdR, DivJ, DivK, RcdA, and SpmX) and mutations in *popZ* result in localization defects of several of these key proteins, including CckA, CpdR, DivJ, and DivK (Ardissone and Viollier 2012; Bowman et al. 2010; Ebersbach et al. 2008); it is likely that mutations in the *Agrobacterium* homolog would produce similar localization defects.

Interestingly, recent data suggests that the *A. tumefaciens* PopZ homolog (PopZ$_{At}$) does not affect the localization of the histidine kinases DivJ, PdhS1, or PdhS2, but does briefly interact with PdhS2 when PopZ$_{At}$ migrates to the new growth pole (Ehrle et al. 2017). Unlike PopZ$_{At}$ in *C. crescentus*, the *A. tumefaciens* PopZ$_{At}$, which is 22.8% identical (32.9% identical plus positives) and nearly twice the length, localizes primarily to the new (growth) pole (Grangeon et al. 2015). Deletion of PodJ$_{At}$ affects the localization of PopZ$_{At}$, causing localization at ectopic poles (Anderson-Furgeson et al. 2016; Howell et al. 2017) while deletion of PopZ$_{At}$ causes aberrant cell division, similar ectopic growth poles, and other growth defects (Ehrle et al. 2017; Grangeon et al. 2017; Howell et al. 2017). As with PodJ$_{At}$, PopZ$_{At}$, is not essential for polar growth; it is, however, required for chromosomal segregation like in *C. crescentus* (Anderson-Furgeson et al. 2016; Ehrle et al. 2017; Grangeon et al. 2015, 2017; Howell et al. 2017). In *Agrobacterium* ParBI (Atu2828), the ParB homolog with highest similarity to the *C. crescentus* homolog, associates with PopZ to properly partition the centromeres and segregate the chromosomes. Deletion of PopZ$_{At}$ causes improper localization of ParBI and failure to translocate chromosome I to the new pole (Ehrle et al. 2017). FtsA and FtsZ are likewise associated with PopZ$_{At}$ and are also mislocalized in the Δ*popZ*$_{At}$ mutant (Howell et al. 2017). These data suggest that PopZ$_{At}$ plays a key role in chromosomal segregation and the transition of peptidoglycan biosynthesis from polar growth to mid-cell synthesis. This relationship with peptidoglycan biosynthesis, however, has not been confirmed to be a direct result of PopZ$_{At}$ mutations and may be a secondary effect (Figueroa-Cuilan and Brown 2018). Perhaps related to the observed functional differences between the *C. crescentus* and *A. tumefaciens* PodJ and PopZ homologs, a homolog of the new pole landmark protein, TipN, was thought to be absent in *A. tumefaciens*. However, recent discovery and characterization of the large GROWTH POLE RING protein (Atu1348) has identified this protein as a possible functional homolog to *C. crescentus* TipN (Zupan et al. 2019).

3 Additional Regulators of Developmental Phenotypes

3.1 Secondary Messenger Cyclic Diguanylate Monophosphate

Cyclic diguanylate monophosphate, c-di-GMP, is one of the universal secondary messengers that plays a profound role in controlling cell signaling pathways in many prokaryotic systems. Initially described by Ross et al. in 1987 as a bis-(3′-5′)-cyclic diguanylic acid responsible for activation of cellulose synthase in *Komagataeibacter xylinus* (formerly *Acetobacter xylinum*) before being identified in *A. tumefaciens* in 1989, c-di-GMP has been implicated in regulating bacterial phenotypes including attachment, motility, virulence, and cell cycle progression in addition to other processes (Amikam and Benziman 1989; Jenal et al. 2017; Römling et al. 2013; Ross et al. 1987).

c-di-GMP synthesis and degradation are the result of diguanylate cyclase (DGC) and c-di-GMP specific phosphodiesterase (PDE) activities. These enzymatic activities are encoded by the GGDEF, EAL, and HD-GYP domains, all of which are conserved between multiple proteins as well as across major bacterial phyla. This conservation establishes c-di-GMP as a universal bacterial secondary messenger important to all bacteria, not just *A. tumefaciens* (Römling et al. 2013). Synthesis of this secondary messenger occurs via the condensation of two molecules of guanosine triphosphate (GTP) by a DGC, resulting in one molecule of c-di-GMP. This reaction is controlled primarily through a GGDEF c-di-GMP synthesizing domain present in many DGCs. Conversely, degradation of c-di-GMP occurs via one of two pathways that is dependent on the active domain of the PDE. PDE activity through an EAL domain results in the hydrolysis of c-di-GMP into a linear form (5′-phosphoguanylyl-(3′- > 5′)-guanosine, pGpG) while hydrolysis via the HD-GYP domain produces GMP, not pGpG (Römling et al. 2013).

Cellular c-di-GMP pools are inversely regulated by an organism's collective DGC and PDE activity resulting in the activation of effector proteins. Many bacteria encode for multiple proteins with the conserved GGDEF and EAL domains. *A. tumefaciens*, for example, possesses numerous proteins that control c-di-GMP regulation, including 16 with a GGDEF domain, 1 each of an EAL or HD-GYP domain, and 13 that include both a GGDEF and EAL domain (Römling et al. 2013).

3.1.1 Effect of c-di-GMP on Biofilm and Motility

c-di-GMP plays a critical role in the regulation of both biofilm formation and motility in *A. tumefaciens*. Generally, *A. tumefaciens* attachment and motility are inversely regulated by c-di-GMP. Low levels of c-di-GMP leads to flagellar motility while high levels lead to attachment and biofilm formation through the production of different exopolysaccharides (EPS). These EPSs include unipolar polysaccharide

(UPP), cellulose, succinoglycan, cyclic β-1, 2-glucans, lipopolysaccharide, and curdlan (Heindl et al. 2014; Xu et al. 2013). UPP and cellulose have been demonstrated to play the primary role in attachment and biofilm formation with UPP being required for biofilm production (Xu et al. 2012). The effect of UPP and cellulose on attachment and motility has been shown to be regulated by c-di-GMP via the DGC protein, PleD (Xu et al. 2012, 2013). In *A. tumefaciens*, overexpression of *pleD* increases c-di-GMP activity by enhancing the DGC activity of the GGDEF domain, thereby increasing the level of both UPP and cellulose (Xu et al. 2013). However, overexpressing *pleD* resulted in branched and elongated cellular morphologies suggesting that multiple processes are affected by c-di-GMP and PleD overexpression may signal cell division initiation (Barnhart et al. 2013).

Cellulose synthesis is controlled by several genes in two adjacent operons and overproduction of this EPS has been shown to result in loose and extensive biofilm formation in *A. tumefaciens* (Heindl et al. 2014; Matthysse et al. 2005; Matthysse et al. 1995). Cellulose synthase (*celA* and *celB*) gene homologs are likely to bind with c-di-GMP, with CelA binding the secondary messenger via a PilZ c-di-GMP binding domain at its carboxy-terminus (Matthysse 2014). This effector domain is a well-studied c-di-GMP binding domain that acts as an adaptor between the signaling of c-di-GMP and the output phenotypes (Jenal et al. 2017). A mutation of the PilZ domain at a conserved serine residue results in decreased binding affinity for c-di-GMP, demonstrating the importance of the PilZ domain of CelA in c-di-GMP interaction (Barnhart et al. 2014).

Furthermore, in most alphaproteobacteria, the PdhS-DivK-CtrA regulatory pathway impacts biofilm formation, motility, and cell cycle regulation. In *A. tumefaciens*, this pathway affects the phosphorylation of the global response regulator CtrA, which then regulates attachment and motility by altering gene expression (Heindl et al. 2019). While *ctrA* regulation is key for these phenotypic variations, c-di-GMP interaction with effector proteins, such as CckA, can lead to similar phenotypes. Binding of c-di-GMP to CckA results in the switch from kinase to phosphatase activity resulting in progression of cell cycle (Lori et al. 2015).

To better understand the genes involved in the motile to sessile transition in *A. tumefaciens*, a transposon mutagenesis study was performed to screen for mutations which increased UPP production. Multiple Tn-mutants were isolated, among which were *visR* and *visN* (Xu et al. 2013). These two LuxR-FixJ transcriptional regulators are vital to the swimming phenotype and were earlier reported to control motility as a global regulator for flagellum in *S. meliloti*. VisN and VisR are predicted to be essential for the expression of the motility genes in *A. tumefaciens* and mutations in either gene resulted in increased biofilm formation and a reciprocal depression of motility as seen in *S. meliloti* due to the loss of flagellar gene expression (Sourjik et al. 2000; Xu et al. 2013). In *S. meliloti*, the activation of motility by *visN* and *visR* occurs via a two-component response regulator, transcription factor, and regulator of exponential growth motility (Rem). However, this interaction is not observed in *A. tumefaciens*, suggesting that control of *visN* and *visR* is independent of Rem. Mutations in either *visN* and *visR* profoundly influence the motile to sessile transition, presumably by affecting the c-di-GMP pools (Xu et al. 2013).

3.1.2 *Diguanylate Cyclases DgcA (Atu1257), DgcB (Atu1691), DgcC (Atu2179), and DcpA (Atu3495)*

DgcA is a putative GGDEF-type DGC homolog with seven transmembrane domains identified during a mutagenesis screen examining the function of swimming motility genes *visN* and *visR*. Dysregulated UPP production and increased attachment of *visNR* mutants were found to be dependent on DgcA, and, while two other GGDEF-type DCGs were shown to be under VisR regulation (DgcB and DgcC), DgcA was shown to be dominant in this activity. Interestingly, VisR does not affect *dgcA* transcription, suggesting that VisR controls *dgcA* indirectly through secondary gene products (Xu et al. 2013).

Like DgcA, DgcB is a GGDEF-type DGC involved in exopolysaccharide production and biofilm formation. DgcB in *C. crescentus* ($DgcB_{Cc}$) activity is inhibited by the PDE PdeA. Once $PdeA_{Cc}$ is degraded by ClpXP, $DgcB_{Cc}$ and $PleD_{Cc}$ initiate the stalked-cell state via increased c-di-GMP levels, driving the degradation of $CtrA_{Cc}$ and leading to pole morphogenesis (Abel et al. 2011). Unlike both DgcA and DgcC, the *A. tumefaciens* DgcB ($DgcB_{At}$) does not contain any transmembrane domains and is predicted to be cytoplasmic (Xu et al. 2013).

A similar homolog to DgcA, DgcC, is a membrane-associated DGC that does not appear to have the same degree of effect on biofilm formation as either DgcA or DgcB. It remains unclear under what conditions DgcC is enzymatically active and it is possible that, despite containing a GGDEF motif, it does not function as a DGC (Xu et al. 2013).

Examination of exopolysaccharide synthesis in *A. tumefaciens* mutants lacking a combination of *visR* and *dgcA,B,* or *C* suggests that DgcA and DgcB are involved in the stimulation of cellulose and UPP production while DcgC plays a lesser role. Additionally, deletion of *dgcA* or *dgcB* resulted in significant depression of biofilm formation compared to wild type, further supporting the role of these proteins in biofilm formation (Xu et al. 2013). Overall swimming motility was not impacted by single mutations or ectopic expression of *dgcA, B,* or *C*, and cellulose-dependent Congo Red staining was only affected by the simultaneous deletion of all three genes (Heindl et al. 2019; Xu et al. 2013). Furthermore, recent data suggests that the ci-di-GMP pools generated via DgcA, DgcB, or PleD increase the overall biofilms generated in a Δ*pdhS2* mutant strain and loss of DgcB abolishes this increase in biofilm formation (Heindl et al. 2019).

The diguanylate cyclase-phosphodiesterase (DGC-PDE) protein DcpA (Atu3495) is a dual domain protein and part of the pterin-mediated signaling pathway that affects these phenotypes in *A. tumefaciens*. DcpA activity is dependent on the presence of an active pteridine reductase PruA (Atu1130), acting as a PDE in a presence of this reductase and as DGC, via a GGDEF motif, in its absence. Additionally, DcpA activity is reliant on pterin-binding protein PruR (Atu3496) and it is assumed that this binding protein acts as the link connecting DcpA's DGC/PDE switch with pterin levels as mediated by PruA activity. In frame deletion of *pruA* profoundly increases biofilm formation and c-di-GMP levels compared to the wild type. Furthermore, staining of UPP using wheat germ agglutinin (WGA) in the

Δ*pruA* strain exhibited increased UPP staining and higher stain concentration on large cellular aggregates, suggesting that synthesis of adhesin is independent from attachment in this background. Similar observations were also seen in *visN* and *visR* mutants (Feirer et al. 2015). These data suggest that the DcpA EAL domain activity is an important regulator in controlling the motile to sessile transition in *A. tumefaciens*.

3.1.3 *R*ough *O*uter *S*urface Repressor, Ros (Atu0916)

The *A. tumefaciens* MucR homolog, Ros (*r*ough *o*uter *s*urface repressor), was the first identified zinc finger domain protein in bacteria (Chou et al. 1998). Ros was initially identified as a repressor of virulence gene expression in *A. tumefaciens* (Close et al. 1987). Ros also represses intracellular c-di-GMP concentrations resulting in altered exopolysaccharide production ultimately reducing biofilm formation. Contrarily, Ros positively regulates motility genes such as the hook protein FlgE and the Rem regulator. Transcriptomic and phenotypic analysis shows that Ros is a global regulator involved in the production of exopolysaccharides, virulence, c-di-GMP turnover, pAT conjugation, Type VI secretion, and motility (Wang 2014). As noted at the beginning of this review, in *Caulobacter* MucR is involved in regulating the entry to G1 by repressing the G1 active genes. This repression is lost as $CtrA_{Cc}$ activates G1 gene transcription (Fumeaux et al. 2014). Based on the phenotypes affected by loss of *ros* in *A. tumefaciens*, and based on its relatively large regulon, it is likely that *Agrobacterium* Ros functions in much the same way as MucR of *C. crescentus* and is a critical regulator of cellular processes during cell cycle progression.

3.2 Environmental Effects on Cell Cycle Regulation Phenotypes and Virulence

A. tumefaciens senses and responds to environmental signals within the rhizosphere. Alterations in the concentration of phosphate, iron, or manganese, for example, all affect biofilm formation (Danhorn et al. 2004; Heindl et al. 2015). Successful infiltration of its plant host through a naturally occurring opening or wound requires a chemotactic response to environmental cues, namely a response to plant-derived chemical signals. Once *A. tumefaciens* has infiltrated its host, the efficiency of transformation via transfer of T-DNA from bacterium to a host cell is determined by the microenvironmental conditions surrounding the host–pathogen interaction. Below we briefly describe the known effects on cell cycle-dependent phenotypes of several environmental factors. The exact mechanism underlying the described regulation of these phenotypes is unclear, but one possible point of regulation may be through the central PdhS-DivK-CtrA regulatory pathway. Indeed, a very recent description of the gene expression changes occurring during *A. tumefaciens* interaction with the plant host demonstrates a remarkable downregulation of activity of this pathway, suggestive of cell cycle arrest (Gonzalez-Mula et al. 2018).

3.2.1 Temperature

Changes to environmental temperature can produce varying cellular responses in *A. tumefaciens*. These temperature fluctuations can impact enzymatic activity and DNA transfer (Biran et al. 2018; Dillen et al. 1997). The transfer mechanism for T-DNA from *A. tumefaciens* to its plant host calls for optimal temperature-regulated environments to maximize the induction rate of tumor formation. Temperatures above 32 °C result in increased sensitivity of signal transduction by the VirA-VirG two-component regulatory system resulting in a loss of efficiency in transfer of the virulence genes and inhibition of *vir* induction in *A. tumefaciens* (Dillen et al. 1997; Jin et al. 1993; Yuan et al. 2008).

Although growth is inhibited at elevated temperatures in *A. tumefaciens*, a heat-shock response cascade can be initiated to prevent cell death and arrest the cell cycle. The heat-shock response involves highly conserved heat-shock proteins (Hsp): the chaperones Hsp60 (GroEL) and Hsp70 (DnaK), and ATP-dependent proteases including HspL and cell-cycle associated serine peptidase Lon protease. Due to its ability to function at elevated temperatures, sigma factor σ^{32} can selectively transcribe these heat-shock proteins when this cascade is activated (Biran et al. 2018).

3.2.2 Effects of Light on *A. tumefaciens*

The *A. tumefaciens* motility phenotype is affected by a wide variety of internal and external factors, including the response to changing light conditions. *A. tumefaciens* cells exposed to increased light conditions express lower motility as compared to those grown in darker environments. Phytochrome genes (Agp1/AtBphP1 and Agp2/AtBphP2) located within *A. tumefaciens* serve as photoreceptors and elicit a phototactic response via light-regulated histidine kinase activity (Lamparter et al. 2002; Oberpichler et al. 2008). Proteomic analysis shows that flagellar proteins FlaA and FlaB respond to various light conditions; in darker conditions, FlaA and FlaB are expressed at higher levels as compared to higher light conditions. These data suggest that flagellar gene expression is regulated via a light-dependent control mechanism. Analysis of this mechanism showed that *flaA*, *flab*, and *flaC* gene expression increases when incubated in a dark environment. Furthermore, there is a quantitative difference in the mean diameters of *A. tumefaciens* colonies in swim motility assays, with colonies grown in darker conditions displaying an approximate 1.8-fold increase in swim-ring diameter. Finally, attachment to root surfaces and virulence are both increased under dark conditions (Oberpichler et al. 2008).

3.2.3 The Role of Oxygen: Aerobic vs. Anaerobic Conditions

Several developmental phenotypes are sensitive to oxygen tension. Maturation of normal biofilm formation in *A. tumefaciens* requires regulation by the FNR-type transcriptional regulator, SinR. Mutation of *fnrN*, an inducer of *sinR* in oxygen-limited cultures, leads to a depletion of SinR expression resulting in a decrease in biofilm maturation. These data suggest a correlation between oxygen sensing and FnrN-SinR regulatory cascade-regulated biofilm formation. FNR-type regulators have previously been shown to be involved in the transition between aerobic and anaerobic metabolism pathways and this link between anaerobic metabolism and biofilm formation can also be observed in *P. aeruginosa*. Biofilms formed by *P. aeruginosa* under anaerobic conditions produces a more robust biofilm compared to aerobic conditions. These findings suggest that some bacteria, like *A. tumefaciens*, may produce a more robust biofilm when cultivated in anaerobic conditions (Ramey et al. 2004). A second transcriptional regulator known to be responsive to local redox conditions is BigR. BigR is believed to regulate detoxification of products of sulfur metabolism, and mutation of *bigR* results in more robust biofilm formation (Barbosa and Benedetti 2007). Finally, the LysR-family transcriptional regulator, VtlR, was recently described to regulate several cell cycle-dependent phenotypes of *A. tumefaciens*. Loss of VtlR activity reduces growth rate, motility, biofilm formation, and tumor-inducing efficiency. VtlR controls expression of a large number of genes, including the small regulatory RNAs AbcR1 and VrsA. The exact mechanism of VtlR regulation of developmental phenotypes is not known, but it is notable that *divK* was identified as a member of the VtlR regulon (Budnick et al. 2020). All three of these transcriptional regulatory pathways (FnrN-SinR, BigR, and VtlR) may sense oxygen tension using redox-reactive disulfide bonds. AbcR1 and VrsA are not the only sRNAs present in *A. tumefaciens*. In addition to AbcR1 and VrsA, several comprehensive studies have identified numerous sRNAs expressed in *A. tumefaciens* and conserved among the Alphaproteobacteria (Reinkensmeier et al. 2011; Robledo et al. 2015; Wilms et al. 2012). Related to cell cycle regulation, the recently described EcpR1 of *S. meliloti* is conserved in *A. tumefaciens* (Robledo et al. 2015).

3.2.4 The Role of Polyamines on Growth, Biofilm Formation, and Virulence

Polyamines are small, positively charged, omnipresent metabolites containing at least two primary amino groups that are essential for growth regulation, surface attachment, and cell proliferation in prokaryotes and eukaryotes (Becerra-Rivera and Dunn 2019; Michael 2018). Spermidine and putrescine are two of the most common polyamines found in bacteria and function to regulate biofilm formation and growth in *A. tumefaciens*. When polyamine levels are altered, changes in *A. tumefaciens* phenotypes can be observed, suggesting that these molecules play an important role in bacterial response to various abiotic and biotic stressors (Wang et al. 2016). Spermidine and putrescine, and other polyamines, are also generated in plant tissue and may affect *A. tumefaciens* biology during host–pathogen interactions (Chen et al. 2018).

A recent study has confirmed that spermidine plays an essential role in the growth of *A. tumefaciens*, however, this function can be performed by any polyamine containing a 1,3-diaminopropane moiety (Kim et al. 2016). In addition to its role in cellular growth, spermidine has been shown to inversely affect biofilm formation in *A. tumefaciens*; limitation or absence of the polyamine leads to an increase in biofilm formation and adherence (Wang et al. 2016). Furthermore, modification of polyamine levels leads to increased virulence induction. Putrescine, spermidine, and spermine (a derivative of spermidine) all affect *vir* gene induction with putrescine and spermidine showing significant increases in *vir* induction (Kumar and Rajam 2005).

3.2.5 Ions: Iron, Manganese, and Phosphate

In addition to the previous biofilm effectors, several ions have been identified as influencers on biofilm formation including iron (Fe^{2+}), manganese (Mn^{2+}), and phosphate ($[HPO_4]^{2-}$, $[H_2PO_4]^{-}$). Varying the concentrations of both metals influences the formation of biofilms, with low levels of each diminishing overall biofilm formation.

Agrobacterium tumefaciens cultivated in iron-deficient conditions show a decrease in biofilm formation and growth (Heindl et al. 2015). Like iron, limited manganese availability results in a reduction of biofilm formation observed in *A. tumefaciens* (Heindl et al. 2015). Although the direct mechanism by which either iron or manganese affects biofilm formation is unclear, one possibility is via the PruA-PruR-DcpA pathway. The PruA-PruR-DcpA pathway impacts the synthesis of UPP through the turnover of c-di-GMP via the activity of diguanylate cyclase-phosphodiesterase DcpA. Under manganese limited conditions, *pruR* is upregulated, possibly resulting in the reduction of surface attachment and biofilm formation (Heindl et al. 2015).

Limitation of phosphate in bacteria, including *A. tumefaciens*, *P. aeruginosa*, and *S. meliloti*, can cause an increased capacity for resistance to oxidative stress. Under oxidative stress conditions, *A. tumefaciens* has been shown to upregulate the bifunctional catalase-peroxidase gene *katA*. Conversely, *katA* expression is reduced in phosphate starvation conditions (Yuan et al. 2005). In environments where there is limited phosphate availability, attachment and biofilm formation are seen to increase. Regulation of phosphate limiting conditions requires the two-component regulatory system PhoB/PhoR. PhoB/PhoR regulates phosphate limited conditions by way of transcription regulation of genes that require phosphate. Phosphorylation of PhoB drives the increase in surface attachment and UPP adhesion. In turn, biofilm formation is increased when limited phosphate conditions are present (Danhorn et al. 2004; Xu et al. 2012).

3.2.6 Response to Plant Generated Compounds

A. tumefaciens pathogenicity can be tied directly to its motility and biofilm formation. Plant-produced small molecules, such as polyamines or phenolics, also affect *A. tumefaciens* phenotypes (Kim et al. 2016). The phenolic acetosyringone is required for full virulence gene activation and also inhibits growth of *A. tumefaciens* (Fortin et al. 1992). Several other plant hormones, including salicylic acid, indole acetic acid, abscisic acid, and gibberellic acid affect growth or attachment (Anand et al. 2008; Liu and Nester 2006; Plyuta et al. 2013; Yuan et al. 2008). The mechanism by which plant signals affect cell growth of the Alphaproteobacteria is relatively underexplored, although recent results on bacteroid differentiation in *S. meliloti* demonstrate that CtrA activity may be involved (Defez et al. 2016; Kobayashi et al. 2009; Pini et al. 2013, 2015).

A. tumefaciens T-DNA is transferred into the host via a type IV secretion system (T4SS), ultimately inducing the process of tumor formation. Infection of a host by *A. tumefaciens* generally occurs in the slightly acidic (pH ~5.5) rhizosphere. This acidic environment induces the two-component *vir* regulon that is comprised of the histidine kinase VirA and its response regulator VirG (Yuan et al. 2008). While *A. tumefaciens* exhibits steady-state exponential growth at this pH, the rate is lower in comparison with the growth rates observed in a neutral pH environment. In response to this low pH, exopolysaccharide production and biofilm formation are regulated via the ExoR regulatory pathway. Acidic pH triggers the proteolysis of ExoR, resulting in activity of the two-component system ChvG-ChvI. Activation of these genes leads to suppression of both motility and biofilm formation (Heckel et al. 2014; Tomlinson and Fuqua 2009; Tomlinson et al. 2010). Activation of this pathway also leads to expression of the type VI secretion system (T6SS) used to compete against other bacteria (Heckel et al. 2014; Wu et al. 2012). The T6SS and the T4SS are both affected by high levels of c-di-GMP at the transcriptional level via endogenous DGC activity (McCarthy et al. 2019). Ultimately, *A. tumefaciens* virulence phenotypes appear to be directly correlated to c-di-GMP levels, with specific outputs requiring specific c-di-GMP thresholds (Dahlstrom and O'Toole 2017; McCarthy et al. 2019; Römling et al. 2013).

4 Conclusion

The *A. tumefaciens* cell cycle is a tightly controlled process regulated by several key two-component systems and phosphorelays that enable expedient response to internal and external environmental changes. As shown here, nearly all the key components of the central cell cycle regulatory pathway share homology with other alphaproteobacterial species. This homology is greatest among the Rhizobiales, but does extend to the model alphaproteobacterium, *C. crescentus*. This conservation of the cell cycle-associated proteins between *A. tumefaciens* and *C. crescentus* allows for strong predictions to be drawn regarding the interactions of the individual proteins and their effect on downstream developmental phenotypes where no data yet exists. The primary histidine kinase cascades provide crucial regulatory control over the cell cycle through transcriptional regulation, dynamic protein localization and degradation, and generation of a c-di-GMP gradient. Each of these aspects ultimately results in downstream effects on developmental phenotypes, including changes to motility, attachment and biofilm formation, and overall cell morphology and viability. However, much of the mechanistic detail associated with these pathways remains elusive. Significant advances in elucidating this information have been made recently (Heindl et al. 2019; Grangeon et al. 2017; Howell et al. 2017; Ehrle et al. 2017), but continued work is necessary to improve the *Agrobacterium* cell cycle model.

Acknowledgements Work on *Agrobacterium tumefaciens* in the Heindl lab is currently supported by the National Institutes of Health/National Institute of General Medical Sciences Academic

Research Enhancement Award (1R15GM135933-01). This work was also supported by the W. W. Smith Charitable Trust (Grant #H1804) and start-up funding provided by the University of the Sciences in Philadelphia. We thank John Pinto for work on early drafts of this manuscript, and all current and past undergraduate trainees for their energy, inquisitiveness, and perseverance. Thanks to Clay Fuqua and Peter Chien for sharing unpublished results.

References

Abel S et al (2011) Regulatory cohesion of cell cycle and cell differentiation through interlinked phosphorylation and second messenger networks. Mol Cell 43:550–560. https://doi.org/10.1016/j.molcel.2011.07.018

Aldridge P, Jenal U (1999) Cell cycle-dependent degradation of a flagellar motor component requires a novel-type response regulator. Mol Microbiol 32:379–391. https://doi.org/10.1046/j.1365-2958.1999.01358.x

Aldridge P, Paul R, Goymer P, Rainey P, Jenal U (2003) Role of the GGDEF regulator PleD in polar development of *Caulobacter crescentus*. Mol Microbiol 47:1695–1708. https://doi.org/10.1046/j.1365-2958.2003.03401.x

Amikam D, Benziman M (1989) Cyclic diguanylic acid and cellulose synthesis in *Agrobacterium tumefaciens*. J Bacteriol 171:6649–6655. https://doi.org/10.1128/jb.171.12.6649-6655.1989

Anand A, Uppalapati SR, Ryu CM, Allen SN, Kang L, Tang Y, Mysore KS (2008) Salicylic acid and systemic acquired resistance play a role in attenuating crown gall disease caused by *Agrobacterium tumefaciens*. Plant Physiol 146:703–715. https://doi.org/10.1104/pp.107.111302

Anderson-Furgeson JC, Zupan JR, Grangeon R, Zambryski PC (2016) Loss of PodJ in *Agrobacterium tumefaciens* leads to rctopic polar growth, branching, and reduced cell division. J Bacteriol 198:1883–1891. https://doi.org/10.1128/JB.00198-16

Angelastro PS, Sliusarenko O, Jacobs-Wagner C (2010) Polar localization of the CckA histidine kinase and cell cycle periodicity of the essential master regulator CtrA in *Caulobacter crescentus*. J Bacteriol 192:539–552. https://doi.org/10.1128/JB.00985-09

Ardissone S, Viollier PH (2012) Developmental and environmental regulatory pathways in alphaproteobacteria. Front Biosci 17:1695–1714. https://doi.org/10.2741/4013

Barbosa RL, Benedetti CE (2007) BigR, a transcriptional repressor from plant-associated bacteria, regulates an operon implicated in biofilm growth. J Bacteriol 189:6185–6194. https://doi.org/10.1128/JB.00331-07

Barnhart DM, Su S, Baccaro BE, Banta LM, Farrand SK (2013) CelR, an ortholog of the diguanylate cyclase PleD of *Caulobacter*, regulates cellulose synthesis in *Agrobacterium tumefaciens*. Appl Environ Microbiol 79:7188–7202. https://doi.org/10.1128/AEM.02148-13

Barnhart DM, Su S, Farrand SK (2014) A signaling pathway involving the diguanylate cyclase CelR and the response regulator DivK controls cellulose synthesis in *Agrobacterium tumefaciens*. J Bacteriol 196:1257–1274. https://doi.org/10.1128/JB.01446-13

Barton IS, Fuqua C, Platt TG (2018) Ecological and evolutionary dynamics of a model facultative pathogen: *Agrobacterium* and crown gall disease of plants. Environ Microbiol 20:16–29. https://doi.org/10.1111/1462-2920.13976

Becerra-Rivera VA, Dunn MF (2019) Polyamine biosynthesis and biological roles in rhizobia. FEMS Microbiol Lett 366. https://doi.org/10.1093/femsle/fnz084

Biondi EG, Reisinger SJ, Skerker JM, Arif M, Perchuk BS, Ryan KR, Laub MT (2006) Regulation of the bacterial cell cycle by an integrated genetic circuit. Nature 444:899–904. https://doi.org/10.1038/nature05321

Biran D, Rotem O, Rosen R, Ron EZ (2018) Coping with high temperature: a unique regulation in *A. tumefaciens*. Curr Top Microbiol Immunol 418:185–194. https://doi.org/10.1007/82_2018_119

Bowman GR et al (2010) *Caulobacter* PopZ forms a polar subdomain dictating sequential changes in pole composition and function. Mol Microbiol 76:173–189. https://doi.org/10.1111/j.1365-2958.2010.07088.x

Bowman GR et al (2008) A polymeric protein anchors the chromosomal origin/ParB complex at a bacterial cell pole. Cell 134:945–955. https://doi.org/10.1016/j.cell.2008.07.015

Breidenstein EB, Bains M, Hancock RE (2012) Involvement of the lon protease in the SOS response triggered by ciprofloxacin in *Pseudomonas aeruginosa* PAO1. Antimicrob Agents Chemother 56:2879–2887. https://doi.org/10.1128/AAC.06014-11

Brilli M, Fondi M, Fani R, Mengoni A, Ferri L, Bazzicalupo M, Biondi EG (2010) The diversity and evolution of cell cycle regulation in alpha-proteobacteria – a comparative genomic analysis. BMC Syst Biol 4:52. https://doi.org/10.1186/1752-0509-4-52

Brown PJ et al (2012) Polar growth in the Alphaproteobacterial order Rhizobiales. Proc Natl Acad Sci USA 109:1697–1701. https://doi.org/10.1073/pnas.1114476109

Brown PJ, Kysela DT, Brun YV (2011) Polarity and the diversity of growth mechanisms in bacteria. Semin Cell Dev Biol 22:790–798. https://doi.org/10.1016/j.semcdb.2011.06.006

Budnick JA, Sheehan LM, Ginder MJ, et al. (2020) A central role for the transcriptional regulator VtlR in small RNA-mediated gene regulation in Agrobacterium tumefaciens. Sci Rep 10:14968. https://doi.org/10.1038/s41598-020-72117-0

Camberg JL, Hoskins JR, Wickner S (2009) ClpXP protease degrades the cytoskeletal protein, FtsZ, and modulates FtsZ polymer dynamics. Proc Natl Acad Sci USA 106:10614–10619. https://doi.org/10.1073/pnas.0904886106

Cameron TA, Anderson-Furgeson J, Zupan JR, Zik JJ, Zambryski PC (2014) Peptidoglycan synthesis machinery in *Agrobacterium tumefaciens* during unipolar growth and cell division. mBio 5:e01219–e01214. https://doi.org/10.1128/mBio.01219-14

Chen D, Shao Q, Yin L, Younis A, Zheng B (2018) Polyamine function in plants: metabolism, regulation on development, and roles in abiotic stress responses. Front Plant Sci 9:1945. https://doi.org/10.3389/fpls.2018.01945

Chen JC, Hottes AK, McAdams HH, McGrath PT, Viollier PH, Shapiro L (2006) Cytokinesis signals truncation of the PodJ polarity factor by a cell cycle-regulated protease. EMBO J 25:377–386. https://doi.org/10.1038/sj.emboj.7600935

Chen YE, Tropini C, Jonas K, Tsokos CG, Huang KC, Laub MT (2011) Spatial gradient of protein phosphorylation underlies replicative asymmetry in a bacterium. Proc Natl Acad Sci USA 108:1052–1057. https://doi.org/10.1073/pnas.1015397108

Chen YE, Tsokos CG, Biondi EG, Perchuk BS, Laub MT (2009) Dynamics of two phosphorelays controlling cell cycle progression in *Caulobacter crescentus*. J Bacteriol 191:7417–7429. https://doi.org/10.1128/JB.00992-09

Chien P, Perchuk BS, Laub MT, Sauer RT, Baker TA (2007) Direct and adaptor-mediated substrate recognition by an essential AAA+ protease. Proc Natl Acad Sci USA 104:6590–6595. https://doi.org/10.1073/pnas.0701776104

Childers WS, Shapiro L (2014) A pseudokinase couples signaling pathways to enable asymmetric cell division in a bacterium. Microbial Cell 2:29–32. https://doi.org/10.15698/mic2015.01.184

Childers WS, Xu Q, Mann TH, Mathews II, Blair JA, Deacon AM, Shapiro L (2014) Cell fate regulation governed by a repurposed bacterial histidine kinase. PLoS Biol 12:e1001979. https://doi.org/10.1371/journal.pbio.1001979

Chou AY, Archdeacon J, Kado CI (1998) *Agrobacterium* transcriptional regulator Ros is a prokaryotic zinc finger protein that regulates the plant oncogene ipt. Proc Natl Acad Sci USA 95:5293–5298

Close TJ, Rogowsky PM, Kado CI, Winans SC, Yanofsky MF, Nester EW (1987) Dual control of *Agrobacterium tumefaciens* Ti plasmid virulence genes. J Bacteriol 169:5113–5118. https://doi.org/10.1128/jb.169.11.5113-5118.1987

Collier J, Murray SR, Shapiro L (2006) DnaA couples DNA replication and the expression of two cell cycle master regulators. EMBO J 25:346–356. https://doi.org/10.1038/sj.emboj.7600927

Costa ED, Chai Y, Winans SC (2012) The quorum-sensing protein TraR of *Agrobacterium tumefaciens* is susceptible to intrinsic and TraM-mediated proteolytic instability. Mol Microbiol 84:807–815. https://doi.org/10.1111/j.1365-2958.2012.08037.x

Crymes WB, Zhang D, Ely B (1999) Regulation of podJ expression during the *Caulobacter crescentus* cell cycle. J Bacteriol 181:3967–3973

Curtis PD, Brun YV (2014) Identification of essential alphaproteobacterial genes reveals operational variability in conserved developmental and cell cycle systems. Mol Microbiol 93:713–735. https://doi.org/10.1111/mmi.12686

Dahlstrom KM, O'Toole GA (2017) A symphony of cyclases: specificity in diguanylate cyclase signaling. Annu Rev Microbiol 71:179–195. https://doi.org/10.1146/annurev-micro-090816-093325

Danhorn T, Hentzer M, Givskov M, Parsek MR, Fuqua C (2004) Phosphorus limitation enhances biofilm formation of the plant pathogen *Agrobacterium tumefaciens* through the PhoR-PhoB regulatory system. J Bacteriol 186:4492–4501. https://doi.org/10.1128/JB.186.14.4492-4501.2004

Defez R, Esposito R, Angelini C, Bianco C (2016) Overproduction of indole-3-acetic acid in free-living Rhizobia induces transcriptional changes resembling those occurring in noduleb. Mol Plant Microbe Interact 29:484–495. https://doi.org/10.1094/MPMI-01-16-0010-R

Deutscher J, Saier MH Jr (2005) Ser/Thr/Tyr protein phosphorylation in bacteria - for long time neglected, now well established. J Mol Microbiol Biotechnol 9:125–131. https://doi.org/10.1159/000089641

Dillen W, De Clercq J, Kapila J, Zambre M, Van Montagu M, Angenon G (1997) The effect of temperature on *Agrobacterium tumefaciens* mediated gene transfer to plants. Plant J 12:1459–1463. https://doi.org/10.1046/j.1365-313x.1997.12061459.x

Domian IJ, Quon KC, Shapiro L (1997) Cell type-specific phosphorylation and proteolysis of a transcriptional regulator controls the G1-to-S transition in a bacterial cell cycle. Cell 90:415–424. https://doi.org/10.1016/S0092-8674(00)80502-4

Ebersbach G, Briegel A, Jensen GJ, Jacobs-Wagner C (2008) A self-associating protein critical for chromosome attachment, division, and polar organization in *Caulobacter*. Cell 134:956–968. https://doi.org/10.1016/j.cell.2008.07.016

Ehrle HM, Guidry JT, Iacovetto R, Salisbury AK, Sandidge DJ, Bowman GR (2017) Polar organizing protein PopZ is required for chromosome segregation in *Agrobacterium tumefaciens*. J Bacteriol 199:e00111–e00117. https://doi.org/10.1128/JB.00111-17

Erbse A et al (2006) ClpS is an essential component of the N-end rule pathway in *Escherichia coli*. Nature 439:753–756. https://doi.org/10.1038/nature04412

Esser D et al (2016) Protein phosphorylation and its role in archaeal signal transduction. FEMS Microbiol Rev 40:625–647. https://doi.org/10.1093/femsre/fuw020

Feirer N et al (2015) A pterin-dependent signaling pathway regulates a dual-function diguanylate cyclase-phosphodiesterase controlling surface attachment in *Agrobacterium tumefaciens*. mBio 6:e00156. https://doi.org/10.1128/mBio.00156-15

Figueroa-Cuilan W, Daniel JJ, Howell M, Sulaiman A, Brown PJ (2016) Mini-Tn7 insertion in an artificial attTn7 site enables depletion of the essential master regulator CtrA in the phytopathogen *Agrobacterium tumefaciens*. Appl Environ Microbiol 82:5015–5025. https://doi.org/10.1128/AEM.01392-16

Figueroa-Cuilan WM, Brown PJB (2018) Cell wall biogenesis during elongation and division in the plant pathogen *Agrobacterium tumefaciens*. Curr Top Microbiol Immunol 418:87–110. https://doi.org/10.1007/82_2018_92

Fioravanti A et al (2013) DNA binding of the cell cycle transcriptional regulator GcrA depends on N6-adenosine methylation in *Caulobacter crescentus* and other Alphaproteobacteria. PLoS Genet 9:e1003541. https://doi.org/10.1371/journal.pgen.1003541

Fortin C, Nester EW, Dion P (1992) Growth inhibition and loss of virulence in cultures of *Agrobacterium tumefaciens* treated with acetosyringone. J Bacteriol 174:5676–5685

Fumeaux C et al (2014) Cell cycle transition from S-phase to G1 in *Caulobacter* is mediated by ancestral virulence regulators. Nat Commun 5:4081. https://doi.org/10.1038/ncomms5081

Gao R, Mukhopadhyay A, Fang F, Lynn DG (2006) Constitutive activation of two-component response regulators: characterization of VirG activation in *Agrobacterium tumefaciens*. J Bacteriol 188:5204–5211. https://doi.org/10.1128/JB.00387-06

Gao R, Stock AM (2017) Quantitative kinetic analyses of shutting off a two-component system. mBio 8. https://doi.org/10.1128/mBio.00412-17

Gonzalez-Mula A et al (2018) Lifestyle of the biotroph *Agrobacterium tumefaciens* in the ecological niche constructed on its host plant. New Phytol 219:350–362. https://doi.org/10.1111/nph.15164

Goodner B et al (2001) Genome sequence of the plant pathogen and biotechnology agent *Agrobacterium tumefaciens* C58. Science 294:2323–2328. https://doi.org/10.1126/science.1066803

Gora KG, Cantin A, Wohlever M, Joshi KK, Perchuk BS, Chien P, Laub MT (2013) Regulated proteolysis of a transcription factor complex is critical to cell cycle progression in *Caulobacter crescentus*. Mol Microbiol 87:1277–1289. https://doi.org/10.1111/mmi.12166

Gora KG, Tsokos CG, Chen YE, Srinivasan BS, Perchuk BS, Laub MT (2010) A cell-type-specific protein-protein interaction modulates transcriptional activity of a master regulator in *Caulobacter crescentus*. Mol Cell 39:455–467. https://doi.org/10.1016/j.molcel.2010.06.024

Gorbatyuk B, Marczynski GT (2005) Regulated degradation of chromosome replication proteins DnaA and CtrA in *Caulobacter crescentus*. Mol Microbiol 55:1233–1245. https://doi.org/10.1111/j.1365-2958.2004.04459.x

Grangeon R, Zupan J, Jeon Y, Zambryski PC (2017) Loss of PopZat activity in *Agrobacterium tumefaciens* by deletion or depletion leads to multiple growth poles, minicells, and growth defects. mBio 8:e01881–e01817. https://doi.org/10.1128/mBio.01881-17

Grangeon R, Zupan JR, Anderson-Furgeson J, Zambryski PC (2015) PopZ identifies the new pole, and PodJ identifies the old pole during polar growth in *Agrobacterium tumefaciens*. Proc Natl Acad Sci USA 112:11666–11671. https://doi.org/10.1073/pnas.1515544112

Greene SE, Brilli M, Biondi EG, Komeili A (2012) Analysis of the CtrA pathway in *Magnetospirillum* reveals an ancestral role in motility in alphaproteobacteria. J Bacteriol 194:2973–2986. https://doi.org/10.1128/JB.00170-12

Gur E, Sauer RT (2009) Degrons in protein substrates program the speed and operating efficiency of the AAA+ Lon proteolytic machine. Proc Natl Acad Sci USA 106:18503–18508. https://doi.org/10.1073/pnas.0910392106

Haakonsen DL, Yuan AH, Laub MT (2015) The bacterial cell cycle regulator GcrA is a σ70 cofactor that drives gene expression from a subset of methylated promoters. Genes Dev 29:2272–2286. https://doi.org/10.1101/gad.270660

Hallez R, Bellefontaine AF, Letesson JJ, De Bolle X (2004) Morphological and functional asymmetry in alpha-proteobacteria. Trends Microbiol 12:361–365. https://doi.org/10.1016/j.tim.2004.06.002

Hallez R et al (2007) The asymmetric distribution of the essential histidine kinase PdhS indicates a differentiation event in *Brucella abortus*. EMBO J 26:1444–1455. https://doi.org/10.1038/sj.emboj.7601577

Hecht GB, Lane T, Ohta N, Sommer JM, Newton A (1995) An essential single domain response regulator required for normal cell division and differentiation in *Caulobacter crescentus*. EMBO J 14:3915–3924

Hecht GB, Newton A (1995) Identification of a novel response regulator required for the swarmer-to-stalked-cell transition in *Caulobacter crescentus*. J Bacteriol 177:6223–6229. https://doi.org/10.1128/jb.177.21.6223-6229.1995

Heckel BC, Tomlinson AD, Morton ER, Choi JH, Fuqua C (2014) *Agrobacterium tumefaciens* exoR controls acid response genes and impacts exopolysaccharide synthesis, horizontal gene transfer, and virulence gene expression. J Bacteriol 196:3221–3233 doi:https://doi.org/10.1128/JB.01751-14

Heindl JE et al (2019) Reciprocal control of motility and biofilm formation by the PdhS2 two-component sensor kinase of *Agrobacterium tumefaciens*. Microbiology 165:146–162. https://doi.org/10.1099/mic.0.000758

Heindl JE, Hibbing ME, Xu J, Natarajan R, Buechlein AM, Fuqua C (2015) Discrete responses to limitation for iron and manganese in *Agrobacterium tumefaciens*: influence on attachment and biofilm formation. J Bacteriol 198:816–829. https://doi.org/10.1128/JB.00668-15

Heindl JE, Wang Y, Heckel BC, Mohari B, Feirer N, Fuqua C (2014) Mechanisms and regulation of surface interactions and biofilm formation in *Agrobacterium*. Front Plant Sci 5:176. https://doi.org/10.3389/fpls.2014.00176

Holtzendorff J, Reinhardt J, Viollier PH (2006) Cell cycle control by oscillating regulatory proteins in *Caulobacter crescentus*. Bioessays 28:355–361. https://doi.org/10.1002/bies.20384

Hordt A et al (2020) Analysis of 1,000+ type-strain genomes substantially improves taxonomic classification of alphaproteobacteria. Front Microbiol 11:468. https://doi.org/10.3389/fmicb.2020.00468

Howard-Flanders P, Simson E, Theriot L (1964) A locus that controls filament formation and sensitivity to radiation in *Escherichia coli* K-12. Genetics 49:237–246

Howell M, Aliashkevich A, Salisbury AK, Cava F, Bowman GR, Brown PJB (2017) Absence of the polar organizing protein PopZ results in reduced and asymmetric cell division in *Agrobacterium tumefaciens* e00101-00117. https://doi.org/10.1128/JB.00101-17

Howell M et al (2019) *Agrobacterium tumefaciens* divisome proteins regulate the transition from polar growth to cell division. Mol Microbiol 111:1074–1092. https://doi.org/10.1111/mmi.14212

Huynh TN, Stewart V (2011) Negative control in two-component signal transduction by transmitter phosphatase activity. Mol Microbiol 82:275–286. https://doi.org/10.1111/j.1365-2958.2011.07829.x

Iniesta AA, Hillson NJ, Shapiro L (2010) Cell pole-specific activation of a critical bacterial cell cycle kinase. Proc Natl Acad Sci USA 107:7012–7017. https://doi.org/10.1073/pnas.1001767107

Iniesta AA, McGrath PT, Reisenauer A, McAdams HH, Shapiro L (2006) A phospho-signaling pathway controls the localization and activity of a protease complex critical for bacterial cell cycle progression. Proc Natl Acad Sci USA 103:10935–10940. https://doi.org/10.1073/pnas.0604554103

Iniesta AA, Shapiro L (2008) A bacterial control circuit integrates polar localization and proteolysis of key regulatory proteins with a phospho-signaling cascade. Proc Natl Acad Sci USA 105:16602–16607. https://doi.org/10.1073/pnas.0808807105

Jacobs C, Domian IJ, Maddock JR, Shapiro L (1999) Cell cycle-dependent polar localization of an essential bacterial histidine kinase that controls DNA replication and cell division. Cell 97:111–120. https://doi.org/10.1016/S0092-8674(00)80719-9

Jacobs C, Hung D, Shapiro L (2001) Dynamic localization of a cytoplasmic signal transduction response regulator controls morphogenesis during the *Caulobacter* cell cycle. Proc Natl Acad Sci USA 98:4095–4100. https://doi.org/10.1073/pnas.051609998

Janakiraman B, Mignolet J, Narayanan S, Viollier PH, Radhakrishnan SK (2016) In-phase oscillation of global regulons is orchestrated by a pole-specific organizer. Proc Natl Acad Sci USA 113:12550–12555. https://doi.org/10.1073/pnas.1610723113

Jenal U, Fuchs T (1998) An essential protease involved in bacterial cell-cycle control. EMBO J 17:5658–5669. https://doi.org/10.1093/emboj/17.19.5658

Jenal U, Reinders A, Lori C (2017) Cyclic di-GMP: second messenger extraordinaire. Nat Rev Microbiol 15:271–284. https://doi.org/10.1038/nrmicro.2016.190

Jin S, Song Y-N, Deng W-Y, Gordon MP, Nester EW (1993) The regulatory VirA protein of *Agrobacterium tumefaciens* does not function at elevated temperatures. J Bacteriol 175:6830–6835. https://doi.org/10.1128/jb.175.21.6830-6835.1993

John RZ, Todd AC, James AF, Patricia CZ (2013) Polar localization of FtsA/FtsZ in Agrobacterium. Proc Natl Acad Sci 110(22):9060–9065. https://doi.org/10.1073/pnas.1307241110

Jonas K, Liu J, Chien P, Laub MT (2013) Proteotoxic stress induces a cell-cycle arrest by stimulating Lon to degrade the replication initiator DnaA. Cell 154:623–636. https://doi.org/10.1016/j.cell.2013.06.034

Kaewnum S, Zheng D, Reid CL, Johnson KL, Gee JC, Burr TJ (2012) A host-specific biological control of grape crown gall by *Agrobacterium vitis* strain F2-5 – its regulation and population dynamics. Phytopathology 103:427–435. https://doi.org/10.1094/PHYTO-07-12-0153-R

Kahng LS, Shapiro L (2001) The CcrM DNA methyltransferase of *Agrobacterium tumefaciens* is essential, and its activity is cell cycle regulated. J Bacteriol 183:3065–3075. https://doi.org/10.1128/JB.183.10.3065-3075.2001

Kahng LS, Shapiro L (2003) Polar localization of replicon origins in the multipartite genomes of *Agrobacterium tumefaciens* and *Sinorhizobium meliloti*. J Bacteriol 185:3384–3391. https://doi.org/10.1128/jb.185.11.3384-3391.2003

Kelly AJ, Sackett MJ, Din N, Quardokus EM, Brun YV (1998) Cell cycle-dependent transcriptional and proteolytic regulation of FtsZ in *Caulobacter*. Genes Dev 12:880–893. https://doi.org/10.1101/gad.12.6.880

Kim J, Heindl JE, Fuqua C (2013) Coordination of division and development influences complex multicellular behavior in *Agrobacterium tumefaciens*. PLoS One 8:e56682. https://doi.org/10.1371/journal.pone.0056682

Kim SH, Wang Y, Khomutov M, Khomutov A, Fuqua C, Michael AJ (2016) The essential role of spermidine in growth of *Agrobacterium tumefaciens* is determined by the 1,3-diaminopropane moiety. ACS Chem Biol 11:491–499. https://doi.org/10.1021/acschembio.5b00893

Kobayashi H, De Nisco NJ, Chien P, Simmons LA, Walker GC (2009) *Sinorhizobium meliloti* CpdR1 is critical for co-ordinating cell cycle progression and the symbiotic chronic infection. Mol Microbiol 73:586–600. https://doi.org/10.1111/j.1365-2958.2009.06794.x

Kobir A, Shi L, Boskovic A, Grangeasse C, Franjevic D, Mijakovic I (2011) Protein phosphorylation in bacterial signal transduction. Biochim Biophys Acta 1810:989–994. https://doi.org/10.1016/j.bbagen.2011.01.006

Kowallis KA, Silfani EM, Kasumu A, Rong G, So V, Childers WS (2020) Synthetic control of signal flow within a bacterial multi-kinase network ACS. Synth Biol. https://doi.org/10.1021/acssynbio.0c00043

Kumar SV, Rajam MV (2005) Polyamines enhance *Agrobacterium tumefaciens* vir gene induction and T-DNA transfer. Plant Sci 168:475–480. https://doi.org/10.1016/j.plantsci.2004.09.018

Kyriakis JM (2014) In the beginning, there was protein phosphorylation. J Biol Chem 289:9460–9462. https://doi.org/10.1074/jbc.R114.557926

Laloux G, Jacobs-Wagner C (2013) Spatiotemporal control of PopZ localization through cell cycle-coupled multimerization. J Cell Biol 201:827–841. https://doi.org/10.1083/jcb.201303036

Lamparter T, Michael N, Mittmann F, Esteban B (2002) Phytochrome from *Agrobacterium tumefaciens* has unusual spectral properties and reveals an N-terminal chromophore attachment site. Proc Natl Acad Sci USA 99:11628–11633. https://doi.org/10.1073/pnas.152263999

Lasker K, Mann TH, Shapiro L (2016) An intracellular compass spatially coordinates cell cycle modules in *Caulobacter crescentus*. Curr Opin Microbiol 33:131–139. https://doi.org/10.1016/j.mib.2016.06.007

Lassalle F et al (2011) Genomic species are ecological species as revealed by comparative genomics in *Agrobacterium tumefaciens*. Genome Biol Evol 3:762–781. https://doi.org/10.1093/gbe/evr070

Laub MT, Chen SL, Shapiro L, McAdams HH (2002) Genes directly controlled by CtrA, a master regulator of the *Caulobacter* cell cycle. Proc Natl Acad Sci USA 99:4632–4637. https://doi.org/10.1073/pnas.062065699

Laub MT, McAdams HH, Feldblyum T, Fraser CM, Shapiro L (2000) Global analysis of the genetic network controlling a bacterial cell cycle. Science 290:2144–2148. https://doi.org/10.1126/science.290.5499.2144

Lawler ML, Larson DE, Hinz AJ, Klein D, Brun YV (2006) Dissection of functional domains of the polar localization factor PodJ in *Caulobacter crescentus*. Mol Microbiol 59:301–316. https://doi.org/10.1111/j.1365-2958.2005.04935.x

Lee K-E, Park D-K, Baek C-H, Hwang W, Kim K-S (2006) repABC-type replicator region of megaplasmid pAtC58 in *Agrobacterium tumefaciens* C58. J Microbiol Biotechnol 16:118–125

Liu P, Nester EW (2006) Indoleacetic acid, a product of transferred DNA, inhibits vir gene expression and growth of *Agrobacterium tumefaciens* C58. Proc Natl Acad Sci USA 103:4658–4662. https://doi.org/10.1073/pnas.0600366103

Lori C et al (2015) Cyclic di-GMP acts as a cell cycle oscillator to drive chromosome replication. Nature 523:236–239. https://doi.org/10.1038/nature14473

Mahmoud SA, Chien P (2018) Regulated proteolysis in bacteria. Annu Rev Biochem 87:677–696. https://doi.org/10.1146/annurev-biochem-062917-012848

Mann TH, Shapiro L (2018) Integration of cell cycle signals by multi-PAS domain kinases. Proc Natl Acad Sci USA 115:E7166–E7173. https://doi.org/10.1073/pnas.1808543115

Marczynski GT, Shapiro L (1992) Cell-cycle control of a cloned chromosomal origin of replication from *Caulobacter crescentus*. J Molec Biol 226:959–977. https://doi.org/10.1016/0022-2836(92)91045-Q

Martin DD, Beauchamp E, Berthiaume LG (2011) Post-translational myristoylation: fat matters in cellular life and death. Biochimie 93:18–31. https://doi.org/10.1016/j.biochi.2010.10.018

Matroule JY, Lam H, Burnette DT, Jacobs-Wagner C (2004) Cytokinesis monitoring during development; rapid pole-to-pole shuttling of a signaling protein by localized kinase and phosphatase in *Caulobacter*. Cell 118:579–590. https://doi.org/10.1016/j.cell.2004.08.019

Matsui H et al (2003) Oral immunization with ATP-dependent protease-deficient mutants protects mice against subsequent oral challenge with virulent *Salmonella enterica* serovar *typhimurium*. Infect Immun 71:30–39. https://doi.org/10.1128/iai.71.1.30-39.2003

Matthysse AG (2014) Attachment of *Agrobacterium* to plant surfaces. Front Plant Sci 5:252. https://doi.org/10.3389/fpls.2014.00252

Matthysse AG (2018) Exopolysaccharides of *Agrobacterium tumefaciens*. In: Ahmed R et al (eds) Current topics in microbiology and immunology, vol 418. Springer, Cham, pp 111–142. https://doi.org/10.1007/978-3-030-03257-9

Matthysse AG, Marry M, Krall L, Kaye M, Ramey BE, Fuqua C, White AR (2005) The effect of cellulose overproduction on binding and biofilm formation on roots by *Agrobacterium tumefaciens*. Mol Plant-Microb Interact 19:1002–1010. https://doi.org/10.1094/MPMI-18-1002

Matthysse AG, White S, Lightfoot R (1995) Genes required for cellulose synthesis in *Agrobacterium tumefaciens*. J Bacteriol 177:1069–1075. https://doi.org/10.1128/jb.177.4.1069-1075.1995

McCarthy RR, Yu M, Eilers K, Wang YC, Lai EM, Filloux A (2019) Cyclic di-GMP inactivates T6SS and T4SS activity in *Agrobacterium tumefaciens*. Mol Microbiol 112:632–648. https://doi.org/10.1111/mmi.14279

McGrath PT, Iniesta AA, Ryan KR, Shapiro L, McAdams HH (2006) A dynamically localized protease complex and a polar specificity factor control a cell cycle master regulator. Cell 124:535–547. https://doi.org/10.1016/j.cell.2005.12.033

Michael AJ (2018) Polyamine function in archaea and bacteria. J Biol Chem 293:18693–18701. https://doi.org/10.1074/jbc.TM118.005670

Mignolet J, Van der Henst C, Nicolas C, Deghelt M, Dotreppe D, Letesson JJ, De Bolle X (2010) PdhS, an old-pole-localized histidine kinase, recruits the fumarase FumC in *Brucella abortus*. J Bacteriol 192:3235–3239. https://doi.org/10.1128/JB.00066-10

Mohari B, Thompson MA, Trinidad JC, Setayeshgar S, Fuqua C (2018) Multiple flagellin proteins have distinct and synergistic roles in *Agrobacterium tumefaciens* motility. J Bacteriol 200:e00327–e00318. https://doi.org/10.1128/JB.00327-18

Oberpichler I, Rosen R, Rasouly A, Vugman M, Ron EZ, Lamparter T (2008) Light affects motility and infectivity of *Agrobacterium tumefaciens*. Environ Microbiol 10:2020–2029. https://doi.org/10.1111/j.1462-2920.2008.01618.x

Ohta N, Lane T, Ninfa EG, Sommer JM, Newton A (1992) A histidine protein kinase homologue required for regulation of bacterial cell division and differentiation. Proc Natl Acad Sci USA 89:10297–10301. https://doi.org/10.1073/pnas.89.21.10297

Olivares AO, Baker TA, Sauer RT (2016) Mechanistic insights into bacterial AAA+ proteases and protein-remodelling machines. Nat Rev Microbiol 14:33–44. https://doi.org/10.1038/nrmicro.2015.4

Panis G, Murray SR, Viollier PH (2015) Versatility of global transcriptional regulators in alpha-Proteobacteria: from essential cell cycle control to ancillary functions. FEMS Microbiol Rev 39:120–133. https://doi.org/10.1093/femsre/fuu002

Pappas KM (2008) Cell-cell signaling and the *Agrobacterium tumefaciens* Ti plasmid copy number fluctuations. Plasmid 60:89–107. https://doi.org/10.1016/j.plasmid.2008.05.003

Paul R et al (2008) Allosteric regulation of histidine kinases by their cognate response regulator determines cell fate. Cell 133:452–461. https://doi.org/10.1016/j.cell.2008.02.045

Paul R, Weiser S, Amiot NC, Chan C, Schirmer T, Giese B, Jenal U (2004) Cell cycle-dependent dynamic localization of a bacterial response regulator with a novel di-guanylate cyclase output domain. Genes Dev 18:715–727. https://doi.org/10.1101/gad.289504

Perez AM, Mann TH, Lasker K, Ahrens DG, Eckart MR, Shapiro L (2017) A localized complex of two protein oligomers controls the orientation of cell polarity. mBio 8. https://doi.org/10.1128/mBio.02238-16

Pierce DL, O'Donnol DS, Allen RC, Javens JW, Quardokus EM, Brun YV (2006) Mutations in DivL and CckA rescue a divJ null mutant of *Caulobacter crescentus* by reducing the activity of CtrA. J Bacteriol 188:2473–2482. https://doi.org/10.1128/JB.188.7.2473-2482.2006

Pini F et al (2015) Cell cycle control by the master regulator CtrA in *Sinorhizobium meliloti*. PLoS Genet 11:e1005232. https://doi.org/10.1371/journal.pgen.1005232

Pini F et al (2013) The DivJ, CbrA and PleC system controls DivK phosphorylation and symbiosis in *Sinorhizobium meliloti*. Mol Microbiol 90:54–71. https://doi.org/10.1111/mmi.12347

Plyuta VA, Lipasova VA, Kuznetsov AE, Khmel IA (2013) Effect of salicylic, indole-3-acetic, gibberellic, and abscisic acids on biofilm formation by *Agrobacterium tumefaciens* C58 and *Pseudomonas aeruginosa* PAO1. Appl Biochem Microbiol 49:706–710. https://doi.org/10.1134/s000368381308005x

Ptacin JL et al (2010) A spindle-like apparatus guides bacterial chromosome segregation. Nat Cell Biol 12:791–798. https://doi.org/10.1038/ncb2083

Quon KC, Marczynski GT, Shapiro L (1996) Cell cycle control by an essential bacterial two-component signal transduction protein. Cell 84:83–93. https://doi.org/10.1016/S0092-8674(00)80995-2

Quon KC, Yang B, Domian IJ, Shapiro L, Marczynski GT (1998) Negative control of bacterial DNA replication by a cell cycle regulatory protein that binds at the chromosome origin. Proc Natl Acad Sci USA 95:120–125. https://doi.org/10.1073/pnas.95.1.120

Radhakrishnan SK, Thanbichler M, Viollier PH (2008) The dynamic interplay between a cell fate determinant and a lysozyme homolog drives the asymmetric division cycle of *Caulobacter crescentus*. Genes Dev 22:212–225. https://doi.org/10.1101/gad.1601808

Ramey BE, Matthysse AG, Fuqua C (2004) The FNR-type transcriptional regulator SinR controls maturation of *Agrobacterium tumefaciens* biofilms. Mol Microbiol 52:1495–1511. https://doi.org/10.1111/j.1365-2958.2004.04079.x

Reinkensmeier J, Schluter JP, Giegerich R, Becker A (2011) Conservation and occurrence of trans-encoded sRNAs in the rhizobiales genes. Basel 2:925–956. https://doi.org/10.3390/genes2040925

Reisenauer A, Shapiro L (2002) DNA methylation affects the cell cycle transcription of the CtrA global regulator in *Caulobacter*. EMBO J 21:4969–4977. https://doi.org/10.1093/emboj/cdf490

Reisinger SJ, Huntwork S, Viollier PH, Ryan KR (2007) DivL performs critical cell cycle functions in *Caulobacter crescentus* independent of kinase activity. J Bacteriol 189:8308–8320. https://doi.org/10.1128/JB.00868-07

Rivera-Rivera I, Roman-Hernandez G, Sauer RT, Baker TA (2014) Remodeling of a delivery complex allows ClpS-mediated degradation of N-degron substrates. Proc Natl Acad Sci USA 111:E3853–E3859. https://doi.org/10.1073/pnas.1414933111

Robledo M, Frage B, Wright PR, Becker A (2015) A stress-induced small RNA modulates alpha-rhizobial cell cycle progression. PLoS Genet 11:e1005153. https://doi.org/10.1371/journal.pgen.1005153

Rogers A, Townsley L, Gallego-Hernandez AL, Beyhan S, Kwuan L, Yildiz FH (2016) The LonA protease regulates biofilm formation, motility, virulence, and the type VI secretion system in *Vibrio cholerae*. J Bacteriol 198:973–985. https://doi.org/10.1128/JB.00741-15

Roman-Hernandez G, Hou JY, Grant RA, Sauer RT, Baker TA (2011) The ClpS adaptor mediates staged delivery of N-end rule substrates to the AAA+ ClpAP protease. Mol Cell 43:217–228. https://doi.org/10.1016/j.molcel.2011.06.009

Römling U, Galperin MY, Gomelsky M (2013) Cyclic di-GMP: the first 25 years of a universal bacterial second messenger. Microbiol Mol Biol Rev 77:1–52. https://doi.org/10.1128/MMBR.00043-12

Ross P et al (1987) Regulation of cellulose synthesis in *Acetobacter xyinum* by cyclic diguanylic acid. Nature 325:279–281. https://doi.org/10.1038/325279a0

Ryan KR (2006) Partners in crime: phosphotransfer profiling identifies a multicomponent phosphorelay. Mol Microbiol 59:361–363. https://doi.org/10.1111/j.1365-2958.2005.04980.x

Sauer RT, Baker TA (2011) AAA+ proteases: ATP-fueled machines of protein destruction. Annu Rev Biochem 80:587–612. https://doi.org/10.1146/annurev-biochem-060408-172623

Schofield WB, Lim HC, Jacobs-Wagner C (2010) Cell cycle coordination and regulation of bacterial chromosome segregation dynamics by polarly localized proteins. EMBO J 29:3068–3081. https://doi.org/10.1038/emboj.2010.207

Smith SC, Joshi KK, Zik JJ, Trinh K, Kamajaya A, Chien P, Ryan KR (2014) Cell cycle-dependent adaptor complex for ClpXP-mediated proteolysis directly integrates phosphorylation and second messenger signals. Proc Natl Acad Sci USA 111:14229–14234. https://doi.org/10.1073/pnas.1407862111

Snider J, Thibault G, Houry WA (2008) The AAA+ superfamily of functionally diverse proteins. Genome Biol 9:216. https://doi.org/10.1186/gb-2008-9-4-216

Sommer JM, Newton A (1991) Pseudoreversion analysis indicates a direct role of cell division genes in polar morphogenesis and differentiation in *Caulobacter crescentus*. Genetics 129:623–630

Sourjik V, Muschler P, Scharf B, Schmitt R (2000) VisN and VisR are global regulators of chemotaxis, flagellar, and motility genes in *Sinorhizobium* (Rhizobium) meliloti. Genetics and Molec Biol 182:782–788. https://doi.org/10.1128/JB.182.3.782-788.2000

Stein BJ, Grant RA, Sauer RT, Baker TA (2016) Structural basis of an N-degron adaptor with more stringent specificity. Structure 24:232–242. https://doi.org/10.1016/j.str.2015.12.008

Stephens C, Reisenauer A, Wright R, Shapiro L (1996) A cell cycle-regulated bacterial DNA methyltransferase is essential for viability. Proc Natl Acad Sci USA 93:1210–1214. https://doi.org/10.1073/pnas.93.3.1210

Stinson BM, Nager AR, Glynn SE, Schmitz KR, Baker TA, Sauer RT (2013) Nucleotide binding and conformational switching in the hexameric ring of a AAA+ machine. Cell 153:628–639. https://doi.org/10.1016/j.cell.2013.03.029

Striebel F, Kress W, Weber-Ban E (2009) Controlled destruction: AAA+ ATPases in protein degradation from bacteria to eukaryotes. Curr Opin Struct Biol 19:209–217. https://doi.org/10.1016/j.sbi.2009.02.006

Su S, Stephens BB, Alexandre G, Farrand SK (2006) Lon protease of the alpha-proteobacterium *Agrobacterium tumefaciens* is required for normal growth, cellular morphology and full virulence. Microbiology 152:1197–1207. https://doi.org/10.1099/mic.0.28657-0

Subramanian K, Paul MR, Tyson JJ (2015) Dynamical localization of DivL and PleC in the asymmetric division cycle of *Caulobacter crescentus*: a theoretical investigation of alternative models. PLoS Comput Biol 11:e1004348. https://doi.org/10.1371/journal.pcbi.1004348

Suzuki K, Iwata K, Yoshida K (2001) Genome analysis of *Agrobacterium tumefaciens*: construction of physical maps for linear and circular chromosomal DNAs, determination of copy number ratio and mapping of chromosomal virulence genes. DNA Res 8:141–152. https://doi.org/10.1093/dnares/8.4.141

Tan MH, Kozdon JB, Shen X, Shapiro L, McAdams HH (2010) An essential transcription factor, SciP, enhances robustness of *Caulobacter* cell cycle regulation. Proc Natl Acad Sci USA 107:18985–18990. https://doi.org/10.1073/pnas.1014395107

Thompson MA, Onyeziri MC, Fuqua C (2018) Function and regulation of *Agrobacterium tumefaciens* cell surface structures that promote attachment. Curr Top Microbiol Immunol 418:143–184. https://doi.org/10.1007/82_2018_96

Tobias JW, Shrader TE, Rocap G, Varshavsky A (1991) The N-end rule in bacteria. Science 254:1374–1377. https://doi.org/10.1126/science.1962196

Tomlinson AD, Fuqua C (2009) Mechanisms and regulation of polar surface attachment in *Agrobacterium tumefaciens*. Curr Opin Microbiol 12:708–714. https://doi.org/10.1016/j.mib.2009.09.014

Tomlinson AD, Ramey-Hartung B, Day TW, Merritt PM, Fuqua C (2010) *Agrobacterium tumefaciens* ExoR represses succinoglycan biosynthesis and is required for biofilm formation and motility Microbiology 156:2670–2681. https://doi.org/10.1099/mic.0.039032-0

Tsokos CG, Perchuk BS, Laub MT (2011) A dynamic complex of signaling proteins uses polar localization to regulate cell-fate asymmetry in *Caulobacter crescentus*. Dev Cell 20:329–341. https://doi.org/10.1016/j.devcel.2011.01.007

van Teeffelen S, Wang S, Furchtgott L, Huang KC, Wingreen NS, Shaevitz JW, Gitai Z (2011) The bacterial actin MreB rotates, and rotation depends on cell-wall assembly. Proc Natl Acad Sci USA 108:15822–15827. https://doi.org/10.1073/pnas.1108999108

Viollier PH, Shapiro L (2003) A lytic transglycosylase homologue, PleA, is required for the assembly of pili and the flagellum at the *Caulobacter crescentus* cell pole. Mol Microbiol 49:331–345. https://doi.org/10.1046/j.1365-2958.2003.03576.x

Viollier PH, Sternheim N, Shapiro L (2002) Identification of a localization factor for the polar positioning of bacterial structural and regulatory proteins. Proc Natl Acad Sci USA 99:13831–13836. https://doi.org/10.1073/pnas.182411999

Wang H, Bowman GR (2019) SpbR overproduction reveals the importance of proteolytic degradation for cell pole development and chromosome segregation in *Caulobacter crescentus*. Mol Microbiol 111:1700–1714. https://doi.org/10.1111/mmi.14247

Wang Q, Lei Y, Xu X, Wang G, Chen LL (2012) Theoretical prediction and experimental verification of protein-coding genes in plant pathogen genome *Agrobacterium tumefaciens* strain C58. PLoS One 7:e43176. https://doi.org/10.1371/journal.pone.0043176

Wang SP, Sharma PL, Schoenlein PV, Ely B (1993) A histidine protein kinase is involved in polar organelle development in *Caulobacter crescentus*. Proc Natl Acad Sci USA 90:630–634. https://doi.org/10.1073/pnas.90.2.630

Wang Y (2014) Functionality and regulation of type IV pili and cellulose during *Agrobacterium tumefaciens* attachment to surfaces. Doctoral thesis, Indiana University

Wang Y et al (2016) Spermidine inversely influences surface interactions and planktonic growth in *agrobacterium tumefaciens*. J Bacteriol 198:2682–2691. https://doi.org/10.1128/JB.00265-16

Wheeler RT, Shapiro L (1999) Differential localization of two histidine kinases controlling bacterial cell differentiation. Mol Cell 4:683–694. https://doi.org/10.1016/s1097-2765(00)80379-2

Wilms I, Overloper A, Nowrousian M, Sharma CM, Narberhaus F (2012) Deep sequencing uncovers numerous small RNAs on all four replicons of the plant pathogen *Agrobacterium tumefaciens*. RNA Biol 9:446–457. https://doi.org/10.4161/rna.17212

Wood DW et al (2001) The genome of the natural genetic engineer *Agrobacterium tumefaciens* C58. Science 294:2317–2323. https://doi.org/10.1126/science.1066804

Wright R, Stephens C, Shapiro L (1997) The CcrM DNA methyltransferase is widespread in the alpha subdivision of proteobacteria, and its essential functions are conserved in *Rhizobium meliloti* and *Caulobacter crescentus*. J Bacteriol 179:5869–5877. https://doi.org/10.1128/jb.179.18.5869-5877.1997

Wright R, Stephens C, Zweiger G, Shapiro L, Alley MRK (1996) *Caulobacter* Lon protease has a critical role in cell-cycle control of DNA methylation Genes Dev 10:1532–1542. https://doi.org/10.1101/gad.10.12.1532

Wu CF, Lin JS, Shaw GC, Lai EM (2012) Acid-induced type VI secretion system is regulated by ExoR-ChvG/ChvI signaling cascade in *Agrobacterium tumefaciens*. PLoS Pathog 8:e1002938. https://doi.org/10.1371/journal.ppat.1002938

Wu J, Ohto N, Zhao J-L, Newton A (1999) A novel bacterial tyrosine kinase essential for cell division and differentiation. Proc Natl Acad Sci USA 96:13068–13073. https://doi.org/10.1073/pnas.96.23.13068

Xu J, Kim J, Danhorn T, Merritt PM, Fuqua C (2012) Phosphorus limitation increases attachment in *Agrobacterium tumefaciens* and reveals a conditional functional redundancy in adhesin biosynthesis. Res Microbiol 163:674–684. https://doi.org/10.1016/j.resmic.2012.10.013

Xu J, Kim J, Koestler BJ, Choi JH, Waters CM, Fuqua C (2013) Genetic analysis of *Agrobacterium tumefaciens* unipolar polysaccharide production reveals complex integrated control of the motile-to-sessile switch. Mol Microbiol 89:929–948. https://doi.org/10.1111/mmi.12321

Young JM, Kerr A, Sawada H (2005) The alpha-, beta-, delta-, and epsilonproteobacteria. In: Brenner DJ, Krieg NR, Staley JTS (eds) The proteobacteria, Bergey's manual of systematic bacteriology, vol 2C, 2nd edn. Bergey's Manual Trust, East Lansing, pp 340–345

Yuan ZC, Liu P, Saenkham P, Kerr K, Nester EW (2008) Transcriptome profiling and functional analysis of *Agrobacterium tumefaciens* reveals a general conserved response to acidic conditions (pH 5.5) and a complex acid-mediated signaling involved in *Agrobacterium*-plant interactions. J Bacteriol 190:494–507. https://doi.org/10.1128/JB.01387-07

Yuan ZC, Zaheer R, Finan TM (2005) Phosphate limitation induces catalase expression in *Sinorhizobium meliloti, Pseudomonas aeruginosa* and *Agrobacterium tumefaciens*. Mol Microbiol 58:877–894. https://doi.org/10.1111/j.1365-2958.2005.04874.x

Zan J, Heindl JE, Liu Y, Fuqua C, Hill RT (2013) The CckA-ChpT-CtrA phosphorelay system is regulated by quorum sensing and controls flagellar motility in the marine sponge symbiont *Ruegeria sp*. KLH11. PLoS One 8:e66346. https://doi.org/10.1371/journal.pone.0066346

Zhou X, Wang J, Herrmann J, Moerner WE, Shapiro L (2019) Asymmetric division yields progeny cells with distinct modes of regulating cell cycle-dependent chromosome methylation. Proc Natl Acad Sci USA 116:15661–15670. https://doi.org/10.1073/pnas.1906119116

Zupan JR, Grangeon R, Robalino-Espinosa JS, Garnica N, Zambryski P (2019) GROWTH POLE RING protein forms a 200-nm-diameter ring structure essential for polar growth and rod shape in *Agrobacterium tumefaciens*. Proc Natl Acad Sci USA 116:10962–10967. https://doi.org/10.1073/pnas.1905900116

Brucella abortus, a Pathogenic Rhizobiale with a Complex Cell Cycle

Mathilde Van der Henst and Xavier De Bolle

Abstract

Brucella abortus in a pathogenic Rhizobiale able to survive and replicate inside host cells. *B. abortus* shares many features of the Rhizobiales, such as an unusual lipopolysaccharide, periplasmic cyclic β-glucans, general stress response, unipolar growth, multipartite genome, asymmetric division, complex cell cycle regulation, and specific recruitment of proteins to the poles. Conserved regulators of cell cycle progression, like CtrA and GcrA, have been characterized in *B. abortus*.

Brucella spp. are intracellular bacteria responsible for Brucellosis. Another classification of the pathogen has emerged as "facultative extracellular, intracellular pathogen" because although this bacterium can grow in extracellular environments such as culture medium, for example, they are mainly described inside cells in natural hosts (Moreno and Moriyon 2006).

With the exception of *Brucella vulpis* (Al Dahouk et al. 2017), *Brucella* spp. are non-motile and present a coccobacilli morphology from 0.5 to 1.5 μm long (Moreno and Moriyon 2006). The genus *Brucella* is commonly divided into different species according to their tropism for host organisms. For decades, six "classical" species were counted in the *Brucella* genus, i.e., *B. abortus, B. melitensis, B. suis, B. canis, B. neotomae*, and *B. ovis* where the first four are known to be pathogenic for humans (Wallach et al. 2004; Young 1995). However, with the improvement of detection methods, several "new" species have been found to belong to the genus *Brucella*, such as *B. microti, B. ceti, B. pinnipedialis, B. papionis*, and *B. vulpis, B. inopinata*,

M. Van der Henst · X. De Bolle (✉)
URBM – University of Namur, Namur, Belgium
e-mail: xavier.debolle@unamur.be

E. Biondi (ed.), *Cell Cycle Regulation and Development in Alphaproteobacteria*,
https://doi.org/10.1007/978-3-030-90621-4_10

reviewed in Olsen and Palmer (2014). More recently, additional motile *Brucella* strains have been isolated from African bullfrogs (Al Dahouk et al. 2017). Despite host preference and several phenotypic variations, *Brucella* species are closely related at the genetic level, displaying more than 98% of sequence identity between strains of different species (Mayer-Scholl et al. 2010).

1 The Genome of *Brucella*

Brucella spp. are part of the 10% of bacterial species having a bipartite genome characterized by a chromosome I (ChrI) of 2.1 Mb, a chromosome II (ChrII) of 1.15 Mb, and the absence of plasmid in natural strains (Jumas-Bilak et al. 1998; Michaux et al. 1993). An exception for *B. suis* biovar 3 has to be mentioned since this strain presents only one large chromosome of 3.1 Mb (Jumas-Bilak et al. 1998). The genomes of the different *Brucella* species are closely related since a genomic microarray study revealed that on 3.198 ORFs analyzed from *B. melitensis*, more than 3.110 were present in the other *Brucella* strains analyzed at that time (Rajashekara et al. 2004).

The segregation of the ChrI is proposed to be mediated through a ParAB system (Van der Henst 2019) and three *parS* centromere-like sequences which are located near the replication origin, in *B. abortus* (Deghelt et al. 2014). The ChrII has been proposed to be the result of an ancestral plasmid domestication. Indeed, this chromosome contains the plasmidic replication and segregation system *repABC*, which is typically found in plasmids that are widely distributed among Rhizobiales (Cevallos et al. 2008). Moreover, a recent Tn-seq study on *B. abortus* reveals a difference in essential genes distribution between both chromosomes (Sternon et al. 2018). Indeed, 19% of the genes present on the ChrI are essential for the growth on rich culture medium plate while the ChrII carries only 5% of essential genes. Moreover, there are clusters of essential genes on ChrII and clusters of non-essential genes in ChrI that would be consistent with recent exchanges between the two chromosomes. Finally, a study of chromosomes replication and segregation highlighted important differences between ChrI and ChrII (Deghelt et al. 2014). The replication and segregation of the origin of ChrI (*oriI*) occur before the corresponding events for ChrII origin (*oriII*), indicating the existence of a mechanism that allows the temporal coordination of both events. Such a mechanism was identified in *Vibrio cholerae* (Val et al. 2016), another pathogen harboring two chromosomes, but remains completely unknown for *B. abortus*. At the end of the replication process, both chromosomes terminate their replication approximately at the same time, like in *V. cholerae* (Deghelt et al. 2014). In addition to being differentially regulated during time, both replication origins show distinct localization patterns. The *oriI* displays an old pole localization before becoming bipolar after the initiation of the replication and segregation process, whereas the replication origin of ChrII does not show any polar attachment (Deghelt et al. 2014), which would be expected for plasmids rather than chromosomes. Taken together, all these evidences support the hypothesis that the ChrII may originate from a megaplasmid acquired during evolution.

2 *Brucella* Are Intracellular Pathogens

Brucella can mainly infect mammalian hosts through ingestion or aerosols. Once inside its host, bacteria can disseminate to specific tissues or organs for which *B. abortus* shows preferential tropism such as the placenta in pregnant females or the reproductive tract. This propagation is mediated by crossing the mucosal epithelium barrier and by entering and surviving inside professional and nonprofessional phagocytic cells, including macrophages and dendritic cells. In this context, different in vitro models for the study of the host cell infection and trafficking have been used such as RAW 264.7 macrophages, THP-1 macrophages, or HeLa epithelial cells. Once internalized in the mammalian cells, *B. abortus* resides in a vacuole named BCV for *Brucella* Containing *V*acuole. This vacuole first acquires early trafficking markers such as Rab5 and EEA-1 by interacting with the endocytic pathway and is therefore named eBCV (Celli 2015; Pizarro-Cerda et al. 2000). This vacuole then maturates and rapidly acquires markers of late endosomal traffic such as the Lysosomal Associated Membrane Protein 1 (LAMP1) (Pizarro-Cerda et al. 1998). Although a fraction of the bacteria is able to avoid lysosomal degradation by preventing the fusion of the BCV with lysosomes (Celli et al. 2003), transient interactions between BCVs and lysosomes have been demonstrated (Starr et al. 2008). The acidification of the BCV is required for the success of the intracellular trafficking (Porte et al. 1999) and it also triggers the expression of the *virB* virulence factor (Starr et al. 2008; Boschiroli et al. 2002). The *virB* operon, which is conserved among *Brucella* species, codes for a type IV secretion system (T4SS) homologous to the well-described VirB system found in *Agrobacterium tumefaciens* (O'Callaghan et al. 1999). The expression of *virB* genes constitutes a crucial step for the *Brucella* trafficking to a compartment derived from the endoplasmic reticulum (ER) where bacteria replicate and establish the proliferation niche therefore named rBCV (Celli 2015; Pizarro-Cerda et al. 1998). Indeed, a mutant for *virB* cannot reach this ER-derived compartment and stays blocked in LAMP1-positive compartment (Comerci et al. 2001; Delrue et al. 2001). Several effectors secreted by the VirB system have been identified such as BspB which can interact with the conserved oligomeric Golgi (COG complex) modifying the Golgi vesicles traffic and allowing the formation of the rBCV (Miller et al. 2017). Interestingly, bacterial growth is initiated in eBCVs while daughter cells are only detected in rBCVs, consistent with the initiation of growth of the *virB* mutant, which never generates daughter cells intracellularly, showing that trafficking and cell cycle progression are intimately connected (Deghelt et al. 2014). Later in the trafficking, after massive proliferation, the bacteria are associated with several atypical autophagic markers, forming the autophagic BCVs or aBCVs (Celli 2015; Starr et al. 2012).

3 *Brucella* Are Rhizobiales

In this part of the chapter, we will describe features that *Brucella* shares with Rhizobiales and that have been characterized at the molecular level. These include the unusual structure of lipopolysaccharide (LPS), general stress response, the role of cyclic beta-glucans, unipolar growth, and asymmetric division.

3.1 Features of *Brucella* Envelope

Like many Rhizobiales but unlike *Escherichia coli*, *Brucella abortus* encodes AcpXL and LpxXL proteins that, respectively, allow synthesis and incorporation of a very long-chain fatty acids (VLCFA) in the lipid A of LPS. These proteins were first characterized in *Rhizobium leguminosarum* (Basu et al. 2002; Vedam et al. 2003) and *Sinorhizobium meliloti* (Sharypova et al. 2003). It was also shown that *B. abortus* LPS contains a VLCFA (Ferguson et al. 2004; Moreno et al. 1990; Velasco et al. 2000). The nonclassical structure of LPS correlates with its poor ability to induce pro-inflammatory molecules and GTPases of the p47 family (Lapaque et al. 2006) and envelope extracts poorly activate the pro-inflammatory response (Barquero-Calvo et al. 2007). The nonclassical feature of LPS, and maybe other components of the envelope as well, could have helped *Brucella* ancestors to establish an infectious cycle, allowing the bacteria to (partially) escape immune surveillance.

At the periplasmic level, *Brucella abortus* was shown to have cyclic beta-glucans (CβG) like other Rhizobiales. These CβG are composed of about 20 β-D-glucopyranosyl residues linked in 1,2 (Bundle et al. 1988). CβGs are synthesized by Cgs (Inon de Iannino et al. 1998), modified by the succinyltransferase Cgm (Roset et al. 2006) and exported to the periplasm by the ABC transporter Cgt (Roset et al. 2004). A *B. abortus cgs* mutant displays surface alterations (Briones et al. 2001) and it is attenuated in cellular and mice infection models (Briones et al. 2001). While virulence attenuation of a mutant affected for its surface is not surprising, it was interestingly reported that CbG is able to extract cholesterol from lipid rafts in host cells (Arellano-Reynoso et al. 2005). Moreover, the treatment of HeLa cells with CβG allows the restoration of the virulence of the *cgs* mutant (Arellano-Reynoso et al. 2005), excluding the possibility that pleiotropic effects resulting from the absence of CβG in the periplasm level would be responsible for the attenuation of the *cgs* mutant in cellular models of infection.

3.2 General Stress Response

Compared to other bacteria, *Brucella* genomes contain only a few σ factors, six for *Brucella melitensis* (Delory et al. 2006). One of them called σ^{E1} is involved in a conserved regulation circuit called general stress response (Fiebig et al. 2015). In *B. abortus*, a sensory histidine kinase named LovhK is able to phosphorylate a response regulator named PhyR (Kim et al. 2014). A small protein named NepR is able to bind phosphorylated PhyR but not unphosphorylated PhyR. Since NepR is acting as an anti-σ factor on σ^{E1}, phosphorylation of PhyR results in the release of active σ^{E1}, that is able to control the expression of genes required for full virulence, such as the *cydA* and *cydB* (encoding cytochrome *bd* ubiquinol oxidase), *pgm* (encoding phosphoglucomutase), and those coding for urease subunits (Kim et al. 2013). In agreement with this, it was shown that PhyR and σ^{E1} are required for chronic infection in mice, suggesting that general stress response is required for long-term infections (Kim et al. 2013). The general stress response is well conserved in alphaproteobacteria, and the genes coding for this system are even syntenic in many Rhizobiales (Fiebig et al. 2015).

3.3 Unipolar Growth

At the level of the envelope, bacterial species have evolved diversified mechanisms leading to incorporation of new cell wall material in order to expand the cell surface and finally to divide to complete their cell cycle. Among these growth mechanisms, we can distinguish two major ways that mediate the addition of new peptidoglycan (PG) and outer membrane components. The lateral growth mode, shared by the bacterial models *E. coli*, *Bacillus subtilis*, or *Caulobacter crescentus*, is defined by dispersed PG synthesis that takes place along the cell axis (Cabeen and Jacobs-Wagner 2005). This canonical cell elongation mode is mediated by a protein complex called *elongasome* that includes the penicillin-binding proteins PBP2 (transpeptidase), the PBP1a (transpeptidase/transglycosylase) as well as structural proteins like MreBCD and RodA. The constriction step of the cell division is achieved through the formation of the Z-ring by redirecting components of the PG synthesis machinery to the midcell, a process mediated by the tubulin homolog FtsZ and the actin homolog FtsA (Lan et al. 2009; Osawa and Erickson 2013). In contrast, the polar growth is characterized by local incorporation of new material that occurs at one or both poles of the cell, and unlike the dispersed elongation, little is known about mechanisms involved in this typical growth. Polar growth was described in bacteria able to form hyphae like *Streptomyces* (Flardh 2010). Unipolar growth, i.e., polar growth occurring at a single pole, was first shown in *A. tumefaciens* using Texas Red Succinimidyl Ester (TRSE) fluorescent labelling, which bind covalently amine groups on the bacterial surface (Brown et al. 2012). In bacteria presenting a dispersed growth mode, the TRSE signal is diluted throughout the entire cell surface whereas the unipolar growth is highlighted by the addition of new unlabelled material at a specific pole of the bacterium. In theory, by growing at only one pole, bacteria can generate two different sibling cells, a mother bacterium inheriting the old envelope material and a daughter cell presenting an envelope made from new precursors, if the envelope material does not disperse too fast, which seems to be the case at least in *B. abortus* (Vassen et al. 2019), and maybe in many Rhizobiales since it is consistent with the observation of unipolar growth using TRSE. The concept of mother cells undergoing deleterious effects of aging and rejuvenating new daughter cells has been proposed to give advantage for bacterial fitness, in asymmetric dividing bacteria (Ackermann et al. 2003) as well as symmetrically dividing bacteria like *E. coli* (Lindner et al. 2008).

It has been suggested that the localization of the PG synthesis machinery governs the mode of growth. In *A. tumefaciens*, the proteins FtsZ, FtsA, PBP1a, and a specific L,D-transpeptidase (Atu0845) are found to be associated to the growing pole as well as to the constriction site when division occurs (Cameron et al. 2014). Unipolar growth is also found in other Rhizobiales like *Sinorhizobium meliloti* and *Ochrobactrum anthropic*, suggesting that this growth mode is probably conserved among the Rhizobiales order (Brown et al. 2012). The *B. melitensis* genome encodes seven penicillin-binding proteins, three of them belonging to the PBP1a family, three from the PBP6 family and one from the PBP2 family (DelVecchio et al. 2002). A Tn-seq study on *B. abortus* indicated that the gene coding for a PBP1a (BAB1_0932) was essential to grow on rich medium (Sternon et al. 2018), as it was already described for its ortholog Atu1341 in *A. tumefaciens* (Curtis and Brun 2014). It is still unknown how Rhizobiales including *B. abortus* directs growth through one pole, new insights into this mechanism could be of great interest for antimicrobial strategies development. Analysis of the insertion of specific components in the envelope of *B. abortus* (Vassen et al. 2019) revealed that immature peptidoglycan, i.e., the one able to incorporate fluorescent D-amino acids, is preferentially located at the new pole, which is growing pole, and at the constriction site during division. It was also found that outer membrane proteins Omp25 and Omp2b are inserted at the same sites on the bacterial surface (Vassen et al. 2019). Interestingly, the Cgs and Cgt proteins, involved in the generation of periplasmic CβG, also display a unipolar localization (Guidolin et al. 2015), possibly preferentially to the growing pole, which remains to be demonstrated. Finally, it was also shown that LPS is incorporated in the outer membrane at the same sites (Vassen et al. 2019). The outer membrane protein was found to be surprisingly spatially stable, since lateral movement of outer membrane proteins and LPS was not detectable during bacterial growth (Vassen et al. 2019).

3.4 Asymmetric Division and Polarity

Symmetric division is thought to be the major way used by bacteria to perform cytokinesis. Indeed, bacterial models like *E. coli* or *V. cholerae* use this mode of division also named binary fission. In this dividing mode, the septum formation occurs at the midcell and gives rise to two daughter cells that are equal in size.

By contrast, some bacterial species present an asymmetric mode of division that is the result of asymmetric localization of the division machinery along the cell length. The two daughter cells generated after cytokinesis have two different sizes and eventually present different morphologies and functions as illustrated with the small swarmer cell and the large stalked cell in *C. crescentus*. Asymmetric division was proposed to be a common feature of alphaproteobacteria's group, as scanning electron micrographs of *S. meliloti, A. tumefaciens*, and *B. abortus* predivisional cells showed two different sizes for daughter cells (Hallez et al. 2004). However, the functional role of this asymmetric division is still unknown and remains to be investigated. This is particularly interesting since many bacteria display morphologies that depart from the classical rod model (Young 2006).

The mechanisms involved in asymmetric division and especially in the correct positioning of the septum are not yet elucidated. Interestingly orthologs of *minCDE*, coding for proteins implicated in the inhibition of the Z-ring formation in *E. coli*, are present in the Rhizobiales *B. abortus, S. meliloti*, and *A. tumefaciens* suggesting the presence of other additional factors that determine the asymmetric position of the septum (Hallez et al. 2004, 2007a; de Boer et al. 1989). Recently, the PopZ protein was proposed to be involved in the control of septum localization in the plant pathogen *A. tumefaciens* (Howell et al. 2017). Although PopZ is relatively well conserved among alphaproteobacteria, its intracellular localization pattern varies between clades showing strict new pole localization in *A. tumefaciens* (Grangeon et al. 2015) and bipolar localization in *C. crescentus* (Bowman et al. 2008; Ebersbach et al. 2008), for example. The deletion of *popZ* in *A. tumefaciens* leads to generation of elongated cells, minicells, and cells presenting ectopic poles (Grangeon et al. 2017). Furthermore, the lack of PopZ was shown to induce the misplacing of the cell division factors FtsA and FtsZ leading to loss of the correct position of the constriction site in *A. tumefaciens* (Howell et al. 2017). *B. abortus* presents a homolog of PopZ*A.tumefaciens* sharing 26.4% of identity, and that is shown to be essential according to a Tn-seq study (Sternon et al. 2018). The role of PopZ in *B. abortus* is not yet defined but still, the protein presents the same localization pattern than in *A. tumefaciens*, i.e., at the new pole of the bacterial cell (Deghelt et al. 2014).

4 Cell Cycle Regulation in *Brucella*

Although cell cycle regulation could have been considered to be rather independent from the pathogenic nature of *Brucella*, a control of cell cycle progression occurs during infection in simple cellular models. The conserved regulators of cell cycle progression CtrA and GcrA, first described in *C. crescentus*, have been characterized in *B. abortus*.

4.1 The Cell Cycle Is Linked to the Virulence

The first two steps of *B. abortus* intracellular trafficking are well illustrated with a counting of colony forming units (CFUs) along cellular infection. In the absence of killing, a first phase characterized by a constant CFUs number where bacteria reside in the eBCV, and a second one showing an exponential increase of CFUs number over time, matching with the establishment of the replicative niche in the ER-derived compartment and the proliferation within this organelle (rBCV) (Sedzicki et al. 2018). Interestingly, during the first step of infection, bacteria do not present any growth (as shown by a uniform TRSE labelling, thus without unlabelled growth zones) and are blocked in G1 phase of the cell cycle (as shown by highlighting *oriI*, the replication origin of ChrI) reflecting the inability of the bacteria to proliferate (Deghelt et al. 2014). Moreover, G1 bacteria are more infectious than S or G2 bacteria suggesting first an active mechanism from *B. abortus* to invade the host cell and secondly that G1 stage seems important for the infection process (Deghelt et al. 2014). The duration of this non-proliferative phase differs according to the host cells model used ranging from 4 h to 6 h, respectively, for RAW 264.7 macrophages and HeLa epithelial cells. Then, bacteria resume their growth and start chromosomes replication just prior to the transition into rBCV, when they are still in Lamp1-positive compartment (Deghelt et al. 2014).

The non-proliferative phase observed at the beginning of the infection and the mechanisms involved in the cell cycle arrest remain to be elucidated. This global cell cycle arrest could represent a widespread strategy shared by pathogenic bacteria to face intracellular stresses such as oxidative/nitrosative stress and alkylating stress (Poncin et al. 2019). Another hypothesis is that starvation could explain the *B. abortus* cell cycle block since mutants of *rsh* in *B. melitensis*, *B. abortus*, and *B. suis*, which are unable to produce the alarmone (p)ppGpp and thus, to adapt to starvation conditions, are strongly attenuated during infection of mammalian cells (Dozot et al. 2006; Kim et al. 2005; Kohler et al. 2002). Recent data suggest that affecting (p)ppGpp levels through the production of enzymes able to deplete or constitutively produce the alarmone effects intracellular proliferation and progression through the cell cycle (Van der Henst et al. 2020).

4.2 The CtrA Regulation Pathway

The PleC/DivJ/DivK pathway that regulates *C. crescentus* cell cycle and differentiation is conserved in *B. abortus* (Hallez et al. 2004; De Bolle et al. 2015; Poncin et al. 2018). In *C. crescentus*, PleC and DivJ histidine kinase homologs control the phosphorylation status of the DivK and PleD response regulators, DivJ acting as a kinase and PleC acting as a phosphatase and a kinase for DivK depending on cell cycle stage. DivK controls the abundance and the activity of CtrA, the central regulator of cell cycle in *C. crescentus*, through DivL, CckA, and ChpT. CtrA is also conserved and functional in *B. abortus* (Bellefontaine et al. 2002;

Francis et al. 2017) although its characterization was limited compared to *C. crescentus*. The functional role of PleC and DivK seems to be conserved between *C. crescentus* and *B. abortus*, as genes coding for these proteins in *B. abortus* are able to heterocomplement *C. crescentus* deletion mutants lacking these genes (Hallez et al. 2007b). As in *C. crescentus*, DivK has been shown to interact with DivJ, PleC, and DivL by yeast two-hybrid experiments in *B. abortus*. Finally, $\text{DivK}_{abortus}$ shows polar localization depending on its phosphorylation state, as observed for $\text{DivK}_{crescentus}$ (Hallez et al. 2007b). Despite these similarities, both bacterial species seem to present a relative plasticity regarding the PleC/DivJ/DivK regulon. For example, $\text{PleC}_{abortus}$ does not present the same localization as in *C. crescentus*, and neither DivJ nor PleC protein seem to control the polar localization of DivK since deletions of the corresponding genes do not impact the polar localization of DivK. This suggest also that DivJ and PleC are not crucially involved in the phosphorylation state of DivK, as it is the case in *C. crescentus* (Hallez et al. 2007b).

Interestingly, a third histidine kinase homologous to PleC and DivJ, named PdhS (*P*leC *D*ivJ *H*omologous *S*ensor), is found in *Brucella* species (Hallez et al. 2007b). The *pdhS* gene was shown to be essential as a thermosensitive mutant (*pdhS*Ts) cannot grow at restrictive temperature and expression of nonfunctional alleles of *pdhS* led to abnormal morphologies typically observed during division defects in Rhizobiales, such as branching (Van der Henst et al. 2012). In *B. abortus*, PdhS was found to interact with the response regulator DivK by yeast two-hybrid assay and to be localized exclusively at the old pole of the bacterium, where it co-localizes with DivK (Hallez et al. 2007b). Moreover, a functional link between PdhS and DivK in *B. abortus* has been suggested since the old pole localization of DivK in the *pdhS*Ts mutant is altered at restrictive temperature (Van der Henst et al. 2012). These results suggest that in *B. abortus* PdhS could play a similar role to DivJ in *C. crescentus* (Hallez et al. 2007b). Because PdhS is exclusively found in the large cell and has to be acquired by the small cell prior to divide, it has been proposed that differentiation event(s) could take place during the cell cycle, and that PdhS constitutes a differentiation marker (Hallez et al. 2007b; Van der Henst et al. 2012). Interestingly, a yeast two-hybrid screen for PdhS partners points out the FumC enzyme and this interaction was consistent with the colocalization of both proteins fused to fluorescent proteins (Mignolet et al. 2010) and with the artificial reconstituted interaction in *E. coli* (Van der Henst et al. 2010). FumC is a fumarate hydratase, which catalyzes the conversion of fumarate into L-malate in the tricarboxylic acid (TCA) cycle, and it is likely that this function is retained in *B. abortus* since deletion of *fumC* is colethal with the deletion of *fumA*, the other unrelated fumarase gene in *B. abortus* (Mignolet et al. 2010). The functional meaning of the PdhS-FumC interaction is still unclear, but these data support a link between cell cycle progression, differentiation event, and metabolism, where PdhS could play an important role. This link is further supported by the observation that GdhZ is required for efficient intracellular replication of *B. abortus* (Beaufay et al. 2016). GdhZ is a catabolic glutamate dehydrogenase feeding the TCA cycle and able to modulate the activity of the cell division protein FtsZ in *C. crescentus* (Beaufay et al. 2015). A *B. abortus gdhZ* mutant displays cell division defects and growth impairment in a glutamate-rich medium, while growth is similar to the wild-type strain in a defined medium with xylose or glucose as a carbon source (Beaufay et al. 2016).

In *C. crescentus*, DivK is regulating CtrA through interaction with DivL and a phosphorelay involving CckA and ChpT. ChpT transfers a phosphate group to CtrA and CpdR, this later enhancing the proteolysis of CtrA when it is not phosphorylated (Iniesta et al. 2010). It was shown that homologs of CckA, ChpT, and CpdR interact similarly in *B. abortus* (Willett et al. 2015), suggesting that the whole DivK-CtrA regulation network initially described in *C. crescentus* is conserved in *B. abortus*. For example, overexpression of the unphosphorylatable allele *cpdR(D52A)* generates impressive cell division defects and a complete blockage of intracellular proliferation (Willett et al. 2015), showing that the phosphorelay is also active inside host cells.

4.3 CtrA Targets in *B. abortus*

In *B. abortus*, the depletion of CtrA or the presence of a thermosensitive allele of *ctrA* induces abnormal morphologies such as branched shapes, that are typically observed during division defects in Rhizobiales (Francis et al. 2017; Willett et al. 2015). This phenotype could be explained in part by the CtrA regulon predicted by a ChIP-seq experiment. Indeed, among the predicted targets found in the reported condition, several genes involved in the division process such as *minCDE* operon (involved in Z ring localization), *ftsQAZ, ftsEX,* and *ftsK* were detected as direct targets of CtrA and are thus proposed to be directly regulated by this transcription factor (Francis et al. 2017).

Some of the first targets found for CtrA are conserved between *C. crescentus* and *B. abortus* such as genes involved in division (*ftsQ*) and DNA methylation (*ccrM*), consistent with the role proposed for CtrA in cell cycle regulation in *B. abortus* (Bellefontaine et al. 2002). In *B. abortus*, CcrM is an essential DNA methyltransferase whose overproduction leads to DNA over-replication and to morphological defects resembling in part those produced by cell division defects (Robertson et al. 2000). In *C. crescentus*, CcrM fully methylates the GAnTC sites generated by semiconservative replication of the methylated chromosomes, and it is crucial for appropriate regulation of transcription by the GcrA protein (Fioravanti et al. 2013) (see below). In *B. abortus* GcrA controls cell division and cell growth, being a pleiotropic regulator able to bind numerous promoters, as suggested by ChIP-seq data (Poncin et al. 2019). The ChIP-seq made with CtrA revealed genomic targets that are part of the CtrA regulation pathway and which are conserved in *C. crescentus* such as *ctrA* itself, *divK* and *divJ*. However, additional targets such as *chpT* and *divL*, were also pointed out by the ChIP-seq, suggesting a more complex transcriptional regulation of CtrA in *B. abortus* than in *C. crescentus* (Poncin et al. 2018; Francis et al. 2017).

An important target of $CtrA_{crescentus}$ is the origin of replication of the chromosome (*oriC*). In *C. crescentus*, CtrA regulates DNA replication by directly binding the *oriC* and by preventing initiation of chromosome replication (Quon et al. 1998). However, CtrA*abortus* was not found to bind to the origin of replication by ChIP-seq experiment, and no CtrA binding box is found in the *oriI* region, suggesting that CtrA could not be implicated in direct inhibition of DNA replication initiation, as it is the case for CtrA*crescentus* (Francis et al. 2017). However, it is not excluded that CtrA*abortus* could act indirectly to control replication initiation. Indeed, the ChIP-seq revealed that $CtrA_{abortus}$ binds the region upstream of the *dnaA* gene which codes for the initiator of the ChrI replication (Francis et al. 2017). In addition, the ChIP-seq pointed out the *repABC* operon as target of $CtrA_{abortus}$, suggesting that CtrA could regulate replication and/or segregation machinery of the ChrII. The identification of other CtrA targets in *B. abortus*, for example, controlling outer membrane composition (Francis et al. 2017) or DNA repair (Poncin et al. 2019), that is not the case in *C. crescentus*, supports the relative plasticity of the CtrA network among these species, as previously suggested (Hallez et al. 2004; Bellefontaine et al. 2002).

4.4 GcrA Function in *B. abortus*

In *C. crescentus*, GcrA is a protein oscillating during cell cycle (Holtzendorff et al. 2004), it is associated with the main σ factor, σ^{70} (Haakonsen et al. 2015) and it is able to discriminate between different states of methylation of the DNA (Fioravanti et al. 2013). As in *C. crescentus*, GcrA is able to bind many promoters in *B. abortus* and these promoters are enriched in GAnTC sites (Poncin et al. 2019). The *gcrA* gene is essential in *B. abortus* (Sternon et al. 2018) and a GcrA depletion strain is impaired for growth and division (Poncin et al. 2019). GcrA depletion renders bacteria very sensitive to the alkylating agent methyl methanesulfonate (Poncin et al. 2019). This is particularly interesting since alkylating stress was found to occur during a cellular infection and genes allowing resistance to alkylating stress are required for a successful infection in a mouse model of intranasal infection (Poncin et al. 2019). It is tempting to speculate that GcrA and CcrM are needed to coordinate gene expression with cell cycle progression along a cellular infection, since alterations of the abundance of these two proteins affect the growth of *B. abortus* in cellular models of infection (Poncin et al. 2019; Robertson et al. 2000).

4.5 CtrA Is Not Crucial for Intracellular Trafficking and G1 Arrest in Endosomal Compartments

Infection experiments with the *B. abortus* CtrA depletion strain indicated that CtrA is not crucially required for the infection process. In host cells, CtrA depletion induces abnormal morphologies, as observed in culture medium, suggesting that bacteria can grow and uptake nutrients in the absence of CtrA (Francis et al. 2017).

In addition, after the cell cycle arrest typically observed during the first hours of the infection, the CtrA depleted bacteria resume their growth approximately at the same time than the WT. The CtrA-depleted bacteria are able to leave the eBCV for the rBCV, indicating that these bacteria are still able to express and produce the *virB* system, which is crucial for this step. These data suggest that the cell cycle arrest observed at the beginning of the infection does not involve CtrA and that other mechanisms should be implicated in this regulation (Francis et al. 2017). The starvation response mediated by (p)ppGpp or adaption to acidic pH constitutes good candidates to explain the cell cycle arrest during the first hours post-infection, but these hypotheses remain to be tested in detail.

References

Ackermann M, Stearns SC, Jenal U (2003) Senescence in a bacterium with asymmetric division. Science 300(5627):1920

Al Dahouk S et al (2017) Brucella spp. of amphibians comprise genomically diverse motile strains competent for replication in macrophages and survival in mammalian hosts. Sci Rep 7:44420

Arellano-Reynoso B et al (2005) Cyclic beta-1,2-glucan is a Brucella virulence factor required for intracellular survival. Nat Immunol 6(6):618–625

Barquero-Calvo E et al (2007) *Brucella abortus* uses a stealthy strategy to avoid activation of the innate immune system during the onset of infection. PLoS One 2(7):e631

Basu SS, Karbarz MJ, Raetz CR (2002) Expression cloning and characterization of the C28 acyltransferase of lipid A biosynthesis in Rhizobium leguminosarum. J Biol Chem 277(32):28959–28971

Beaufay F et al (2015) A NAD-dependent glutamate dehydrogenase coordinates metabolism with cell division in *Caulobacter crescentus*. EMBO J 34(13):1786–1800

Beaufay F, De Bolle X, Hallez R (2016) Metabolic control of cell division in alpha-proteobacteria by a NAD-dependent glutamate dehydrogenase. Commun Integr Biol 9(1):e1125052

Bellefontaine AF et al (2002) Plasticity of a transcriptional regulation network among alpha-proteobacteria is supported by the identification of CtrA targets in *Brucella abortus*. Mol Microbiol 43(4):945–960

Boschiroli ML et al (2002) The *Brucella suis* virB operon is induced intracellularly in macrophages. Proc Natl Acad Sci USA 99(3):1544–1549

Bowman GR et al (2008) A polymeric protein anchors the chromosomal origin/ParB complex at a bacterial cell pole. Cell 134(6):945–955

Briones G et al (2001) *Brucella abortus* cyclic beta-1,2-glucan mutants have reduced virulence in mice and are defective in intracellular replication in HeLa cells. Infect Immun 69(7):4528–4535

Brown PJ et al (2012) Polar growth in the Alphaproteobacterial order Rhizobiales. Proc Natl Acad Sci USA 109(5):1697–1701

Bundle DR, Cherwonogrodzky JW, Perry MB (1988) Characterization of Brucella polysaccharide B. Infect Immun 56(5):1101–1106

Cabeen MT, Jacobs-Wagner C (2005) Bacterial cell shape. Nat Rev Microbiol 3(8):601–610

Cameron TA et al (2014) Peptidoglycan synthesis machinery in *Agrobacterium tumefaciens* during unipolar growth and cell division. MBio 5(3):e01219–e01214

Celli J (2015) The changing nature of the Brucella-containing vacuole. Cell Microbiol 17(7):951–958

Celli J et al (2003) Brucella evades macrophage killing via VirB-dependent sustained interactions with the endoplasmic reticulum. J Exp Med 198(4):545–556

Cevallos MA, Cervantes-Rivera R, Gutierrez-Rios RM (2008) The repABC plasmid family. Plasmid 60(1):19–37

Comerci DJ et al (2001) Essential role of the VirB machinery in the maturation of the *Brucella abortus*-containing vacuole. Cell Microbiol 3(3):159–168

Curtis PD, Brun YV (2014) Identification of essential alphaproteobacterial genes reveals operational variability in conserved developmental and cell cycle systems. Mol Microbiol 93(4):713–735

de Boer PA, Crossley RE, Rothfield LI (1989) A division inhibitor and a topological specificity factor coded for by the minicell locus determine proper placement of the division septum in *E. coli*. Cell 56(4):641–649

De Bolle X et al (2015) *Brucella abortus* cell cycle and infection are coordinated. Trends Microbiol 23(12):812–821

Deghelt M et al (2014) The newborn *Brucella abortus* blocked at the G1 stage of its cell cycle is the major infectious bacterial subpopulation. Nat Commun 5:4366

Delory M et al (2006) An RpoH-like heat shock sigma factor is involved in stress response and virulence in *Brucella melitensis* 16M. J Bacteriol 188(21):7707–7710

Delrue RM et al (2001) Identification of Brucella spp. genes involved in intracellular trafficking. Cell Microbiol 3(7):487–497

DelVecchio VG et al (2002) The genome sequence of the facultative intracellular pathogen *Brucella melitensis*. Proc Natl Acad Sci USA 99(1):443–448

Dozot M et al (2006) The stringent response mediator Rsh is required for *Brucella melitensis* and *Brucella suis* virulence, and for expression of the type IV secretion system virB. Cell Microbiol 8(11):1791–1802

Ebersbach G et al (2008) A self-associating protein critical for chromosome attachment, division, and polar organization in caulobacter. Cell 134(6):956–968

Ferguson GP et al (2004) Similarity to peroxisomal-membrane protein family reveals that Sinorhizobium and Brucella BacA affect lipid-A fatty acids. Proc Natl Acad Sci USA 101(14):5012–5017

Fiebig A et al (2015) General stress signaling in the Alphaproteobacteria. Annu Rev Genet 49:603–625

Fioravanti A et al (2013) DNA binding of the cell cycle transcriptional regulator GcrA depends on N6-adenosine methylation in *Caulobacter crescentus* and other Alphaproteobacteria. PLoS Genet 9(5):e1003541

Flardh K (2010) Cell polarity and the control of apical growth in Streptomyces. Curr Opin Microbiol 13(6):758–765

Francis N et al (2017) CtrA controls cell division and outer membrane composition of the pathogen *Brucella abortus*. Mol Microbiol 103(5):780–797

Grangeon R et al (2015) PopZ identifies the new pole, and PodJ identifies the old pole during polar growth in *Agrobacterium tumefaciens*. Proc Natl Acad Sci USA 112(37):11666–11671

Grangeon R et al (2017) Loss of PopZ At activity in *Agrobacterium tumefaciens* by deletion or depletion leads to multiple growth poles, minicells, and growth defects. MBio 8(6)

Guidolin LS et al (2015) Interaction network and localization of *Brucella abortus* membrane proteins involved in the synthesis, transport, and succinylation of cyclic beta-1,2-glucans. J Bacteriol 197(9):1640–1648

Haakonsen DL, Yuan AH, Laub MT (2015) The bacterial cell cycle regulator GcrA is a sigma70 cofactor that drives gene expression from a subset of methylated promoters. Genes Dev 29(21):2272–2286

Hallez R et al (2004) Morphological and functional asymmetry in alpha-proteobacteria. Trends Microbiol 12(8):361–365

Hallez R et al (2007a) Gateway-based destination vectors for functional analyses of bacterial ORFeomes: application to the Min system in *Brucella abortus*. Appl Environ Microbiol 73(4):1375–1379

Hallez R et al (2007b) The asymmetric distribution of the essential histidine kinase PdhS indicates a differentiation event in *Brucella abortus*. EMBO J 26(5):1444–1455

Holtzendorff J et al (2004) Oscillating global regulators control the genetic circuit driving a bacterial cell cycle. Science 304(5673):983–987

Howell M et al (2017) Absence of the polar organizing protein PopZ results in reduced and asymmetric cell division in *Agrobacterium tumefaciens*. J Bacteriol 199(17)

Iniesta AA, Hillson NJ, Shapiro L (2010) Cell pole-specific activation of a critical bacterial cell cycle kinase. Proc Natl Acad Sci USA 107(15):7012–7017

Inon de Iannino N et al (1998) Molecular cloning and characterization of cgs, the *Brucella abortus* cyclic beta(1-2) glucan synthetase gene: genetic complementation of *Rhizobium meliloti* ndvB and *Agrobacterium tumefaciens* chvB mutants. J Bacteriol 180(17):4392–4400

Jumas-Bilak E et al (1998) Unconventional genomic organization in the alpha subgroup of the Proteobacteria. J Bacteriol 180(10):2749–2755

Kim S et al (2005) Roles of *Brucella abortus* SpoT in morphological differentiation and intramacrophagic replication. Microbiology 151(Pt 5):1607–1617

Kim HS et al (2013) The *Brucella abortus* general stress response system regulates chronic mammalian infection and is controlled by phosphorylation and proteolysis. J Biol Chem 288(19):13906–13916

Kim HS et al (2014) The *Brucella abortus* virulence regulator, LovhK, is a sensor kinase in the general stress response signalling pathway. Mol Microbiol 94(4):913–925

Kohler S et al (2002) The analysis of the intramacrophagic virulome of *Brucella suis* deciphers the environment encountered by the pathogen inside the macrophage host cell. Proc Natl Acad Sci USA 99(24):15711–15716

Lan G et al (2009) Condensation of FtsZ filaments can drive bacterial cell division. Proc Natl Acad Sci USA 106(1):121–126

Lapaque N et al (2006) Differential inductions of TNF-alpha and IGTP, IIGP by structurally diverse classic and non-classic lipopolysaccharides. Cell Microbiol 8(3):401–413

Lindner AB et al (2008) Asymmetric segregation of protein aggregates is associated with cellular aging and rejuvenation. Proc Natl Acad Sci USA 105(8):3076–3081

Mayer-Scholl A et al (2010) Advancement of a multiplex PCR for the differentiation of all currently described Brucella species. J Microbiol Methods 80(1):112–114

Michaux S et al (1993) Presence of two independent chromosomes in the *Brucella melitensis* 16M genome. J Bacteriol 175(3):701–705

Mignolet J et al (2010) PdhS, an old-pole-localized histidine kinase, recruits the fumarase FumC in Brucella abortus. J Bacteriol 192(12):3235–3239

Miller CN et al (2017) A Brucella type IV effector targets the COG tethering complex to remodel host secretory traffic and promote intracellular replication. Cell Host Microbe 22(3):317–329. e7

Moreno E, Moriyon I (2006) The genus Brucella. PRO 5:315–456

Moreno E et al (1990) *Brucella abortus* 16S rRNA and lipid A reveal a phylogenetic relationship with members of the alpha-2 subdivision of the class Proteobacteria. J Bacteriol 172(7):3569–3576

O'Callaghan D et al (1999) A homologue of the Agrobacterium tumefaciens VirB and *Bordetella pertussis* Ptl type IV secretion systems is essential for intracellular survival of *Brucella suis*. Mol Microbiol 33(6):1210–1220

Olsen SC, Palmer MV (2014) Advancement of knowledge of Brucella over the past 50 years. Vet Pathol 51(6):1076–1089

Osawa M, Erickson HP (2013) Liposome division by a simple bacterial division machinery. Proc Natl Acad Sci USA 110(27):11000–11004

Pizarro-Cerda J et al (1998) Brucella abortus transits through the autophagic pathway and replicates in the endoplasmic reticulum of nonprofessional phagocytes. Infect Immun 66(12):5711–5724

Pizarro-Cerda J, Moreno E, Gorvel JP (2000) Invasion and intracellular trafficking of *Brucella abortus* in nonphagocytic cells. Microbes Infect 2(7):829–835

Poncin K, Gillet S, De Bolle X (2018) Learning from the master: targets and functions of the CtrA response regulator in *Brucella abortus* and other alpha-proteobacteria. FEMS Microbiol Rev 42(4):500–513

Poncin K et al (2019) Occurrence and repair of alkylating stress in the intracellular pathogen *Brucella abortus*. Nat Commun 10(1):4847

Porte F, Liautard JP, Kohler S (1999) Early acidification of phagosomes containing *Brucella suis* is essential for intracellular survival in murine macrophages. Infect Immun 67(8):4041–4047

Quon K et al (1998) Negative control of bacterial DNA replication by a cell cycle regulatory protein that binds at the chromosome origin. Proc Natl Acad Sci USA 95(1):120–125

Rajashekara G et al (2004) Comparative whole-genome hybridization reveals genomic islands in Brucella species. J Bacteriol 186(15):5040–5051

Robertson GT et al (2000) The Brucella abortus CcrM DNA methyltransferase is essential for viability, and its overexpression attenuates intracellular replication in murine macrophages. J Bacteriol 182(12):3482–3489

Roset MS et al (2004) Molecular cloning and characterization of cgt, the Brucella abortus cyclic beta-1,2-glucan transporter gene, and its role in virulence. Infect Immun 72(4):2263–2271

Roset MS et al (2006) The *Brucella abortus* cyclic beta-1,2-glucan virulence factor is substituted with O-ester-linked succinyl residues. J Bacteriol 188(14):5003–5013

Sedzicki J et al (2018) 3D correlative electron microscopy reveals continuity of Brucella-containing vacuoles with the endoplasmic reticulum. J Cell Sci 131(4)

Sharypova LA et al (2003) Sinorhizobium meliloti acpXL mutant lacks the C28 hydroxylated fatty acid moiety of lipid A and does not express a slow migrating form of lipopolysaccharide. J Biol Chem 278(15):12946–12954

Starr T et al (2008) Brucella intracellular replication requires trafficking through the late endosomal/lysosomal compartment. Traffic 9(5):678–694

Starr T et al (2012) Selective subversion of autophagy complexes facilitates completion of the Brucella intracellular cycle. Cell Host Microbe 11(1):33–45

Sternon JF et al (2018) Transposon sequencing of *Brucella abortus* uncovers essential genes for growth in vitro and inside macrophages. Infect Immun 86(8)

Val ME et al (2016) A checkpoint control orchestrates the replication of the two chromosomes of *Vibrio cholerae*. Sci Adv 2(4):e1501914

Van der Henst M (2019) Investigation of the replication and the role of (p)ppGpp in the pathogen *Brucella abortus*, in PhD thesis – Department of Biology. University of Namur: Namur (Belgium)

Van der Henst C et al (2010) Overproduced Brucella abortus PdhS-mCherry forms soluble aggregates in *Escherichia coli*, partially associating with mobile foci of IbpA-YFP. BMC Microbiol 10:248

Van der Henst C et al (2012) The histidine kinase PdhS controls cell cycle progression of the pathogenic alphaproteobacterium *Brucella abortus*. J Bacteriol 194(19):5305–5314

Van der Henst M, Carlier E, De Bolle X (2020) Intracellular growth and cell cycle progression are dependent on (p)ppGpp synthetase/hydrolase in *Brucella abortus*. Pathogens 9(7)

Vassen V et al (2019) Localized incorporation of outer membrane components in the pathogen *Brucella abortus*. EMBO J 38(5):e100323

Vedam V et al (2003) A Rhizobium leguminosarum AcpXL mutant produces lipopolysaccharide lacking 27-hydroxyoctacosanoic acid. J Bacteriol 185(6):1841–1850

Velasco J et al (2000) *Brucella abortus* and its closest phylogenetic relative, Ochrobactrum spp., differ in outer membrane permeability and cationic peptide resistance. Infect Immun 68(6):3210–3218

Wallach JC et al (2004) Human infection with M- strain of *Brucella canis*. Emerg Infect Dis 10(1):146–148

Willett JW et al (2015) Structural asymmetry in a conserved signaling system that regulates division, replication, and virulence of an intracellular pathogen. Proc Natl Acad Sci USA 112(28):E3709–E3718

Young EJ (1995) An overview of human brucellosis. Clin Infect Dis 21(2):283–289. quiz 290

Young KD (2006) The selective value of bacterial shape. Microbiol Mol Biol Rev 70(3):660–703

The manufacturer's authorised representative in the EU is Springer Nature Customer Service Centre GmbH, Europaplatz 3, 69115 Heidelberg, Germany. If you have any concerns regarding our products, please contact ProductSafety@springernature.com

Printed and bound by CPI Group (UK) Ltd, Croydon, CR0 4YY
28/09/2026
02224093-0006